The British Dam

Maintaining the Safety of our Dams and Reservoirs

Proceedings of the 18th Biennial Conference of the British Dam Society at Queen's University, Belfast, from 3–6 September 2014

Edited by Andrew Pepper

Cover photograph by Esler Crawford Photography, Belfast: Altnaheglish Dam, Banagher, Northern Ireland

Conference organised by the British Dam Society

www.britishdams.org

Organising committee: Stephen Orr (Chairman); Dr Andy Hughes; Andrew Pepper; Barbara Sharp; Sharon Doyley

Published by ICE Publishing, One Great George Street, Westminster, London SW1P 3AA.

Full details of ICE Publishing sales representatives and distributors can be found at:

www.icevirtuallibrary.com/info/printbooksales

Other titles by ICE Publishing:

Dams: Engineering in a Social and Environmental Context (17th British Dam Society Conference)
British Dam Society. ISBN 978-0-7277-5799-9
A Guide to the Reservoirs Act 1975 (Second edition)
Institution of Civil Engineers/Defra. ISBN 978-0-7277-5769-2
Floods and Reservoir Safety (Fourth edition)
Institution of Civil Engineers/Defra. ISBN 978-0-7277-6006-7

www.icevirtuallibrary.com

A catalogue record for this book is available from the British Library

ISBN 978-0-7277-6034-0

© Thomas Telford Limited 2014

ICE Publishing is a division of Thomas Telford Ltd, a wholly-owned subsidiary of the Institution of Civil Engineers (ICE).

All rights, including translation, reserved. Except as permitted by the Copyright, Designs and Patents Act 1988, no part of this publication may be reproduced, stored in a retrieval system or transmitted in any form or by any means, electronic, mechanical, photocopying or otherwise, without the prior written permission of the Publisher, ICE Publishing, One Great George Street, Westminster, London SW1P 3AA.

This book is published on the understanding that the author is solely responsible for the statements made and opinions expressed in it and that its publication does not necessarily imply that such statements and/or opinions are or reflect the views or opinions of the publishers. Whilst every effort has been made to ensure that the statements made and the opinions expressed in this publication provide a safe and accurate guide, no liability or responsibility can be accepted in this respect by the author or publishers.

Whilst every reasonable effort has been undertaken by the author and the publisher to acknowledge copyright on material reproduced, if there has been an oversight please contact the publisher and we will endeavour to correct this upon a reprint.

Printed and bound in Great Britain by Ashford Colour Press, Gosport, Hants PO13 0FW

Preface

This book contains the proceedings of the 18th Biennial Conference of the British Dam Society, *Maintaining the Safety of our Dams and Reservoirs*, held at Queen's University, Belfast, in September 2014.

The 44 papers of these proceedings are grouped into seven chapters, although many papers cover more than one chapter topic area. The papers have been written by professionals who have investigated, designed, and overseen a wide variety of work on dams and reservoirs in the UK and around the world.

Recently introduced updates to legislation mean different regulations for each of the four countries that comprise the UK. Papers covering these changes and the variations across the UK form essential reading for dam engineers practising in these countries.

Throughout the papers there are references to the environmental framework within which dam engineers work, whether it be discontinuing reservoirs, raising retained water levels, installing additional spillway capacity, or carrying out repairs in the interest of safety.

Earth embankment dams, many over 100 years old, are common throughout the UK, and a number of papers address the geotechnical issues of maintaining and repairing such dams, including various means of locating and dealing with leakage. But leakage is just one of a number of potential causes of dam failure, and a chapter covers the analysis of risk and actions that can be taken to reduce such risks – risks not only of the escape of water but risks to the public from the water retained in the reservoir.

Many reservoirs rely on gates and other mechanical devices to function safely, and papers covering the repair and replacement of gates and valves show how such works can be installed in existing civil engineering structures.

Several papers refer to the use of modelling – both physical and mathematical – that has assisted in optimising designs for new works such as spillways, where increased capacity has been required on an existing dam while complying with all the associated site constraints.

Overseas dam construction is also covered, with papers on the construction of new hydropower dams in Vietnam and Georgia, and a comprehensive monitoring system for two Portuguese dams being described.

The conference included the presentation of the biennial Geoffrey Binnie Lecture by Alan Cooper OBE, entitled '*The Heritage of Dams in Northern Ireland*'. This 2014 lecture is published in the Society's technical journal *Dams and Reservoirs*.

Contents

1. Legislation and Environmental Challenges

Changes to the Reservoirs Act 1975 – the enforcement authority's perspective in England 3
R. I. LEWIS, A. DEAKIN AND S. RUNDLE

Implementation of the Reservoirs (Scotland) Act 2011 16
J. R. ASHWORTH AND H. V. H. THOMAS

The Reservoirs Bill for Northern Ireland 24
D. N. PORTER

Challenges of planning significant reservoir safety improvements within an historic estate 35
D. E. NEEVE, S. PORTER AND L. MARCHANT

How overcoming environmental challenges shaped the design of a new reservoir in North West Ireland 48
D. A. McKILLEN AND G. A. COOPER

Chingford Pond Restoration 59
T. R. WANNER AND H. COUTTS

2. Geotechnical Issues of Dam Construction and Maintenance

International Levee Handbook - new guidance and a UK approach to the use of Eurocodes 75
J. SIMM, L. CLARKE, R. SANDHAM AND A. PICKLES

The Nearly Perfect Nineteenth Century Embankment Dam 88
P. J. RIGBY, A. N. THOMPSON AND D. E. JONES

The successful rehabilitation of Heapey Embankment, Anglezarke Reservoir 101
C. D. PARKS, M. EDMONDSON, G. MULREID AND D. THOMSON

Ground Engineering aspects of the Grane Valley Reservoir Safety Improvements 113
D. THOMSON, H. TAYLOR AND M. EDDLESTON

Remedial Grouting at Shon Sheffrey Dam, Wales 125
D. A. BRUGGEMANN AND O. J. FRANCIS

Bransholme Lagoon: Problems in design and implementation of a sheet pile solution to remedial works 135
A. PETERS, S. A. PRYCE, A. K. HUGHES AND M. WHEELER

Skhalta Dam – Design of a hardfill dam founded on deep alluvium and lacustrine deposits 148
J. R. PAWSON AND E. RUSSELL

Didachara Dam – Site Constraints in Dam Hydraulic Design 161
P. J. HARVEY

3. Mechanical Components of Dams

Refurbishment of the Portora Sluice Gates, Enniskillen Northern Ireland 175
K. J. McCUSKER AND G. A. COOPER

Dunalastair Dam Floodgate Replacement 186
R. J. DIGBY AND M. A. NOBLE

Rudyard Reservoir (Staffordshire) Safety Related Works 195
D. M. WINDSOR AND M. COOMBS

Sliplining Bottom Draw-Offs 207
J. P. WALKER AND M. T. TIETAVAINEN

4. Risk Analysis and Reduction Measures

Improving serviceability through Portfolio Risk Assessment 221
O. J. CHESTERTON, I. M. HOPE, T. J. HILL AND R. L. GAULDIE

A Practical Application of UK Guidelines for the Public Acceptability of the Risk of Dam Failure 234
K. D. GARDINER AND C. BROWN

Emergency planning for mining waste facilities in England 243
M. CAMBRIDGE, T. J. HILL AND P. HARVEY

Quantitative risk assessment applied to sludge lagoon embankments 257
M. EDDLESTON, C. ROSE, E. GALLAGHER, I. HOPE AND P. SUGDEN

Setting standards for draw-down capability at Scottish Water's Reservoirs 271
R. MANN, J. MALIA AND S. LOCKETT

Public Safety at Dams – A Canadian Perspective 283
G. J. SAUNDERS

Enhancements in reservoir flood risk mapping: example application for Ulley 295
A. D. SMITH, C. A. GOFF AND M. PANZERI

5. Design and Construction Case Studies

Planning a new Water Resource Development - Cheddar Reservoir Two 309
P. KELHAM, R. GROSFILS, M. BROWN AND M. ATYEO

Construction challenges at a roller compacted concrete dam in Vietnam 322
A. M. KIRBY AND T. BLOWER

The design and construction of an enlargement scheme for Black Esk reservoir (Scotland) 336
J. C. ACKERS, D. A. GETHIN, G. KARUNARATNE, S. A. PRYCE, T. A. SCOTT, J. TUDHOPE AND M. WHEELER

Eller Beck Flood Storage Reservoir – the challenges of low impact flood storage design 348
P. BRINDED, R. GILBERT, P. KELHAM AND A. PETERS

Shon Sheffrey Reservoir – Labyrinth overflow and replacement of masonry spillway 361
R. J. TERRELL, D. M. PRISK AND J. C. ACKERS

The issues associated with the discontinuance of impounding reservoirs 369
G. PICKLES AND D. REBOLLO

Construction of a pre-cast concrete service reservoir using BIM 381
S. RUSSELL AND T. JACKSON

6. Monitoring and Incidents at Dams

Emergencies, monitoring and surveillance, and asset management – a new approach 391
A. K. HUGHES

The monitoring and performance of rock anchors at Seathwaite Tarn Reservoir, Cumbria 402
D. E. JONES AND C. D. PARKS

Liquefaction failure in a Derbyshire fluorspar tailings dam 414
M. CAMBRIDGE

Improving the overtopping resistance of existing flood detention reservoirs 426
J. D. GOSDEN, T. AMBLER AND A. P. COURTNADGE

The Rhymney Bridge Incident 438
A. K. HUGHES AND T. WILLIAMSON

Safe access to reservoirs in winter 450
A. W. D. ROSS

The monitoring systems of the dams of the Baixo Sabor Hydroelectric Development 457
G. MOURA AND D. S. MATOS

7. Repairs to Dams

Rehabilitation of an 800 year old masonry dam 469
A. J. BROWN AND A. J. ELDER

An investigation into the impact of a 50yr old discrepancy on the safety of Tittesworth Reservoir 480
O. J. CHESTERTON, I. M. HOPE, A. M. KIRBY AND J. R. CLAYDON

Planning for emergencies and rehabilitation to improve operational safety at Spelga Dam, NI 494
J. R. BRADSHAW, K. J. McCUSKER AND D. A. McKILLEN

Refurbishment of Woodburn Reservoirs Eduction Towers and Scour Pipework 504
G. BRIGGS, G. A. COOPER AND D. BELL

Design of a new grout curtain for Wimbleball Dam 517
J. G. PENMAN, M. J. PALMER, A. C. MORISON, D. K. MASON AND J. J. WELBANK

Author index 530

SECTION 1:
LEGISLATION AND ENVIRONMENTAL CHALLENGES

Maintaining the Safety of our Dams and Reservoirs
ISBN 978-0-7277-6034-0

ICE Publishing: All rights reserved
doi: 10.1680/mdam.60340.003

Changes to the Reservoirs Act 1975 – the enforcement authority's perspective in England

R I LEWIS, Environment Agency, Exeter, UK
A DEAKIN, Environment Agency, Sheffield, UK
S RUNDLE, Environment Agency, Exeter, UK

SYNOPSIS   The Environment Agency has been the enforcement authority for the Reservoirs Act 1975 ('the Act') in England since 2004, when it took over the role from 140 local authorities. The Act places a duty on the enforcement authorities to ensure that reservoir undertakers observe and comply with its requirements. In July 2013 the Act was amended by the commencement of the amendments in Schedule 4 of the Flood and Water Management Act 2010. This paper discusses the changing legal requirements and how they are being enforced in England.

INTRODUCTION
The purpose of the Reservoirs Act 1975 ('the Act') is to 'make provision against escapes of water from large raised reservoirs or from lakes or lochs artificially created or enlarged'. It built upon the Reservoir (Safety Provisions) Act 1930, which was enacted following a number of dam and reservoir failures in Britain in the 19th and early 20th centuries that resulted in loss of human life. The Act was amended by the Water Act 2003 and the Flood and Water Management Act 2010. The latter amendments were brought into effect in July 2013.

The Environment Agency took over the enforcement authority role from 140 local authorities in 2004, when the amendments brought in by the Water Act 2003 were given effect. The Act was amended again in April 2013, when Natural Resources Wales took over from the Environment Agency as the enforcement authority for Wales. The Act places a duty on the enforcement authorities to ensure that reservoir undertakers observe and comply with its requirements.

This paper discusses the amendments and some of the implications in England for reservoir undertakers, panel engineers and the enforcement authority. It assumes some prior knowledge of the Act and does not purport to explain the provisions that remain unaltered. Neither does it discuss details of the effects of any changes in Wales and Scotland, where the Act also applies, albeit with some differences.

REASONS FOR CHANGE
In summer 2007 there was widespread flooding in England. During the floods, Ulley dam near Rotherham came close to failure, leading to the closure of the M1 motorway and the evacuation of 1,000 people from downstream properties (Environment Agency, 2007). Subsequently, Sir Michael Pitt carried out a comprehensive review of the floods. The subsequent Pitt Report (2008) recommended a number of improvements to the Act, following discussions with the reservoir industry and based on operational experience over the 25 years since it was commenced, in order to make the provisions of the Act more risk-based and effective.

THE AMENDMENTS
Following consultation the subsequent amendments, some of which had been suggested by the Environment Agency (2009) in one of its biennial reports to Defra, were included in schedule 4 of the Flood and Water Management Act 2010. Much of schedule 4 was commenced between 28 and 30 July 2013, through four statutory instruments (2013). The main changes to the Act that were commenced in July 2013 are listed in Table 1 and briefly discussed in the following paragraphs.

Table 1. Main changes to the Reservoirs Act 1975 in force in England from July 2013

Section	Statutory Instrument 2013 No.	Description
A1	1590 regulation 4(1), 1677 regulation 3 and 1896 regulation 3	Definition of a large raised reservoir. Note that the change in qualifying capacity for a reservoir to come within the ambit of the Act remains at 25,000m^3 and has not yet been changed to 10,000m^3.
2(2B)	1677 regulation 7	Undertaker must register a large raised reservoir, unless it was already registered before 30 July 2013
2(2C-2E)	1677 regulations 4 and 5	Regulations about registration and requirements for undertaker to provide up to date information to the enforcement authority

Section	Statutory Instrument 2013 No.	Description
2A-2D		The Environment Agency shall make risk designations for registered large raised reservoirs
2E	1896 regulation 4	Undertakers may appeal to the First-tier Tribunal against confirmed high-risk designation notices
10	1896 regulation 7	Until its risk designation is confirmed, undertakers shall have any existing large raised reservoir inspected and carry recommended measures to be taken in the interests of safety as soon as practicable. Thereafter, amended section 10 only applies to designated high-risk reservoirs
10(1)		Undertakers shall have designated high-risk reservoirs inspected
10(2)	1896 regulation 6	Additional inspection requirements
10(3)(b)		Inspecting engineers of designated high-risk reservoirs shall include maintenance recommendations in their reports
10(3)(c)		Inspecting engineers of designated high-risk reservoirs shall include in their reports the period within which measures required in the interests of safety must be taken
10(3A)		If the inspecting engineer has not provided his report on a designated high-risk reservoir within 6 months of the inspection, the engineer must notify the enforcement authority and explain why
10(5A)		Undertakers must comply with maintenance recommendations for designated high-risk reservoirs

Section	Statutory Instrument 2013 No.	Description
10(6)		Undertakers must carry recommendations as to measures to be taken in the interests of safety into effect for designated high-risk reservoirs within the period specified in the inspection report
10(6A)		Inspecting engineers of designated high-risk reservoirs must include in inspection reports statements and explanations about safety measures recommended in previous reports and not yet taken
11	1590 regulation 4(2), 1677 regulation 9	Until its risk designation is confirmed, undertakers of every existing large raised reservoir shall keep records in the prescribed form. Thereafter, this requirement only applies to designated high-risk reservoirs.
12	1590 regulation 4(2)	Until its risk designation is confirmed and at all times when not under construction, undertakers of every existing large raised reservoir shall employ a supervising engineer to supervise the reservoir. Thereafter, this requirement only applies to designated high-risk reservoirs.
12(2A), 12(2B)		At least once every 12 months, supervising engineers of designated high-risk reservoirs must provide a written statement of steps taken to maintain the reservoir
12(6)		Supervising engineers of designated high-risk reservoirs may direct undertakers to carry out visual inspections
12(7)		Undertakers of designated high-risk reservoirs must notify supervising engineers of each visual inspection carried out and anything noticed in the course of it
12A(1A)		Definition of a flood plan
12AA		Flood plan requirements (when directed by the Secretary of State)

Section	Statutory Instrument 2013 No.	Description
13(1A)		Interim certificates for large raised reservoirs being discontinued
13(5)		The Environment Agency may serve notice on undertaker to appoint a discontinuance engineer
19A	1896 regulation 5	Undertakers may appeal to the First-tier Tribunal against enforcement notices
20(1)	1677 regulations 10, 11 and 12	New prescribed forms for certificates, reports and directions
20(4)(b)		Inspecting engineers to copy to the enforcement authority any inspection report or abandonment report, whether or not it includes measures to be taken in the interests of safety
20(4)(f)		Supervising engineers to copy to the enforcement authority any written statement on matters to be watched or steps taken to maintain the reservoir
20(4)(g)		Supervising engineers to copy to the enforcement authority any undertaker direction to carry out visual inspections
20A		The Minister may by regulations make provision for the assessment of the quality of reports and written statements
21A		The Environment Agency may serve notice on an undertaker to provide information
21B	1677 regulation 14	Undertakers must report incidents to the Environment Agency
22(A1)		Offence to fail to register a large raised reservoir or to comply with regulatory requirements
22(1)		Strict liability offences for failure to comply with the requirements of the Act or to comply with an enforcement notice or direction

Section	Statutory Instrument 2013 No.	Description
22C		Undertakers to pay the Environment Agency expenses reasonably incurred in consulting qualified civil engineers for times to be specified in enforcement notices

DEFINITION OF A LARGE RAISED RESERVOIR

A large raised reservoir is a large raised structure, raised lake or other area capable of storing water and artificially created or enlarged. The arguably superfluous words 'as such' in the original Act no longer apply to 'water'.

Mine and quarry tips, canals, inland navigations, road and railway embankments (unless artificially designed to store water) are expressly excluded from the ambit of the Act.

The issue of how siltation affects the escapable volume of water can only be considered at the time when a panel engineer issues a final certificate or a discontinuance certificate. Reservoir undertakers must calculate the capacity of their reservoir whenever a panel engineer issues a final certificate or discontinuance certificate. The method for calculating the capacity is defined in the new regulations as the maximum volume of water capable of being stored above reservoir bed level and between its toe (defined as the point where the downstream side of any structure or dam meets lowest natural ground level) and top water level. It excludes any allowance for silt that is judged by the panel engineer to be incapable of flowing out of the reservoir in the event of an uncontrolled release of water.

For the time being, the qualifying capacity for a reservoir to come within the ambit of the Act remains unchanged at 25,000m^3 above lowest surrounding ground level, as before.

REGISTRATION REQUIREMENTS

It is now the reservoir undertakers' responsibility to register their large raised reservoirs with the Environment Agency. Failure to do so is a criminal offence under the Act. There are three new pieces of information not previously required to be on the public register of large raised reservoirs: dam or reservoir top level; top water level and whether the reservoir is designated as high-risk.

Reservoir information already on the Environment Agency's English register before 30 July 2013 is deemed to have been provided as required. For new or altered reservoirs, the information must be registered within 28 days of the issue of the final certificate.

This is after and in addition to the issue of a section 21 notice of intention to construct, alter or bring back into use a large raised reservoir. Much of the prescribed information in a section 21 notice is the same as that in the list of registration requirements. By implication, the information in the original section 21 notice must be reviewed and supplemented as necessary when the final certificate is issued.

Undertakers must provide any changed or additional information to the Environment Agency within 28 days of the change or addition occurring. This includes:

- The appointment of a construction, supervising or inspecting engineer
- The cessation of an appointment of a construction or supervising engineer
- Any change of undertaker and the date it will take effect

RISK DESIGNATIONS

For all registered large raised reservoirs, as soon as reasonably practicable after registration, the Environment Agency must consider whether to designate them as 'high-risk' or 'not high-risk'. The term 'high-risk' is defined in the new section 2C of the Act. A large raised reservoir may be designated high-risk if the Environment Agency thinks that, in the event of an uncontrolled release of water from the reservoir, human life could be endangered.

The Environment Agency has the discretion to consider the probability of reservoir failure in its decision to designate. However, there are many varying factors which can influence the probability of failure, which will inevitably change over time. Therefore for designation purposes, until and unless a different approach is agreed, the Environment Agency assumes a conditional probability of failure of unity for each reservoir.

If the probability of failure at a certain time was considered to be low and a reservoir were to be designated as not high-risk, then the supervision and inspection requirements of the Act would no longer apply, and the condition of the reservoir could deteriorate in the absence of expert engineering surveillance. Therefore it is undesirable from a public safety point of view for the risk designation process to take into account the probability of failure, which would vary over time. The consequence of failure is the over-riding consideration.

The Environment Agency has published the methodology (2013) it uses to carry out the risk designation process. It expects to complete the initial risk designation process for the 1982 existing reservoirs on the English register by April 2015. The following considerations are taken into account:

- The proximity of infrastructure and residential, business and recreational areas downstream of the reservoir
- Advice contained in section 10 inspection reports
- Reservoir flood maps showing the potential impact of a dam or reservoir failure
- Local site knowledge, maps and photographs
- The recommendations of an All Reservoirs Panel Engineer

Undertakers have an opportunity to make representations to the Environment Agency against provisional high-risk designations and appeals to the First-tier Tribunal against confirmed high-risk designations for their reservoirs. Both the provisional high-risk designation and confirmed high-risk designations are made by sending a notice to the relevant undertaker. Where the Environment Agency has assessed a reservoir as not high-risk, the undertaker is informed of this decision by letter. Only the undertaker can make representations after receiving a notice of provisional high-risk designation.

Risk designations must be reviewed if the Environment Agency thinks it appropriate. Any party, including members of the general public, can ask for a review at any time.

Under the transitional provisions in the new regulations, until the Environment Agency has notified the undertakers of the confirmed risk designation for their reservoir, undertakers of existing large raised reservoirs that were in operation on 30 July 2013 must continue to keep the prescribed records, have the reservoir supervised and inspected and carry into effect safety measures as soon as practicable, as before.

Once the Environment Agency has designated a reservoir as not high-risk, the supervision and inspection requirements of the Act will no longer apply to that reservoir. However, not high-risk large raised reservoirs will remain on the public register and the construction, discontinuance and incident reporting provisions of the Act will continue to apply to them. It is important to remember that the common law liabilities of the undertaker will continue, even if a reservoir is assessed as not high-risk. It therefore may be prudent to continue to manage the reservoir as if it were still under the Act.

INSPECTIONS
Inspecting engineers must provide their inspection reports for designated high-risk reservoirs within six months of the date of completion of the inspection, or else provide a written statement to the Environment Agency of the reasons why the report has not been provided.

Inspecting engineers must include in their inspection reports for designated high-risk reservoirs both the period for taking any recommended safety measures and a status update on any previously recommended safety measures from the previous inspection report. They must also include any maintenance recommendations in the reports. These maintenance recommendations will be monitored by the supervising engineer and have the force of law.

For designated high-risk reservoirs, the requirement for undertakers to carry recommended measures to be taken in the interests of safety into effect as soon as practicable has been replaced by the requirement to carry recommended measures into effect within the period specified in the inspection report. All reports must contain dates by which the measures must be completed. Any report not containing such dates can be viewed as not complying with the provisions of the Act and would therefore be invalid.

Undertakers if aggrieved by any safety measure or maintenance recommendation can refer their complaint to a referee.

SUPERVISION

At least once in every twelve month period, the supervising engineer must provide a written statement of any steps taken to maintain a designated high-risk reservoir in accordance with the recommendations of the latest inspection report. By contrast, the required frequency of the statement on matters to be watched remains unchanged at not less often than once a year, which is often interpreted to mean once in every calendar year.

The supervising engineer may direct undertakers to carry out visual inspections of a designated high-risk reservoir. The undertaker must notify the supervising engineer of each visual inspection and anything noticed in the course of it.

FLOOD PLANS

A flood plan is defined in the amended Act as a document specifying the action undertakers would take to prevent, control or mitigate the effects of a reservoir flood, and information about the areas that may be flooded as a consequence of a reservoir flood.

The Secretary of State has not yet directed any undertakers to prepare flood plans for their reservoirs. However, many undertakers laudably have already prepared such plans either voluntarily or on the recommendation of an inspecting engineer, as a measure to be taken in the interests of safety.

DISCONTINUANCE

A panel engineer supervising the discontinuance of a large raised reservoir to take it out of the Act may issue an interim certificate, requiring the water level in the reservoir to be reduced before it is discontinued. The Environment Agency may serve an enforcement notice on an undertaker requiring them to appoint a panel engineer to supervise discontinuance, if it appears that such an engineer has not been employed as required.

APPEALS

Undertakers have a right of appeal to the First-tier Tribunal against a confirmed high-risk designation notice, an enforcement notice requiring them to appoint a construction, inspection, supervising or discontinuance engineer, or an enforcement notice requiring them to carry into effect measures to be taken in the interests of safety.

CERTIFICATES AND REPORTS

There are new prescribed forms for certificates, inspection reports, supervising engineers' directions, section 11 records and section 21 notices.

Inspecting engineers must send a copy to the enforcement authority of any report they have made, whether or not it includes a recommendation as to measures to be taken in the interests of safety. Previously, only inspection reports with measures to be taken in the interests of safety were required to be copied to the enforcement authority.

Supervising engineers must send a copy to the enforcement authority of any written statement they give the undertakers on matters to be watched or steps taken to maintain the reservoir, and any direction given to the undertakers to carry out a visual inspection.

POWER TO REQUIRE INFORMATION

For the purposes of carrying out its functions under the Act, the Environment Agency can serve notice on an undertaker to provide information. It is a new criminal offence not to provide such information within the time specified in the notice, which must be at least 28 days after the notice is issued.

INCIDENT REPORTS

It is now a legal requirement for the undertakers of all large raised reservoirs (whether or not they have been designated high-risk) to report any 'incident' at their reservoir to the Environment Agency. Up until July 2013, such incidents were only reported on a voluntary basis.

A reportable incident is defined in the regulations as an event which results, or could result, in the uncontrolled release of water from a large raised reservoir, and in respect of which emergency measures have been taken, to prevent an uncontrolled release of water and to minimise the danger to human life.

The undertaker must send the Environment Agency a preliminary report of the incident as soon as practicable after it occurs. A final report must follow within a year of the incident. The template (available on gov.uk at https://brand.environment-agency.gov.uk/mb/C8XPiB) previously available to report voluntary incidents to the Environment Agency can continue to be used as before for this purpose.

It is now a criminal offence for an undertaker not to report such incidents in the prescribed manner to the Environment Agency.

CRIMINAL LIABILITY
The words 'by the wilful default of the undertakers' and 'unless there is reasonable excuse for the default or failure' have been deleted from the Act. Hence, failure to comply with the requirements of the Act and failure to comply with an enforcement notice served by the Environment Agency have both become strict liability offences, similar to pollution offences. This means that if there is non-compliance with a requirement of the Act a criminal offence has been committed, whatever the knowledge or intention of the offender may have been.

The same sections of the Act as before can be enforced by criminal sanctions. In addition, there are new offences in relation to failure to register new or changed information about any large raised reservoir. For designated high-risk reservoirs, further new offences are failure to comply with maintenance recommendations in section 10 inspection reports and failure without reasonable excuse to comply with the directions of a supervising engineer.

As already mentioned, there are new offences of failure to comply with a notice to provide information to the Environment Agency and failure to report an incident at a large raised reservoir.

The Environment Agency views enforcement as any action it takes where it suspects that an offence has occurred, or in some cases is about to occur. These enforcement actions may range from providing advice and guidance, serving enforcement notices, formal warning letters and formal cautions through to prosecution, or any combination that best achieves the desired outcome.

The Environment Agency follows the Regulators' Compliance Code (2007) of better regulation, which aims to ensure that regulation is risk-based, proportionate and targeted. Under the Code, regulators should ensure that their sanctions and penalties policies are consistent with the principles set out in the Macrory Review (2006). This means that sanctions and penalties should:
- aim to change the behaviour of the offender

- aim to eliminate any financial gain or benefit from non-compliance
- be responsive and consider what is appropriate for the particular offender
- be proportionate to the nature of the offence and the harm caused
- aim to restore the harm caused by regulatory non-compliance, where appropriate
- aim to deter future non-compliance

Underlying the Environment Agency's commitment to firm but fair regulation are the principles of proportionality, consistency of approach, transparency and accountability for any enforcement action it may take.

FURTHER CHANGES NOT YET COMMENCED
At the time of writing, the following changes to the Act in schedule 4 of the Flood and Water Management Act 2010 have not yet been commenced:

- the reduction in the threshold capacity for a reservoir to come within the Act from 25,000m³ to 10,000m³ (or some other volume substituted by the Minister)
- the consideration of smaller reservoirs in cascade as a criteria for them to come within the ambit of the Act
- provision for the Environment Agency to charge reservoir undertakers as a means of recovering costs incurred in performing its functions under the Act

CONCLUSIONS
The views expressed in this paper are the legal interpretations and opinions of the authors and do not purport to represent Environment Agency policy or have the force of law. There is little case law for the Act. Legal interpretations and case law precedents will develop over the coming years, arising from the variety of operational situations that may arise in seeking to correctly and accurately apply the spirit and letter of the law.

REFERENCES
BERR (2007). *Regulators' Compliance Code.* Better Regulation Executive, London. http://www.berr.gov.uk.

Environment Agency (2007). *Reservoir safety – learning from Ulley*

(2007 Summer floods case study). http://www.environment-agency.gov.uk/static/documents/Research/reservoirscasestudy_1917484.pdf

Environment Agency (2009). *Biennial report on reservoir safety 1 April 2007 to 31 March 2009.*

https://publications.environment-agency.gov.uk/skeleton/publications

Environment Agency (2013). *Reservoir risk designation guidance.*

https://publications.environment-agency.gov.uk/skeleton/publications

HMSO (1975). *Reservoirs Act 1975.* HMSO, London.

HMSO (2003). *Water Act 2003.* HMSO, London.

HMSO (2010). *Flood and Water Management Act 2010.* HMSO, London.

HMSO (2013). *The Natural Resources Body for Wales (Functions) Order 2013.* SI 2013 No. 755 (W.90), HMSO, London.

HMSO (2013). *The Flood and Water Management Act 2010 (Commencement No. 2, Transitional and Savings Provisions)(England) Order.* SI 2013 No. 1590 (C. 64), HMSO, London.

HMSO (2013). *The Reservoirs Act 1975 (Referees)(Appointment, Procedure and Costs)(England) Rules.* SI 2013 No. 1676, HMSO, London.

HMSO (2013). *The Reservoirs Act 1975 (Capacity, Registration, Presecribed Forms, etc.) (England) Regulations.* SI 2013 No. 1677, HMSO, London.

HMSO (2013). *The Reservoirs Act 1975 (Exemptions, Appeals and Inspections)(England) Regulations.* SI 2013 No. 1896, HMSO, London.

Macrory, R.B. (2006) *Regulatory Justice: Making sanctions effective.* Better Regulation Executive, London. http://www.bis.gov.uk/files/file44593.pdf

Pitt, M (2008). *The Pitt Review - learning lessons from the 2007 floods.* Cabinet Office, London.

Maintaining the Safety of our Dams and Reservoirs
ISBN 978-0-7277-6034-0

ICE Publishing: All rights reserved
doi: 10.1680/mdam.60340.016

Implementation of the Reservoirs (Scotland) Act 2011

J R ASHWORTH, Scottish Environment Protection Agency (SEPA), Stirling, UK
H V H THOMAS, Scottish Environment Protection Agency (SEPA), Edinburgh, UK

SYNOPSIS   The regulatory regime for reservoir safety in Scotland is undergoing major change through introduction of the Reservoirs (Scotland) Act 2011. Following phased implementation of the new legislation, due to begin in 2015, the Scottish Environment Protection Agency (SEPA) will take over responsibility for regulation and enforcement of reservoir safety from the 32 Scottish local authorities. This paper describes the main legislative changes taking place and the benefits to the reservoir industry from the new regulatory regime. The paper explains the new risk designation process being introduced and describes SEPA's indicative information requirements for reservoir registration. When the new legislation is fully implemented, the registration of smaller reservoirs with a capacity of between 10,000m^3 and 25,000m^3 above the natural level of the surrounding land will become a requirement for the first time and will present major challenges for SEPA. Successful implementation of the new legislation will rely on early and effective communication with reservoir owners regarding their reservoir safety responsibilities and associated regulatory requirements.

INTRODUCTION
The regulatory landscape for reservoir safety in Scotland is changing with the phased implementation of new legislation. The existing regime under the Reservoirs Act 1975 was in need of modernisation. As a result, the Scottish Government has introduced new, risk-based legislation, namely the Reservoirs (Scotland) Act 2011, parts of which are due to be commenced in 2015.

LEGISLATIVE CHANGES
The 1975 Act is currently still in effect in Scotland and is enforced by 32 different local authorities. The existing legislation requires registration of all 'large raised reservoirs' designed to hold, or capable of holding, more than 25,000m^3 of water above the natural level of the adjoining land.

These reservoirs must comply with the existing legislation which requires the appointment of engineers to undertake monitoring, carry out inspections and produce reports detailing works that must be undertaken by the reservoir undertaker to ensure the continued structural integrity of the reservoir.

The Reservoirs (Scotland) Act 2011 provides a new regulatory framework which will be supported by a significant amount of secondary legislation in the form of regulations. The first regulations are due to be introduced in 2015 and these will require reservoirs regulated under the 1975 Act to be registered with the Scottish Environment Protection Agency (SEPA).

The transitional programme has commenced with key steps in this process already underway. However, until the transitional programme is fully implemented, the regulation of reservoir safety in Scotland will remain with the local authorities under the Reservoirs Act 1975. Once the transitional period is complete, SEPA will become the single regulatory authority for reservoir safety in Scotland.

Summary of key legislative changes and benefits
Through implementation of the new Act, a number of key changes and benefits to the reservoir industry will be introduced. Table 1 summarises the key differences between the existing and new reservoirs legislation and the changes to be brought into force in Scotland.

The 2011 Act is an example of risk based legislation which provides a proportionate, flexible and targeted approach to improving reservoir safety in Scotland. Other benefits include improved consistency with complete and accurate records on controlled reservoirs held by one regulatory authority (SEPA). This will also allow an improved customer interface providing reservoir managers with one, accessible point of contact.

Improved consistency of reporting and recording will also be achieved as copies of all supervising engineers' annual statements and inspection reports from inspecting engineers must be provided to SEPA. There is a further requirement for an enhanced public register which will include copies of safety reports, inspecting engineer inspection reports, supervising engineer written statements and a reservoir inundation map.

The 2011 Act, when fully implemented, will reduce the volumetric threshold of reservoirs which fall under regulation from >25,000m^3 to 10,000m^3. Furthermore, calculation of this reduced threshold will include the combination of smaller reservoirs (individually under 10,000m^3) between which water could flow and which could therefore give rise to an uncontrolled release of 10,000m^3 or more of water from the combination (i.e. cascade systems).

Table 1. Key differences between the Reservoirs Act 1975 and the Reservoirs (Scotland) Act 2011.

Topic	Reservoirs Act 1975	Reservoirs (Scotland) Act 2011
Registration Threshold	>25,000m^3	≥10,000m^3
Regulatory Authority	32 Scottish Local Authorities	Single Body - SEPA
Risk Designation	No accounting for risk	Risk designation (High, Medium or Low) assigned to all 'controlled reservoirs'
Inspecting Engineer	Required at all sites. No time limit for reports to be sent to reservoir manager. Reports only to be sent to regulatory authority when containing 'measures in the interest of safety'	Required at all High risk sites. Only required at Medium risk sites where requested by the Supervising Engineer. Reports to be sent to reservoir manager within 9 months. Copies of all reports to be sent to SEPA within 28 days
Supervising Engineer	Required at all sites. Supervising Engineer statements not required to be sent to regulatory authority	Only required at High & Medium risk sites. Supervising Engineer written statements sent to reservoir manager at least every 12 months; copied to SEPA within 28 days
Monitoring	Reservoir manager required to keep the 'Prescribed Record' incl. recording of such matters (monitoring) as directed	Visual inspection by reservoir manager at intervals specified by the Supervising Engineer and keeping of written record
Inundation Maps	Not required	Required
Flood (On-site) Plans	Not required	Contains options for mandatory Flood Plans
Incident Reporting	Not required but undertaken on voluntary basis	Contains options for incident reporting as statutory requirement
Public Register	Yes - limited	Yes – expanded requirements incl. risk designation, inundation map, inspection reports, certificates & statements
Enforcement Powers	Contains option for criminal sanctions only	Options for criminal & civil* sanctions (* via related legislation)

SEPA will also be required to assign a risk designation to each controlled reservoir in which Medium and Low risk reservoirs will be subject to less regulation than those deemed as High risk.

THE RISK DESIGNATION PROCESS

The 2011 Act requires SEPA to assign a risk designation of High, Medium or Low to each controlled reservoir. The risk designation will take into account the reservoir inundation map which will show the areas of land likely to be flooded by an uncontrolled release of water from each dam associated with a reservoir. These maps will be used to determine the impact of the water on a number of receptors as required by the 2011 Act:

- Human health
- The environment
- Cultural heritage
- Medical facilities
- Power supplies
- Transport
- The supply of water for consumption
- Other social or economic interests
- Such other potential damage as SEPA considers relevant

These receptors are aligned to those considered in SEPA's National Flood Risk Assessment (NFRA), first published in December 2011 as a requirement of the Flood Risk Management (Scotland) Act 2009 (FRM Act). In future revisions of the NFRA, reservoirs information can be included, providing a comprehensive picture of flood risk in Scotland.

For the purposes of the FRM Act and the 2011 Act, SEPA has grouped the risk receptors into seven categories:

- Human health (A) - People
- Human health (B) - Community
- Economic activity (A) – Business
- Economic activity (B) – Transport
- Economic activity (C) – Agriculture
- Environment
- Cultural Heritage

Within each of these receptor groups sits a suite of indicators which will be assessed for potential impact. This assessment will take into account the reservoir inundation maps that SEPA has now produced for each reservoir. To date, these have only been produced for reservoirs covered by the 1975 Act.

Following the assessment of potential impacts, SEPA will be able to assign a provisional risk designation. Risk designations must be reviewed by SEPA at any time it considers the designation may have ceased to be appropriate or by the end of a period of six years.

Reservoir managers may also make a representation to SEPA within a statutory timescale of two months, should they disagree with the provisional risk designation assigned to their reservoir.

Guidance on the risk designation process will be published by SEPA before it takes on its regulatory function.

Impact of risk designation
The risk designation assigned to a reservoir will inform the level of statutory monitoring and inspection required. The impact of the risk designation, relative to current requirements under the 1975 Act is summarised in Table 2.

Table 2. Impact of Risk Designation on regulatory requirements

Reservoirs Act 1975	Reservoirs (Scotland) Act 2011	
All sites are required to appoint a Supervising Engineer. All sites required to appoint an Inspecting Engineer at least once every 10 years or when recommended in an inspection report or by a Supervising Engineer.	High risk	All High risk sites are required to appoint a Supervising Engineer.
		All High risk sites required to appoint an Inspecting Engineer at least once every 10 years or more frequently when recommended in an inspection report or when requested by a Supervising Engineer.
	Medium risk	Required to appoint a Supervising Engineer at all times. Only required to appoint an Inspecting Engineer when recommended by Supervising Engineer.
	Low risk	No statutory requirement to appoint either a Supervising or Inspecting Engineer

If a reservoir is considered 'High risk', the reservoir manager will be required to appoint a supervising engineer to supervise their site in accordance with the legislation. They will also be required to appoint an inspecting engineer at any time recommended by the supervising engineer, in an inspection report or before the end of the period of 10 years beginning with the date of the latest inspection.

For sites that are 'Medium risk', the reservoir manager will again be required to appoint a supervising engineer to supervise their site in accordance with the legislation. With regard to inspecting engineers, the reservoir manager will only be required to appoint one when recommended by the supervising engineer. There will be no mandatory requirement for a statutory 10 year inspection.

Low risk sites are not required to have a supervising engineer appointed to supervise them nor an inspecting engineer to undertake inspections.

THE REGISTRATION PROCESS

Through the introduction of new regulations, SEPA expects to begin, in 2015, the registration of existing sites which currently fall under the Reservoirs Act 1975. Relevant sections of the 1975 Act as they apply to Scotland will therefore be repealed.

For existing sites registered under the 1975 Act, SEPA is already in possession of the information which was needed in order to produce the inundation maps for use in the risk designation process.

Prior to registration, SEPA plans to make available to the reservoir manager the existing information already held for a particular site. As part of the registration process, SEPA will seek confirmation from the reservoir manager that the information held is still valid or to indicate any changes to the data as required.

As previously mentioned, the inundation maps for existing (1975 Act) sites have already been produced. This amounted to some 750 maps covering 665 reservoirs including sites where multiple dams or dam breach scenarios were modelled. Following completion of the registration phase, the next stage will be to assign a risk designation to each reservoir.

For any newly constructed or altered sites which fall under regulation, a reservoir manager will need to provide certain key information and will also be required to submit an inundation map produced in accordance with SEPA's methodology.

Table 3 provides an indication of the basic information requirements for registration of existing and new sites. For new sites, this information is needed to enable SEPA to assess the inundation map submitted at registration. Note however that for new sites some of this information, such as supervising and inspecting engineer details, will not be available at the point of registration but will be required to be provided subsequently by the reservoir manager.

Table 3. Indicative information requirements for reservoir registration (covering existing 1975 Act and/or new sites)

Reservoir Manager/Owner	Name, address and contact details
Construction Engineer	Name, address and contact details
Supervising & Inspecting Engineer	Name, address, contact details and date of appointment
Inspections	Date of last/next Section 10 inspection(s), measures in the interest of safety
Reservoir and Dam	
Name and NGR of reservoir/dam (Year Dam Construction Completed)	Status of reservoir (In operation, abandoned)
Reservoir Type (Impounding / non-impounding / service reservoir)	Primary purpose of stored water
Surface water area at Top Water Level (m²)	Surface Water Area at Dam Crest Level (m²) (if known)
Top Water Level (mAOD)	Escapable volume of water at top water level or dam crest level (m3)
Cubic capacity of reservoir at Top Water Level (m³)	Cubic capacity of reservoir at Dam Crest Level (m³) (if known)
Dam Type (arch /gravity / buttress / embankment /other)	Principal material used in construction of dam (concrete / earth / masonry / rockfill / other)
Maximum height of dam (m)	Bottom level of dam at ground level (mAOD)
Dam Crest Level (mAOD) and Crest Length (m)	Fetch length and direction

FUTURE IMPLEMENTATION PHASE

When the Reservoirs (Scotland) Act 2011 is fully implemented, reservoirs between 10,000m³ and 25,000m³ capacity (including 'cascades' as referred to previously) will also require to be registered with SEPA.

It is estimated that approximately 800-900 existing reservoirs in Scotland will fall under regulation for the first time due to the reduction in registration threshold.

Identifying the owners of these smaller sites, facilitating the submission of essential data used in inundation mapping and managing the registration process will be major challenges for SEPA.

It is expected that many of the smaller sites will be owned or operated by individuals, small organisations or associations with little understanding and experience of reservoir safety or how the 2011 Act will impact on them.

SEPA will engage with the reservoir industry and those newly affected by legislation through ongoing liaison, in order to raise awareness of the responsibilities reservoir managers have under the 2011 Act.

Early and effective communication regarding the new regulatory requirements will be key to successful implementation. For example, SEPA submitted an article to the online Journal of the Law Society of Scotland earlier this year as part of the strategy for raising awareness amongst existing reservoir owners or potential future owners.

Finally, SEPA will help to support the reservoir industry through a suite of guidance documents that offer advice including good practice on how to fulfil the requirements of the legislation.

REFERENCES

1. Ashworth, J.R. (2014). Reservoir safety regulation: a changing landscape *The Journal of the Law Society of Scotland, March 2014*

www.journalonline.co.uk/Magazine/59-3/1013717.aspx

2. Reservoirs (Scotland) Act 2011, HMSO

3. Reservoirs Act 1975, HMSO

Maintaining the Safety of our Dams and Reservoirs
ISBN 978-0-7277-6034-0

ICE Publishing: All rights reserved
doi: 10.1680/mdam.60340.024

The Reservoirs Bill for Northern Ireland

D N PORTER, Rivers Agency, Belfast, Northern Ireland

SYNOPSIS   The 1975 Reservoir Act[i] does not extend to Northern Ireland (NI) and there is currently no regulation of reservoir safety, which means that there is no requirement for routine inspection or maintenance of the 151 reservoirs identified as holding 10,000m^3 or more of water. In early 2011 DARD Rivers Agency, the flood risk and drainage authority for Northern Ireland, established a Bill Team to take forward reservoir safety policy development. The draft legislation was approved by the NI Executive in November 2013 and was subsequently introduced in the NI Assembly in January 2014 as the 'The Reservoirs Bill'. The Bill is currently being scrutinised by the Agriculture and Rural Development Committee and is expected to be on the statute book towards the end of 2014, with commencement in early 2015.

BACKGROUND
The vulnerability of impoundments has been dramatically demonstrated throughout their history and the need for regulation of these structures within Great Britain was identified by the jury at the inquest into the Dale Dyke Dam failure, which occurred 150 years ago, in 1864. They concluded that *'the Legislature ought to take such action as would result in frequent, regular, and sufficient inspection of all reservoirs of that character'*[ii]. Even though Hansard recorded that *'265 lives were sacrificed, property was destroyed to the amount of about £1,000,000, and 20,000 persons were reduced to destitution'*[iii], no legislation was brought forward at that time. Over sixty years passed and again the issue of dam safety was brought to the attention of the British public as there were two catastrophic structural failures which occurred within a short space of time in 1925. A cascade failure, involving two dams, *'caused a flood that swamped the village of Dolgarrog in North Wales, killing 16 people'*[iv], and a dam failure in Skelmorlie, North Ayrshire, Scotland killed 5 people[v]. These failures led to the introduction of the Reservoirs (Safety Provisions) Act in 1930[vi], which ensured that reservoirs with a capacity in excess of five million gallons (about 22,700m^3) of water *'would be inspected at least every 10 years by a qualified engineer (Inspecting Engineer).'*[vii]

Reservoir incidents involving loss of life across Europe triggered a review of reservoir safety in Great Britain, which resulted in the Reservoirs Act 1975. This updated the 1930s legislation and introduced a requirement to appoint a supervising engineer for reservoirs with a capacity greater than 25,000m³ of water above the natural ground level. The 1975 Act has subsequently been amended by various pieces of legislation but it still provides the legislative basis for reservoir safety in England, Scotland and Wales. None of the Great Britain legislation applies in Northern Ireland, as Section 30 of the Reservoir Act 1975 is very clear, saying '*this Act shall not extend to Northern Ireland*'[viii].

There are two references in the Northern Ireland Statute book to reservoir safety. One is contained in Article 33 of the Drainage (Northern Ireland) Order 1973[ix]. However this legislation can only be used for '*the purpose of preventing or arresting injury to land*' and it cannot be used '*in relation to any dam or sluice which is vested in or controlled by any other government department, any harbour authority, any district council or the Northern Ireland Electricity Service*'[x]. The other reference to reservoir safety is in Article 297 of the Water and Sewerage Services (Northern Ireland) Order 2006[xi], which repealed an earlier version of this legislation, dating from 1973. This legislation enables the making of '*regulations with respect to the construction, inspection, maintenance and repair of reservoirs and dams*'[xii], however, no such regulations have ever been brought forward. The Health and Safety at Work (Northern Ireland) Order 1978[xiii] may apply to those impoundments which are part of a workplace but there is no explicit reference in this legislation to reservoirs.

There is, therefore, no regulation of reservoir safety in Northern Ireland, which means that it is left to the discretion of owners and operators under common law (that is law developed by judges as the outcome of legal cases). This was established in the Ryland v Fletcher decision by the House of Lords in 1868 when Justice Blackburn stated: "*we think that the true rule of law is, that the person who for his own purposes brings on his lands and collects and keeps there anything likely to do mischief if it escapes, must keep it in at his peril, and, if he does not do so, is prima facie answerable for all the damage which is a natural consequence of its escape.*"[xiv]

Anecdotal evidence shows that this common law liability is not well understood among reservoir owners or operators, as many are choosing not to carry out inspections or maintenance to their structures. This clearly is a gap which cannot be allowed to persist and it was highlighted in the Government response to the independent flood management policy review entitled '*Living with Rivers and the Sea*'[xv], published in September 2008. One of the recommendations in this document was that '*appropriate legislation will be proposed to provide for regulatory control of reservoir safety in Northern Ireland by Rivers Agency*'.

DARD Rivers Agency is the competent authority for implementation of the EU Floods Directive[xvi] and the Preliminary Flood Risk Assessment (PFRA)[xvii], completed in December 2010, identified reservoir failure as a potentially significant flood source within Northern Ireland. The lack of regulation means that the likelihood of dam failure is impossible to determine, as there is no overall view of the condition of the reservoir stock. Rivers Agency, therefore, began the process of understanding this risk by first quantifying the potential impacts of total dam failure. Figure 1 shows an example of the dam breach inundation mapping produced. The PFRA identified that approximately 66,000 people who reside in reservoir breach inundation areas would be impacted if a total failure of the 156 reservoirs were to occur. While it is accepted that this is a hugely unrealistic situation it was the only way of quantifying the potential hazard, given that no condition data was available. In the absence of reservoir regulation and, with the information on the potential impact of failure, the NI Executive agreed in late 2010 that the Minister for Agriculture and Rural Development should bring forward reservoir safety legislation.

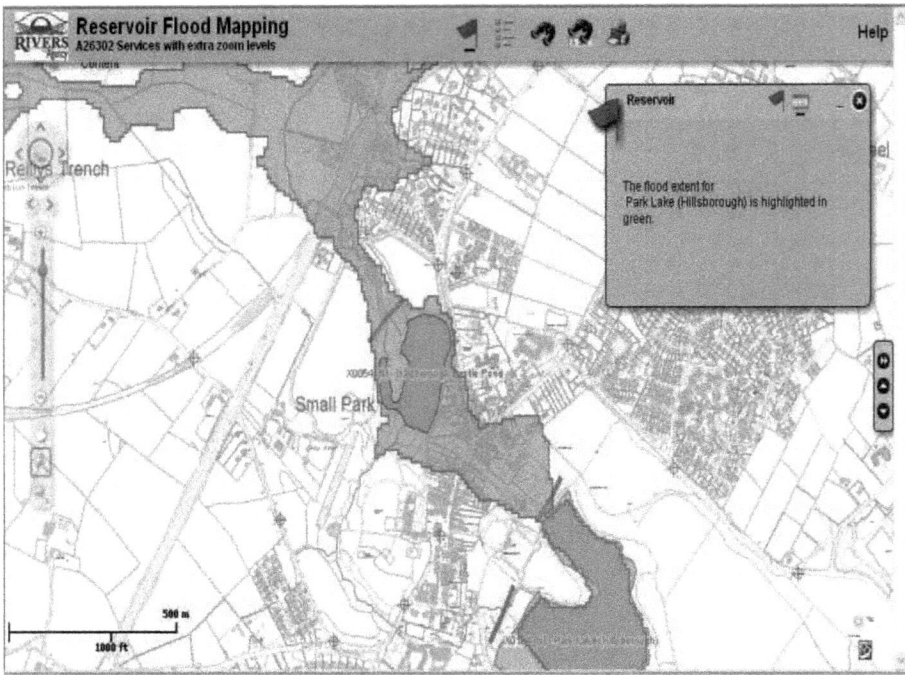

Figure 1. Dam breach inundation map. © Crown Copyright

The 156 figure has subsequently been reduced to 151, as the definition of a reservoir has been refined as part of the legislative process. Forty five of the 151 reservoirs identified are the responsibility of Northern Ireland Water, the body which delivers public water and sewerage services. Although not regulated, these structures are generally maintained within the spirit of the 1975 Act. Further detail on the major upgrade programme of these impoundment structures is available in Alan Cooper's 1987 paper[xviii]. Table 1 details the ownership of all the reservoirs, based on current information.

Table 1. Reservoir ownership in Northern Ireland

Impounding Reservoirs	
Public Sector	77
Third Sector (Not for profit organisations)	9
Private company or individual	59
Unknown	6
Service Reservoirs	
Public Sector	40

While there have been no recorded deaths caused by reservoir failure in Northern Ireland, there have been a number of dam incidents. The earliest uncovered by this research occurred on the 19 March 1822 at Clea Lake, Keady, Co Armagh. HD Gribbon records in his book that *'following a cloudburst, the dam gave way and a flood swept down the valley damaging mills and water wheels nearly as far as Armagh.'* [xix]

In December 1876 a newspaper reported that Wilson's Dam in Carrickfergus *'as usual after heavy rain overflowed its banks'*, flooding many houses and injuring property.

In September 1902, a dam burst on the Springfield Road in Belfast and discharged into the nearby Blackstaff River, causing it to burst its banks. It was reported that due to this uncontrolled release of water, a nearby street flooded to a depth in excess of four metres.

On the 4th/5th August 1971, 63.4 mm of rain fell in 78 minutes, a probable frequency return period of one in 150 or 200 years, causing the earth embankment of Creggan Upper Reservoir to fail due to overtopping. The impounded water was, however, stored in the middle reservoir, which happened to be drawn down at the time. The upper reservoir remains abandoned to this day.

In the 1980s, Church Dam in Hillsborough was overtopped. It was reported that the water from this reservoir caused erosion which exposed old coffins in the adjacent graveyard. The clay core of this structure was subsequently upgraded by sheet piling.

A breach in the bank of Wolfhill Millrace in 1995 resulted in a number of dwellings being flooded.

Flooding in the Doagh area in 1998 was attributed to the collapse of a spill weir at Tildarg Dam and resulted in a number of houses being flooded.

POLICY

The purpose of the policy is to create a legal and administrative framework for regulating reservoir safety. The key features of the policy proposals, which were developed through engagement and communication with key stakeholders, are:

- That all reservoirs with a capacity of 10,000m^3 or more would be regarded as controlled reservoirs and would be registered with the Reservoirs Authority;
- DARD would act as the Reservoir Authority and would be responsible for enforcing the provisions under the legislation;
- The reservoir manager is responsible for reservoir safety;
- A risk based approach should be introduced for all controlled reservoirs. Initially this would primarily be an impact designation, based on the consequence of reservoir failure;
- Each reservoir which falls within the scope of a controlled reservoir would be assigned a risk designation according to whether it poses a threat to human life, the environment, cultural heritage and economic activity;
- Each controlled reservoir would be subject to a proportionate supervision and inspection regime, depending on its risk designation;
- Construction and alteration of controlled reservoirs should be undertaken under the supervision of appropriately experienced and qualified engineers;
- Independent, qualified Civil Engineers, drawn from a panel appointed by the Department, would provide technical expertise when undertaking construction, supervision and inspection roles under the framework;
- There should be provisions for the review of risk designation decision, independent appeal, dispute resolution and enforcement, including offences and penalties; and
- There should be a number of miscellaneous provisions, including emergency powers and powers of entry.

The policy proposals were subject to a twelve week public consultation which finished on the 1 June 2012.

WHICH RESERVOIRS?

The proposal is to regulate 'controlled reservoirs', which is any structure or area capable of holding 10,000m³ or more of water above the natural level of any part of the surrounding land. This will include any structure or area designed or used for collecting and storing water, as well as a lake which was created or enlarged in a manner which raises water above the natural ground level. The term 'controlled reservoir' encompasses the reservoir basin and all apparatus, including its spillway, valves, value towers, pumping stations, pipes and sluices or any other thing which is integral to the functioning and operation of the reservoir. Service reservoirs and other similar structures or areas that meet the minimum 10,000m³ threshold are within the scope of the proposed legislation. Whilst these structures and areas are not impoundments, the proposals aim to reduce the risk of flooding resulting from structural failure.

It is also proposed that structures or areas that individually do not meet the volume threshold but water does or can flow between them, and where there could be an uncontrolled release of 10,000m³ or more of water as a result of the combined capacity, will be treated as controlled reservoirs. This is to ensure that cascades or reservoirs that are, or could become, interlinked in a breach situation and have the potential to cause a similar impact as individual larger reservoirs, are subject to control.

It is proposed that structures or areas are excluded from regulation if these are already subject to other legislation, for example ash, sludge, power station and mining lagoons, as these are regulated by the Mines Act (Northern Ireland) 1969[xx] as amended, and the Quarries (Northern Ireland) Order 1983[xxi].

The Department proposes to take the power to be able to treat, by regulation, any structure or area or combination of any structures or areas that do not meet the criteria for a controlled reservoir but that is assessed by the Department as posing a potential significant risk to human health, the environment, cultural heritage and economic activity as a controlled reservoir. There is no plan to implement this power at the early stages of commencement, but it is included within the legislation so that should a need arise in the future the primary legislation will not need to be amended.

WHO WILL BE RESPONSIBLE FOR RESERVOIR SAFETY?

The person with responsibility for reservoir safety will be known as the 'reservoir manager'. A reservoir manager is the person or organisation that manages or operates the reservoir and, by default, the owner of the reservoir will be the reservoir manager. The following principles would apply:-

- Where all, or part, of a controlled reservoir is managed or operated by a water or sewerage undertaker, then that organisation is the reservoir manager for the whole, or part.

- Any person who manages or operates a controlled reservoir in whole, or in part, will be the reservoir manager for the whole, or part.

- Where all, or part, of the controlled reservoir is not managed or operated by a water or sewerage undertaker or other person, then the owner of the controlled reservoir will be the reservoir manager for the whole, or part, of the reservoir.

It is proposed that reservoir managers will have a duty to co-operate where more than one person or organisation is identified as being responsible for a controlled reservoir. Recreational users of controlled reservoirs, such as fishing or sailing clubs, will not be responsible for reservoir safety unless they own or have management responsibility for the controlled reservoir.

IMPLEMENTATION OPTIONS
The following implementation options were considered during the policy development and where subject to public consultation:

Option 1: Do Nothing, Self Regulation
Under this option reservoir safety would continue to be left to the discretion of reservoir owners and managers. The legal basis for the safety of reservoirs under self regulation would remain common law or the Health and Safety at Work legislation, where it applies.

Option 2: Reservoir Licensing System
Under this option all reservoirs would be registered, usually for a fee. An inspection would be undertaken and then a formal license agreement that sets out the conditions to be met would be provided. Licenses would usually be issued for fixed period of time and would need to be renewed or periodically reviewed to update the license conditions and to reflect legal and policy changes.

Option 3: Panel Engineer System
The Panel Engineer System would introduce a management regime whereby the reservoir manager would commission qualified Civil Engineers to provide advice on the safety of reservoirs and to inspect, supervise and construct 'controlled reservoirs'. This would introduce a risk based approach to management, where the level of inspection and supervision would increase for high risk structures.

The preferred option following consideration of the public consultation responses is Option 3 and this was endorsed by the NI Executive in November 2012.

DESIGNATION

Risk is a function of impact of a given hazard or threat and the likelihood that it will occur. The likelihood of dam failure is generally considered to be low but it is difficult to reliably predict. It is determined by factors such as age, construction materials used, maintenance regime and the condition of structure or dam. The draft legislation includes a risk-based approach that accounts for both the consequence or impact of failure and its likelihood or probability. However, until there is an agreed methodology for determining likelihood, the strategic risk assessment of each reservoir will be primarily based on the impact of a reservoir breach, considering the four indicators within the EU Floods Directive of human health, the economy, the environment and cultural heritage.

Figure 2 illustrates the designation process, including reviews and appeals, the inspection regime and the 10 yearly reassessment of designation. Controlled reservoirs designated as high risk will be subject to higher levels of regulation than those designated as medium risk. Owners of low risk reservoirs will have to take very little action, as there will be no requirement to undertake any formal inspection of these structures. Table 2 sets out the designation criteria. This proportionate approach is necessary in order to ensure that the legislative requirements are not overly burdensome, particularly on private owners or third sector organisations. However, by the simple action of requiring reservoir structures to be subject to routine inspection and maintenance, the likelihood of failure is, as far as is reasonably practicable, managed; therefore the legislation is key in reducing this risk. It should be noted that the options considered for implementation do not fully follow the same processes as England, Wales and Scotland.

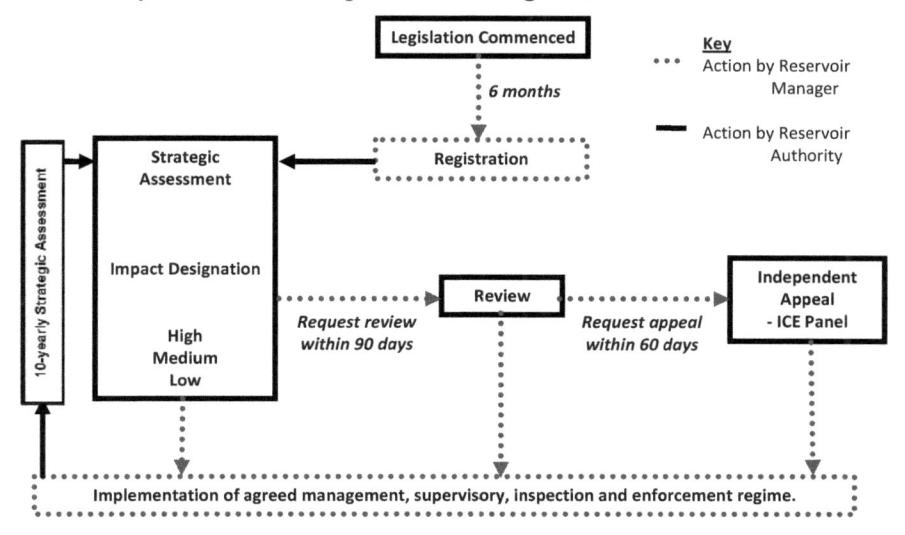

Figure 2. Flow chart of the designation process

Table 2. Designation criteria

HIGH	Where a reservoir failure could cause the loss of one or more life, or result in significant impact on economic activity.
MEDIUM	Where a reservoir failure could impact on people but where no loss of life can be reasonably foreseen, or where significant damage to the environment and cultural heritage could occur.
LOW	Where no loss of life can be reasonably foreseen as a result of a reservoir failure and limited flood damage would be caused.

ROLE OF SUPERVISING ENGINEER

High or medium risk controlled reservoirs must be under the supervision of a supervising engineer 'at all times'. A supervising engineer must be commissioned within six months of the risk designation taking effect and a reservoir manager is required to give notice to the Department within 28 days of the commissioning. A supervising engineer must be an appropriate panel engineer. Unlike other reservoir engineers under the Reservoirs Bill, there is nothing to prevent a supervising engineer from being an employee of the reservoir manager or having previously been a construction engineer or inspecting engineer in relation to the reservoir.

ROLE OF INSPECTING ENGINEER

Reservoir managers of high risk reservoirs are required to commission an inspecting engineer's report every ten years, unless one is recommended by the supervising engineer. Inspecting engineers also supervise the measures in the interest of safety as directed in an inspection report or recommended as a pre-commencement safety recommendation. The reservoir manager must notify the Department within 28 days of the commissioning of an inspecting engineer.

As inspecting engineers are intended to have a degree of independence from the reservoir manager there are criteria for their appointment. They will be disqualified from being commissioned if they are employed by the reservoir manager, or had previously been a construction engineer for the reservoir.

ROLE OF CONSTRUCTION ENGINEER

A construction engineer supervises the relevant works and the safety of the reservoir until a final certificate for the works is given to the Department. The construction engineer inspects the reservoir, designs the construction or alteration and prepares and provides the reservoir manager with safety reports.

LEGISLATIVE PROGRESS

The Reservoirs Bill was introduced in the NI Assembly on the 20 January 2014, and moved to Second Stage on the 4 February 2014. The Second

Stage provides the opportunity for the general principles of the Bill to be debated, however, the detail is not discussed, nor can amendments be introduced. Following the debate, the Assembly voted on the Bill, which then moved to the Committee Stage. The Agriculture and Rural Development Committee began scrutiny of Bill in February and intend to hear evidence from the Institution of Civil Engineers, NI Water, private owners, fishing clubs and the Department.

CONCLUSION

The necessity of a robust reservoir safety regime is beginning to be understood amongst those who own or operate these structures in Northern Ireland. The proposed legislation will enable assurance to be provided to the wider community that this risk of a reservoir failure is being properly managed.

The tried and tested approach of using independent qualified Civil Engineers to inspect and oversee the construction and maintenance of reservoirs is the preferred option for implementation in Northern Ireland. The approach to risk is, however, different from that adopted or proposed in other parts of the United Kingdom.

The current position in Northern Ireland is that reservoir safety is the responsibility of the owners or managers of reservoirs under common law. The new legislation will not change this fundamental principle but it will introduce a regulated system of proportionate management that should ensure structures are in reasonable condition and, therefore in the event of a failure, this may control or limit reservoir managers' liability.

ACKNOWLEDGEMENTS

The work described in this paper is the result of close cooperation between officials of DARD Rivers Agency and the Institution of Civil Engineers (ICE) Northern Ireland reservoir safety advisory panel. From Rivers Agency I would like to thank Kieran Brazier, Mary McKeown, Quinton Campbell and David Henderson for their input. The ICE panel consisted of William Gowdy, Alan Cooper, David McKillen, Stephen Orr, Donal Ryan and Richard Kirk.

REFERENCES

[i] http://www.legislation.gov.uk/ukpga/1975/23

[ii] http://hansard.millbanksystems.com/commons/1864/jul/11/papers-moved-for

[iii] http://hansard.millbanksystems.com/commons/1864/jul/11/papers-moved-for

[iv] http://www.britishdams.org/about_dams/safety.htm

[v] http://www.scotland.gov.uk/Publications/2010/01/22134722/6

[vi] http://www.legislation.gov.uk/all?title=Reservoirs%20%28Safety%20Provisions%29%20Act%2C%201930

[vii] Environment Agency (2010) *The Owner's Guide to Reservoir Safety*

[viii] http://www.legislation.gov.uk/ukpga/1975/23/section/30

[ix] http://www.legislation.gov.uk/nisi/1973/69/contents

[x] http://www.legislation.gov.uk/nisi/1973/69/article/33

[xi] http://www.legislation.gov.uk/nisi/2006/3336/contents

[xii] http://www.legislation.gov.uk/nisi/2006/3336/article/297

[xiii] http://www.legislation.gov.uk/nisi/1978/1039

[xiv] http://www.bailii.org/uk/cases/UKHL/1868/1.html

[xv] http://www.dardni.gov.uk/riversagency/living-with-rivers-and-the-sea.09.097_rivers_agency_-_living_with_rivers_and_the_sea_report_-_published_august_2008.pdf.pdf

[xvi] http://eur-lex.europa.eu/LexUriServ/LexUriServ.do?uri=OJ:L:2007:288:0027:0034:EN:PDF

[xvii] http://www.dardni.gov.uk/riversagency/index/eu-floods-directive/the_european_floods_directive_downloads/the_european_floods_directive_pfra_and_methodology_sfra.htm

[xviii] Cooper GA (1987), *The Reservoir Safety Programme in Northern Ireland*, IWES Summer Conference

[xix] Gribbon HD (1969), *The History of Water Power in Ulster*, Newton Abbot, David & Charles

[xx] http://www.legislation.gov.uk/apni/1969/6/contents

[xxi] http://www.legislation.gov.uk/nisi/1983/150/contents

Maintaining the Safety of our Dams and Reservoirs
ISBN 978-0-7277-6034-0

ICE Publishing: All rights reserved
doi: 10.1680/mdam.60340.035

Challenges of planning significant reservoir safety improvements within an historic estate

D E NEEVE, Arup, Leeds, UK.
S PORTER, Chatsworth House Trust, Chatsworth House, UK.
L MARCHANT, Arup, Leeds, UK.

SYNOPSIS   Chatsworth House is undertaking reservoir safety and improvement works on a series of historic reservoirs located within the Grade I Registered Historic Park. Two of the four reservoirs fall under the Reservoirs Act and have safety recommendations pertaining to the size of the spillways and the discharge of the design events.

The Cavendish family purchased the land Chatsworth House now occupies in 1549 and has been modifying the house and landscape ever since. The earliest surviving reservoir, Swiss Lake, dates from 1710. The four reservoirs are still actively used to provide water for the famous fountains and garden water features along with the fire fighting system, toilets and hydroturbine.

Arup is working closely with the estate to develop low impact, cost effective solutions to achieve the improvements required, whilst minimising the impact on the historic and ecologically diverse landscape within the estate.

This paper covers some of the challenges of working with an historic estate, delivering major reservoir improvements with low impact solutions, whilst ensuring minimal construction cost.

INTRODUCTION
Chatsworth House, a grand stately home in the Peak District National Park, has two reservoirs that fall under the Reservoirs Act 1975 within the Grade I registered historic park and garden.

In late 2011 inspections under section 10 of the Reservoirs Act 1975 were undertaken on two reservoirs, Swiss Lake and Emperor Lake. The findings of the inspections confirmed that the reservoirs were Category A and that they did not meet the guidance set out within Floods and Reservoir Safety (ICE, 1996). Chatsworth House Trust was given a deadline by the Inspecting Engineer to complete the matters in the interest of safety.

35

Swiss Lake and Emperor Lake are set on a valley terrace above a steep escarpment approximately 500m to the east of Chatsworth House. Adjacent to the lakes are two ponds, Mud Pond and Ring Pond, that do not currently fall under the Act's definition of a large raised reservoir as their volumes above natural ground are both around 10,000m^3. The four reservoirs are surrounded by a mixture of woodland and agricultural fields.

The client was faced with two Category A reservoirs which had inadequate freeboard and insufficient overflow capacity to safely convey the design flood. The two adjacent ponds, currently outside the Act, would also require similar improvement works, should Phase 2 of the proposed changes to the Reservoirs Act, as set out in the Flood and Water Management Act 2010, be enacted. The works had to be designed; planning permission obtained; funding secured and the works constructed within three years.

To assist the client with understanding the scale and cost of the required works the Inspecting Engineer provided an outline hydraulic design which was priced by a contractor. Arup was successful in winning a tender to assist Chatsworth House Trust in the design, planning and construction supervision of the works.

Figure 1. Chatsworth House with the reservoirs above the house on the edge of the treeline. (© The Devonshire collection, Chatsworth)

RESERVOIR AND ESTATE HISTORY

Sir William Cavendish, Treasurer of the King's Chamber and the husband of Bess of Hardwick bought the estate in 1549 from the Leche family who created the first enclosed park at Chatsworth in the 15th century.

Bess began to build the new house in 1553 on the site of the present main block with the same quadrangle layout and large central courtyard. Her elder son, after inheriting the estate, sold it to his younger brother William Cavendish, 1st Earl of Devonshire, for £10,000.

William was requested to retire to Chatsworth by James II, and this called for a redesign of the house, starting in 1687. William retained the increasingly unfashionable Elizabethan courtyard plan as initially only works to the south wing were proposed. William Cavendish was honoured with the title of Duke in 1694.

In 1710, the first documentary reference to any of the reservoirs appears for a stone 'head wall' at the 'The Great Pond in the Park'. Further work to the pond included excavation and lining with clay up until 1717. A few years later, in 1719, a catchwater was constructed from 'Humberley Spring to the pond in ye parke' known as Umberley sick. Shortly afterwards stone was carted to the site 'for buttresses to support the pond wall at the top of ye parke'.

The Great Pond is probably Swiss Lake, which received this name in 1839 following the 6th Duke's continental holiday. However, Morton's Pond could also be a candidate as it was known as the Great Fountain Pond in the 19th century. The function of the new pond was to provide water for the canal fountain and for which, in 1727, £10 was paid for enlarging the pond 'Wch plays the fountain in the canal.' References to payments for 'work done' appear in 1743 for another supply reservoir, the 'little pond on the top of ye parke'.

Sometime between the production of the estate maps in 1824 and 1831 the pond now known as the Mud Pond came into existence. Ring Pond was constructed just prior to 1831.

After visiting Russia in 1826 for the coronation of Tsar Nicholas I the 6th Duke wanted a larger fountain than the one at Peterhof to impress the Tsar during his visit in 1844. Joseph Paxton was engaged and created Emperor Lake and its 2.5 mile catchwater Emperor Stream to feed the fountain. The works were completed in just six months over the winter of 1843-44.

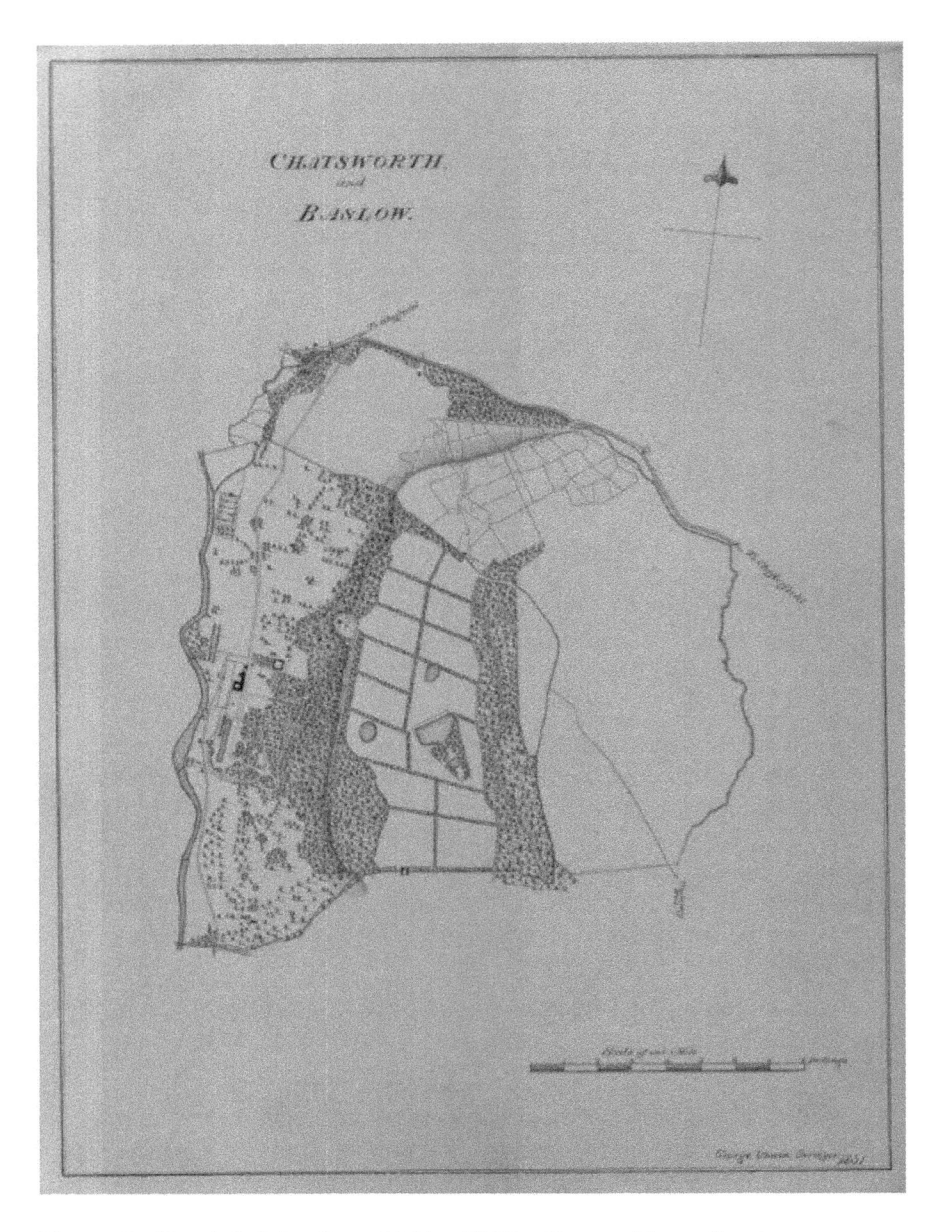

Figure 2. Unwin's Map from 1831. (© The Devonshire collection, Chatsworth)

Figure 3 - Emperor Fountain in front of Chatsworth House. (© The Devonshire collection, Chatsworth)

CLIENT CHALLENGES

Statutory undertakers with large portfolios of dams typically have a well managed system for the operation, maintenance, inspection and implementation of significant remedial works when required. For private owners or undertakers of one or maybe two reservoirs "recommendations in the interest of safety" come with great unknown challenges.

The obvious and initial conundrum is to understand what and why significant works are required to satisfy the Act, i.e. from a legislative and technical perspective. In this particular example, the reservoirs have been operated and maintained by the estate management for hundreds of years. Hence, even when the estate management team understand the requirements of complying with any matters raised in a statutory inspection under the Reservoirs Act, 1975, they still have to convince the larger organisation. This can be difficult when the design parameters include a flood that has around a 1×10^{-6} chance of occurrence in any one year. The risk associated with such an event can seem very remote, especially if other areas of the estate require planned works to protect public safety and are deemed much more likely, or even certain, to occur before the imposed deadline.

Combining the legislative and technical drivers with the financial implications adds to the challenge of convincing an organisation that the works are required, especially when their main remit is the management of the property and estate, rather than specifically the reservoir group. There is inevitably a robust challenge from the organisation to the news that significant additional funding is required at relatively short notice, as this diverts financial resources away from the core business or it is simply not available. The financial strain is not only felt in the short term but its unplanned legacy impacts the medium and long term. In the short and medium term difficult decisions are required to withdraw or seriously postpone agreed works to pay for unforeseen reservoir improvements.

The setting of the project within an historic estate dating back over 500 years creates additional challenges, to convince those within and outside the organisation that works are required and that they will take into consideration the visual impact and historic importance of the structures and landscape.

Therefore these challenges set up a conflict with the traditional cost, time, quality model. The scheme must be delivered on time to meet the compliance deadline; the quality must ensure minimal visual impact; the scheme must satisfy the Qualified Civil Engineer or Construction Engineer, and the cost must be as low as practicable.

A key role for the client is ensuring that the project team fully understands the business drivers to allow meaningful innovation. For example, the obvious notional scheme was unaffordable and had some undesirable visual elements. By understanding the drivers the project team, which includes the client, was able to significantly reduce both the cost and the visual impact (by removing channels, pipelines, concrete weirs and spillways) whilst maintaining compliance with the Reservoirs Act.

There are many potential pitfalls along the way, particularly when commencing the project as the client may not be fully aware of the problem or how it should be solved. Hence they may have a steep learning curve and require support to assist in decision making, which not only extends to engineers who understand the requirements of the Act, but ecologists, planners, heritage advisors, landscape architects and so on.

During normal operation the reservoirs require very little estate management consideration or time. An unknown consequence of large scale reservoir works is the increased workload to understand and make decisions on issues such as the legislative requirements, design options, the planning application and the impact of construction. A significant challenge has been fitting these activities into an already busy full time workload.

PLANNING CHALLENGES
The historic nature and significant profile of Chatsworth House within the Peak District National Park means it is crucial to carefully consider the views of the numerous internal and external stakeholders to produce an acceptable scheme and therefore planning application.

The planning authority and statutory consultees were engaged within the option development stage to deliberate the environmental and heritage impacts of options. The discussion included the form and material used to meet the requirements and how they should be in keeping with the historic estate and surrounding area.

The purpose of a National Park, and to a degree the Chatsworth House Estate, is to conserve and enhance the natural beauty, wildlife and cultural heritage of the area whilst promoting opportunity for understanding and enjoyment by the public. These aims result in the requirement to balance economic and environmental drivers within the National Park.

The Peak Park Planning Team clearly understood the need to provide a balanced view as the works were required by legislation and had to be affordable. However, this did not mean the works could avoid the ecology impacts (e.g. great crested newts), design standards or material specification required by the National Park.

The estate contains a wide diversity of ecology and includes populations of great crested newts and white clawed crayfish, which have been known to exist on the site for many years. The planning authority had previously undertaken ecological surveys to confirm the presence and understand the likely population size of these protected species. Unfortunately, these surveys did not meet the strict requirement for a planning application and were repeated along with a Phase 1 habitat survey and badger, white clawed crayfish and water vole surveys.

Through the early engagement the requirement to submit an Environmental Impact Assessment as part of the planning application was avoided and the detailed response from the formal Screening Opinion assisted in shaping the design and the documents required for the planning application.

Assisting an historic estate with a planning application is very different to working with a developer or statutory organisation, which typically has dedicated teams solely focused on the project delivery. In this case, the reservoir improvement works are a necessary distraction from the successful operation of the Chatsworth House Estate. Therefore obtaining information for the planning application can be time consuming. For example, even though the estate have used the same firm of lawyers for over 300 years obtaining formal proof of ownership from 1549 is not that straight forward.

TECHNICAL CONSTRAINTS
The location, age of the structures and natural environment have created challenges in the development of an acceptable design solution. Some of these challenges are outlined in the following sections.

Historical dam records
As can been observed from the information available on the history of the reservoirs, design or construction records are effectively non-existent. This is perhaps not uncommon with structures that were built some 170 to 300 years ago. However, this creates challenges when designing remedial works. For example, the ground investigation has to provide the first understanding of the embankment construction rather than confirming what is shown on the record drawings. In this particular case the ground investigation had to be very targeted to provide a full understanding of the dams and potential borrow pit areas whilst being cost efficient

The ground investigation revealed some interesting findings, for instance, it appears the grass and topsoil was not removed before Swiss Lake construction commenced, thereby creating a layer of soft material and a potential slip plane and/or leakage path at the location exposed to maximum embankment and hydraulic loading. The embankments appear to be constructed from locally won weathered bedrock, a mixture of sandstone and mudstone.

Flood Conveyance
The existing reservoirs have an inherent problem as the difference between the weir level and the embankment crest is insufficient resulting in embankment overtopping; this is especially true for the auxiliary weirs. In the case of Emperor Lake the embankment starts to overtop in less than a 1 in 5 year event. This example helps to demonstrate how standards and understanding of hydrology and reservoir safety have changed significantly since Swiss Lake was constructed in the early 1700s.

The increasing demand over the centuries for water within the gardens and the house has created a group of reservoirs with direct catchments and a complex arrangement of channels, culverts and sluices to obtain water from indirect catchments. In addition, the reservoirs are interconnected above and below ground and therefore have to be viewed as a system rather than individually.

To establish a viable solution and gain further understanding of the existing arrangement the Inspecting Engineer commissioned a flood study, extracts of which are presented below.

Land use around reservoirs
As well as the visual impact of the final solution and the practicalities of construction, the location and route of the discharge channels from the reservoir have to be considered. In addition to estate access tracks the surrounding fields are used for livestock grazing or given over to mature woodland. Therefore what might be considered the most appropriate in engineering terms is often not the best solution for the adjacent land use.

Careful consideration of all the constraints is required to balance the needs of the organisation's stakeholders and the environment.

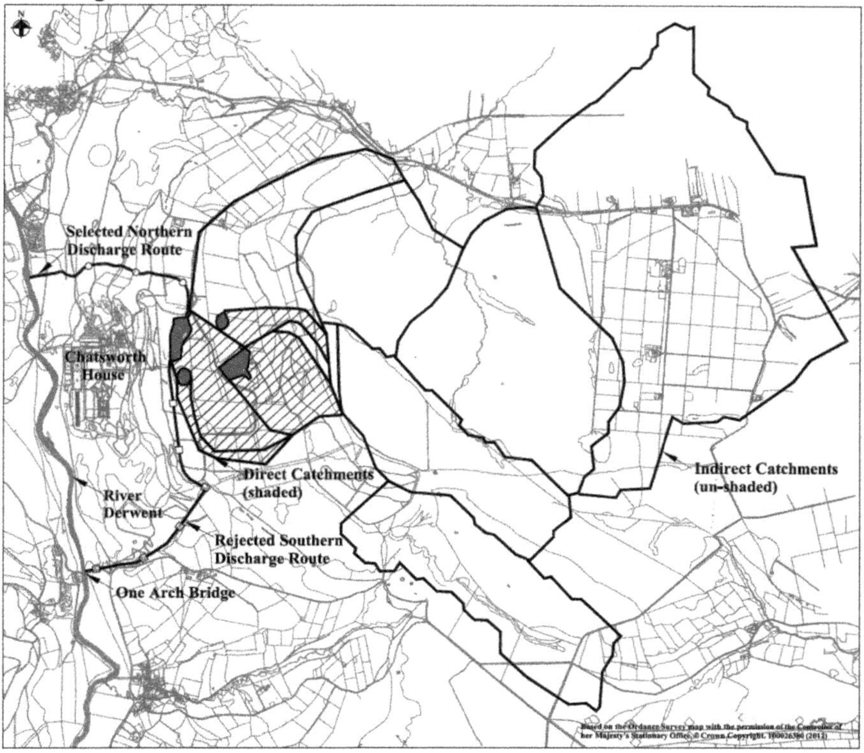

Figure 4. Reservoir direct and indirect catchments, based on AECOM Flood Study, January 2013

Discharge to a suitable watercourse
The design flood could be routed through the reservoirs to flow either north or south and discharge down the escarpment and across the valley floor to the River Derwent. The Inspecting Engineer's notional design followed an historic stream alignment to the north. An estate gardener suggested constructing a 1200m conduit along the top of the escarpment to utilise a ravine on the border of the Old Park to the south of the main House. This has several advantages of avoiding the Grade I registered parkland, the cricket pitch and showground/events area. Avoiding these areas removes disruption and disturbance to the visiting public during construction and concern that the pipeline reinstatement would remain visible within the grassland.

As the benefits were considerable the southern route was investigated as it appeared the preferred option. The route posed a number of manageable challenges including very mature oak trees within the Old Park, potential damage to 6th Duke's Carriage Drive and a discharge point into the River Derwent directed at the footings of the historic One Arch Bridge.

Figure 5. One Arch Bridge. (© The Devonshire collection, Chatsworth)

An outline survey demonstrated there was sufficient fall between the reservoir weir and the top of the ravine, known as Blakes Brook. However, the mature woodland masked a raised section along the escarpment and when a trial pit revealed sandstone bedrock around one metre below ground level this option was rendered economically unviable.

This resulted in the northern option being considered further. This route included concern over construction disturbance and disruption and the potential for a permanent reminder within the grassland of the pipeline location. To avoid the cost and visual impact of the pipeline, consideration was given to allowing flood waters to pass over the ground surface using the natural channel created by the ravine (during the historic landscaping of the parkland the channel had been partial filled to soften its appearance).

Further ground investigation by the client's team resulted in an existing culvert being discovered, which conveyed water from the base of the ravine to the river, presumably constructed during the parkland landscaping to remove surface water allowing uninterrupted grassland.

The client's preference was to utilise the culvert for low order flood events and permit overland flow once its capacity is exceeded. This approach comes with the risk of damage to the culvert and the parkland surface during flood events. However, the client would rather repair future damage than the certain impact on the Grade 1 listed parkland by creating a dedicated pipeline.

Embankments heavily wooded

At some point since construction of the reservoirs, the embankments were allowed or perhaps encouraged to vegetate to screen the water bodies. The now heavily wooded embankments contain substantial trees including sycamore, oak and ash thought to be up to around 60 years old. Such large trees pose concern for embankment stability and the decision to retain and manage the trees, or remove them, is difficult.

Trees on the crest were felled in early 2012 to create a five metre clear access strip on the crest, following a requirement of the Section 10 inspection. The trees were cut just above ground level and the stumps retained.

Sympathetic visual impact with robust engineering

As discussed earlier in the planning section a key challenge was to ensure the requirements of the Reservoirs Act were met, whilst balancing the final solution with the minimal visual impact aspirations of the National Park, English Heritage and the client. Through considerate design and green engineering techniques, the engineered solution will become an integrated part of the existing landscape.

SOLUTION

The purpose of the works is to safely convey the probable maximum flood through the reservoirs and discharge it safely towards the River Derwent. Where possible the existing arrangements for draw-off and top water level regulation have been avoided to maintain the historic integrity of the structures.

The two smaller reservoirs, Mud Pond and Ring Pond, do not currently fall under the provisions of the Act. However, the client has a duty of care and would like to ensure their structural integrity. They also aspired to minimise any risk to estate staff, who have to check the reservoirs for overtopping during storms, even if this is during the night.

The works to all four reservoirs include new auxiliary weirs and associated spillways and discharge channels. In addition, the works on Swiss Lake and Emperor Lake will include embankment raising of approximately 400mm and 600mm respectively to contain the flood rise and wave allowance.

Reinforced grass auxiliary spillways have been designed to minimise visual impact at Mud Pond, Ring Pond and Swiss Lake. To integrate the auxiliary spillway with the landscape at Emperor Lake the existing estate road has been modified to create a lower section that operates as the auxiliary weir.

The design solution located the auxiliary weir on Swiss Lake to utilise the existing topography. This ensured major earthworks were avoided through the open sections of fields, which contain historic ridge and furrow, from Swiss Lake to Mud Pond and then subsequently onto Emperor Stream.

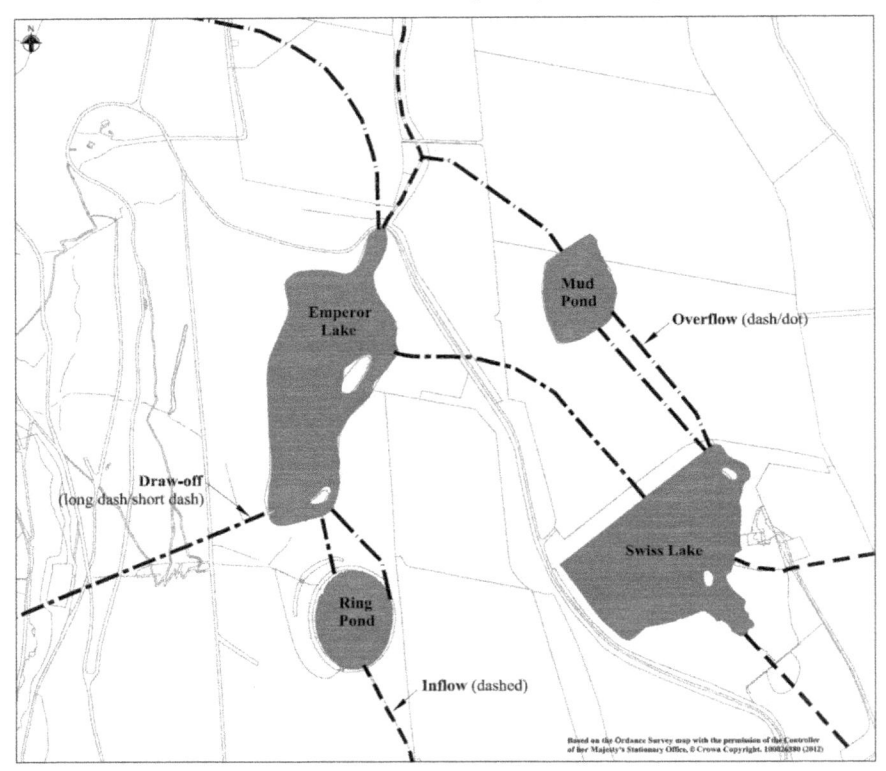

Figure 6. Plan of reservoirs with associated channels

To retain the character of the setting at Mud Pond the required spillway capacity will be split across two distinct sections. This will enable a mature oak tree located immediately downstream of the crest to be retained. As mentioned above the decision to leave or remove large trees on embankment dams has its pros and cons. To minimise damage to the tree, and thereby reduce the risk of breach, the embankment has been raised to prevent overtopping in this location and the top water level weir and spillway relocated. The existing informal spillway has eroded the downstream embankment and exposed large tree roots.

Through close working relationships between the design team, the Estate and external stakeholders the project team has achieved the aim of developing a solution which ensures compliance with the Reservoirs Act, whilst being sympathetic to the landscape and providing a minimal cost solution.

REFERENCES

Institution of Civil Engineers, 1996. *Floods and Reservoir Safety, 3rd Edition*. Thomas Telford, London, UK.

HMSO (1975), *Reservoirs Act, 1975 (1975)* Elizabeth II. Chapter 23. Her Majesty's Stationery Office, London, UK

Barnett & Williamson (2005), *Chatsworth, a Landscape History*, Windgather Press, Macclesfield, UK

Maintaining the Safety of our Dams and Reservoirs
ISBN 978-0-7277-6034-0

ICE Publishing: All rights reserved
doi: 10.1680/mdam.60340.048

How overcoming environmental challenges shaped the design of a new reservoir in North West Ireland

D A McKILLEN, URS Belfast, UK
G A COOPER, URS Belfast, UK

SYNOPSIS   Lough Mourne is a natural upland lake situated in the Blue Stack mountain range to the north-east of Donegal Town in County Donegal, Ireland. The lake has been a water supply source since the 1960s and was identified in the 1980s as having significant development potential. However, as it is situated within and adjacent to a number of sites of high environmental value, any development was believed to have significant planning obstacles. A proposed major development in the mid 1990s, involving the raising in top water level of 6.5m, was abandoned at pre-planning stage as it was considered that the environmental impacts could not be appropriately addressed.

In 2001 Donegal County Council appointed Nicholas O'Dwyer Ltd and Ferguson McIlveen LLP (now URS) to reopen investigations into the Lough's development potential by preparing a detailed feasibility report on a range of options and an associated Environmental Impact Statement (EIS). Following various appropriate studies, the EIS was published in 2005 and the scheme was granted full Planning Approval in 2006 following a public inquiry. The scheme has progressed to full contract document stage and is awaiting funding approval from Irish Water to progress to Tender.

This paper will outline how the range of environmental challenges associated with the scheme were dealt with during the feasibility, outline and detailed design stages of the project and how they have shaped the design of the impoundment, its dams and ancillary features.

INTRODUCTION

The Need for the Scheme
Donegal County Council had identified the need for additional water supplies for the Lough Mourne supply area and adjoining parts of County Donegal, including the rapidly developing town of Letterkenny.

Lough Mourne is a natural upland lake some 96ha in area situated in the Blue Stack mountain range to the North East of Donegal Town in County

Donegal, Ireland (Figure 1). The current available supply from the reservoir, which was developed in the 1960s, is abstracted by the use of a pumped intake located at the north-eastern shore. With no control structure at the natural outlet to the Mourne Beg River, the source depended on lake drawdown and the limited catchment area to provide a reliable yield of only 7.3Ml/d.

The feasibility study carried out by URS confirmed that an output of 31.3Ml/d was required to meet the anticipated supply area demand by 2030. Various options were considered, including the maximum economical sustainable development of the Lough Mourne headworks, augmented by additional adjacent catchments.

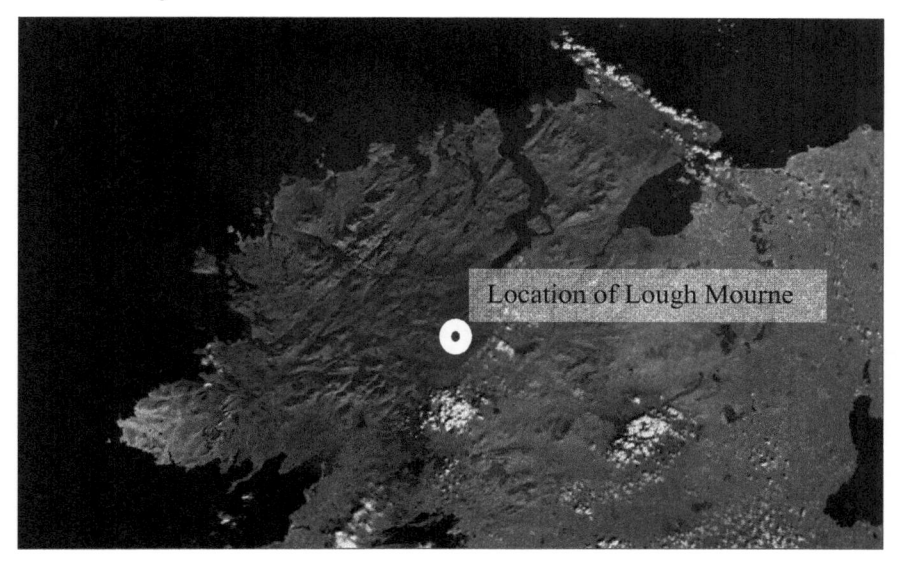

Figure 1. Location Plan (Copyright © Ordnance Survey Ireland. Licence number 2014/02ccma / Donegal County Council)

Engineering Report Recommendations
The feasibility study into sustainably developing the natural lake needed to take account of the constraints imposed by the various environmental factors which included several environmental designations, archaeological features, fish habitat and movement, the effect on adjacent bog land and the proximity of the N15 trunk road.

The study confirmed that, within these constraints, it would be possible to develop a reliable yield of 31.8Ml/d by raising the natural level of the lake by 4.5m. This raising would provide 3,600Ml of additional storage above existing lake level and would require the construction of a 14m high and 380m long dam across the outlet to the Mournebeg River and a 10m high and 230m long dam at the north-eastern shore. Figure 2 shows the locations of the proposed dams. Figures 3 and 4 provide a more detailed view of the Southern and Northern Dams respectively.

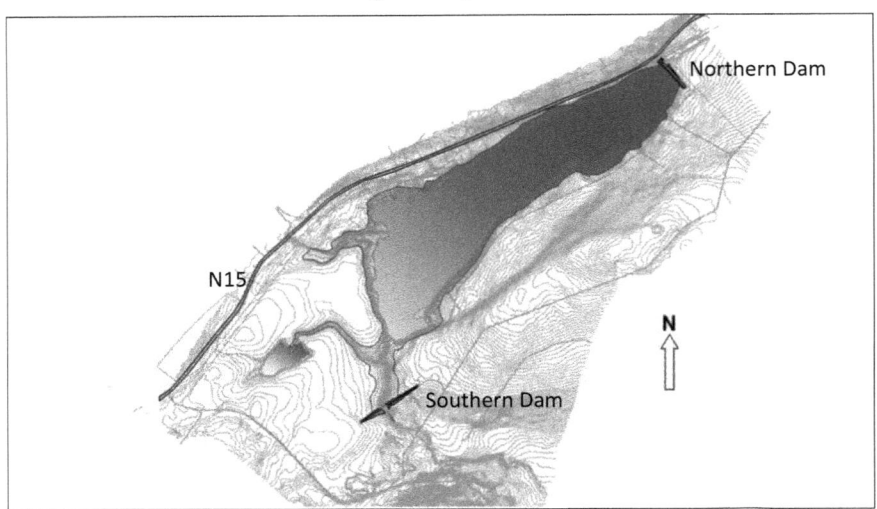

Figure 2. Proposed Dams

The enlarged reservoir would have its direct catchment extended from 862ha by diverting a sustainable percentage of natural stream flow from 500ha of the adjoining Bunadowen catchment via a gravity aqueduct.

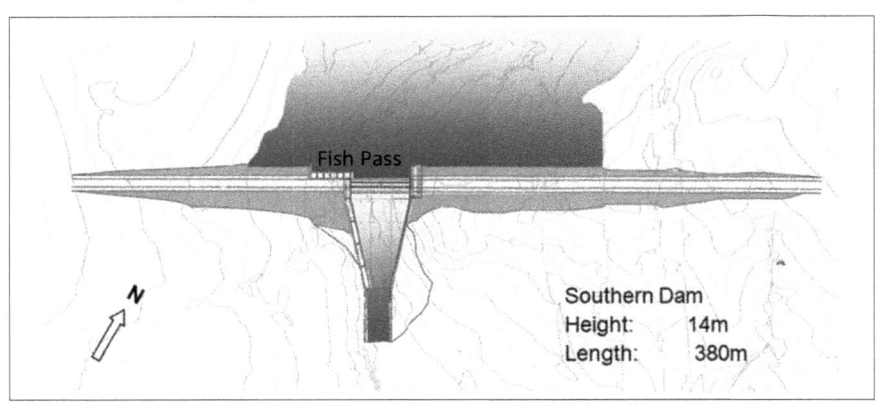

Figure 3. Southern Dam

The Southern Dam would incorporate the spillway, fishpass measuring flume and emergency sluice arrangement. The Northern Dam would incorporate the eduction tower and raw water outlet gallery. The drawoff would be fed only by gravity, since the new bottom water level would correspond to the existing normal water level in the lake.

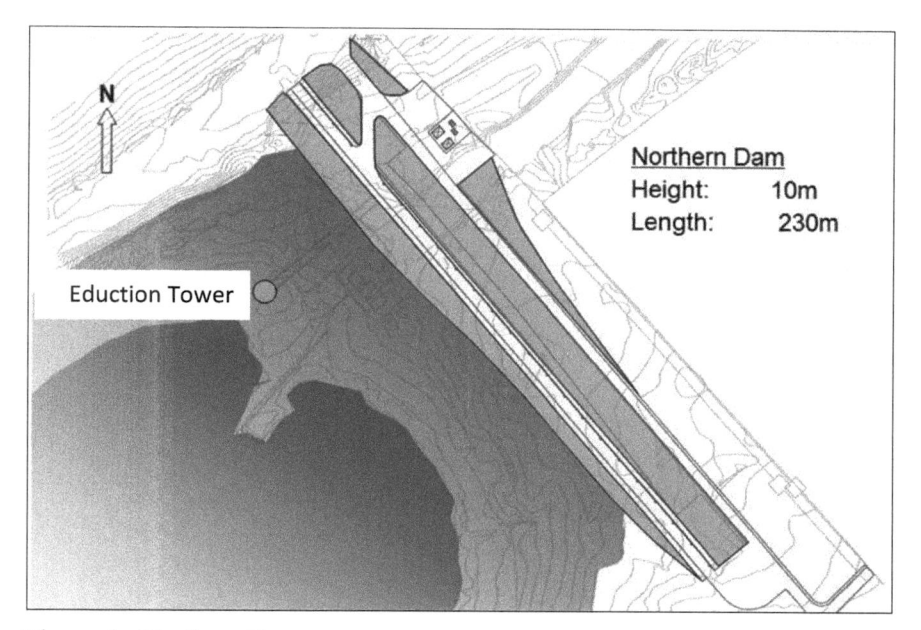

Figure 4. Northern Dam

The Southern Dam would incorporate the spillway, fishpass measuring flume and emergency sluice arrangement. The Northern Dam would incorporate the eduction tower and raw water outlet gallery. The drawoff would be fed only by gravity, since the new bottom water level would correspond to the existing normal water level in the lake.

The proposed raising in water level would not require any raising of the level of the N15 trunk road. However, the abandoned narrow gauge railway embankment, running parallel to the N15, would be partially submerged but would remain intact as a heritage feature.

The development would require the permanent acquisition and appropriate fencing of land around the perimeter of the extended reservoir area. The provision of accommodation works to enable existing landowners, who have turbary rights (the rights to cut peat for fuel), to access adjacent areas for peat abstraction would also need to be incorporated into the design.

ENVIRONMENTAL CONTEXT
Lough Mourne sits within an area of very high environmental significance. There are several designations affecting the site including multiple Special Areas of Conservation (SAC), Natural Habitat Areas (NHA) and Landscape Character Areas (LCA), which presented a number of environmental challenges throughout the entire project. These environmental designations are highlighted in Figure 5.

Several of the environmental challenges were specifically prescribed within the conditions for planning approval and as a result, have shaped the design in some way.

The three main environmental challenges identified in the planning conditions are archaeological testing of the Northern and Southern Dam sites, impact on the fisheries and the impacts on the SAC peat bog located on the south-western shoreline of the lough. A fourth significant challenge, although not a planning condition, surrounded security of supply in terms of water quality and possible spillages from the adjacent N15 primary route.

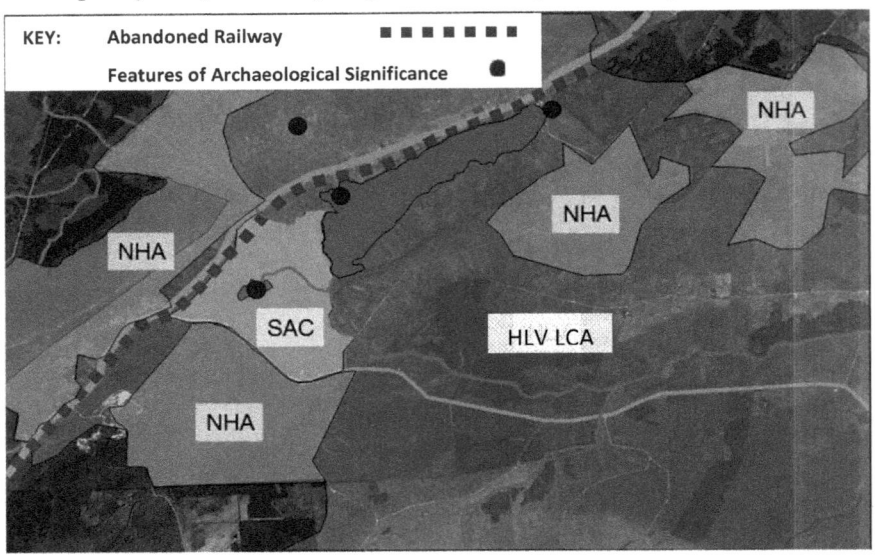

Figure 5. Environmental Designations

ARCHAEOLOGY

Two significant archaeological issues were identified. The first was the need to carry out a watching brief on any excavation carried out at the Northern and Southern Dam sites to identify Neolithic field boundaries that may be buried within the peat. The peat on both sites is in excess of 4m deep in places. This made the requirement to carry out a watching brief very onerous due to the high level of unknowns and the safety risks associated with such deep excavations in peat.

The second issue surrounded the presence of a megalithic tomb at the Northern Dam (Figures 6 and 7). At preliminary design stage, the Department of Environment, Heritage and Local Government (DOEHLG) was consulted regarding the megalithic tomb and identified that the recorded location of the tomb was marked by standing stones on the shoreline of Lough Mourne. During the preparation of the EIS, a targeted appropriate geophysical survey was carried out in order to verify this location. This survey identified that the tomb, known locally as the 'Giant's Grave', and the standing stones were not in fact on the shoreline of Lough Mourne but immediately under the proposed Northern Dam embankment. At the public inquiry, the inspector concluded that the megalithic tomb should be fully excavated, researched and recorded prior to commencing construction of any permanent works on the site.

Figure 6. Megalithic tomb looking south-west

Both of these archaeological issues, when considered in relation to the required form of construction contract, added significant risks if left to be included within the main civil engineering construction contact, both in terms of archaeological costs and main project programme and costs. Accordingly, at an early stage in the design process, a risk management approach was agreed with the client, designed to mitigate any archaeological and overall project risks as early and fully as possible.

An archaeological project design document was produced setting out the methodologies to be used in resolving the archaeology issues on site prior to construction. The report set out the mechanisms for reporting and, most importantly, outlined the mitigation procedures to be used if any archaeology was found on site that was not recorded within the EIS. This document was agreed with Donegal County Council and DOEHLG. It included a specification for the advance archaeological contract providing better control of potential costs and timescales and reduced main contract risk.

Figure 7. Excavated tomb structure facing east.

Both of these archaeological issues, when considered in relation to the required form of construction contract, added significant risks if left to be included within the main civil engineering construction contact, both in terms of archaeological costs and main project programme and costs. Accordingly, at an early stage in the design process, a risk management approach was agreed with the client, designed to mitigate any archaeological and overall project risks as early and fully as possible.

An archaeological project design document was produced setting out the methodologies to be used in resolving the archaeology issues on site prior to construction. The report set out the mechanisms for reporting and, most importantly, outlined the mitigation procedures to be used if any archaeology was found on site that was not recorded within the EIS. This document was agreed with Donegal County Council and DOEHLG. It included a specification for the advance archaeological contract providing better control of potential costs and timescales and reduced main contract risk.

The contract was procured under the Public Works Contract and was separated into two distinct phases. The first phase focussed on the archaeological testing, the trial trenching and ground penetrating radar surveying. This was assessed as a lump sum contract with a fixed price and would form the basis of the second phase works. This phase of the works was completed in the autumn of 2010.

The trial trenches covered both dam sites and included some trial trenches specifically through the megalithic tomb site to help identify the scope and extent of the works there. These works, combined with the geophysical works on both dam sites, have allowed large areas of both sites to be removed from any further monitoring during construction works, resulting in significant risk reduction.

They also allowed the scope of the second phase contract to be identified and defined. The remaining megalithic tomb elements comprised the foundation of the original megalithic tomb (Figure 8). It is likely that the larger stones that were used to construct the tomb have long since been used or crushed and used to make access roads around the site.

Figure 8. Overhead view of the exposed tomb

FISHERIES IMPACT

The effect of water abstraction on fish and freedom of movement into and out of the Lough was investigated. Specialists in fishery aspects concluded that the upper reaches of the Mourne Beg River did not provide a good salmonid habitat and an existing gauging station on the outlet of Lough Mourne had prevented any upstream fish movement between Mourne Beg River and Lough Mourne since the 1960s.

However, it was identified at an early stage that a fish pass should be integrated within the scheme and that approval for the fish pass should be sought from the Loughs Agency as the fishery authority. The fish pass was used to focus the compensation flows required to improve low flow conditions in the upper reaches of the Mourne Beg River and to improve the salmonid habitat. It would also have the additional advantage of improving low flow conditions for an existing fish hatchery further downstream. At the commencement of the design Paul Johnston Associates was appointed to carry out a number of fish surveys and the environmental requirements for the fish pass structure were agreed with the Loughs Agency. The structure was required to be suitable for salmonids, which included sea trout, brown trout and salmon. It also had to operate over the top 2.5m of the reservoir drawdown and maintain a specified minimum compensation flow requirement at all times. Following discussion of options with the Council it was agreed that the fish pass should be a low maintenance structure controlled by an automated system and fundamentally should be within the land take area procured for the scheme.

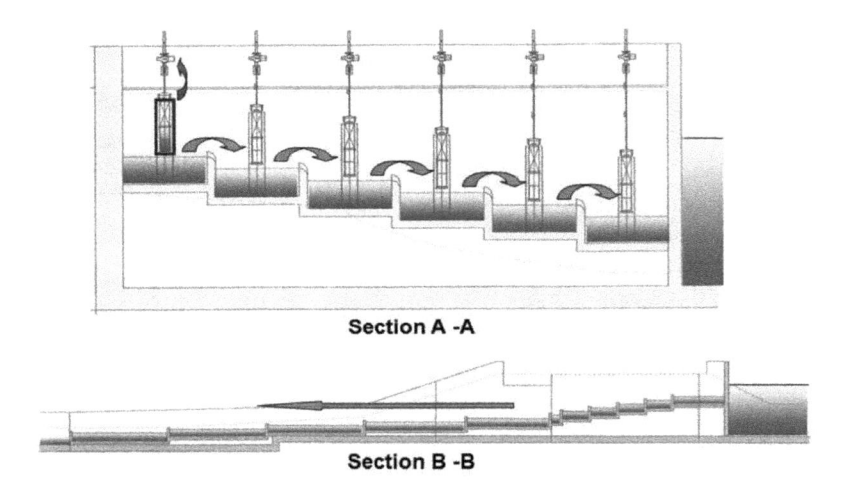

Figure 9. Fish Pass Sluices and Ladder

A number of reports were prepared for the Loughs Agency to investigate the various fish pass types that may be suitable for the site. It was concluded that a pool and traverse fish pass comprising variable inlet levels was the most suitable structure for the site. On the upstream side of the fish pass there are six penstocks at various levels. As the level of the reservoir lowers the appropriate pen stock opens allowing flow into the upper pool and subsequently down the fish pass cascade. As the reservoir drops the next penstock opens continuing the flow until the reservoir reaches a drawdown of 2.5m. The penstocks are the 900mm deep by 350mm wide off-seating, downward opening type (Figure 9). The downward opening allows an accurate flow above the penstock at all times and a sufficient air void for the fish to jump through. The fish pass penstocks will be automatically controlled and will be linked by telemetry to the water level sensors around the lough. Defining the appropriate penstock opening level will be subject to control of the upstream wave action and the fish pass is located in a sheltered part of the lake to minimise issues with this.

EFFECT ON CROAGHONAGH BOG

Croaghonagh Bog, Special Area of Conservation (SAC) is an Annex I habitat type. It is a raised blanket bog and has a number of priority species. The actual flooding of the site and the construction of the dam has a significant impact on the bog and, when the scheme is finished, approximately an additional seven hectares of this habitat will be flooded. At EIS stage this was identified as a very significant constraint on the overall project and as a result Dr John Feenan of University College Dublin was engaged as a recognised authority on peatlands. He identified that the north-eastern shoreline of the peat bog had gradually been drying out, over a number of decades as opposed to a number of years, resulting in tears and collapses. It was concluded that the drying out of the shoreline was a significant risk to the long term integrity of the overall bog structure.

Dr Feenan confirmed that raising the top water level to 4.5m to 170m above datum would in fact re-saturate this area of bog providing stability and protection. While it will flood an additional seven hectares of the SAC, it was flooding the lesser-value components of the special area of conservation.

The tears in the bog fabric had also been worsened by the natural movement of the groundwater through the peat material towards the drying edges. The raised water level would help prevent this deterioration but it was also important that the flow would not realign to discharge downstream of the dam structure. Options to mitigate this were considered and sheet piling was provided to a defined level designed in a plan layout to manage the rate and direction of groundwater movement.

SECURITY OF SUPPLY

The site is in a relatively remote unpopulated location and the only significant infrastructure adjacent to it is the N15 trunk road. When raised, Lough Mourne will become one of the largest water supply sources in Donegal County. Traffic on the N15 Road poses a considerable risk to the site in terms of drainage pollution into the Lough.

The raising of the lough will bring the shoreline to within approximately 10m of the N15 carriageway. There is an ongoing risk of vehicles crashing and their contents entering into the lough. To deal with this as part of the overall works a Sustainable Drainage System (SuDS) comprising a number of retention ponds and petrol interceptors has been specified. In addition, a range of vehicle containment structures will be provided along the N15.

A second security supply issue identified surrounds the potential pollution risk to the site during construction, especially since the existing intake is adjacent to the site of the Northern Dam. The reservoir will remain an active water supply source during the construction period and will be the only water supply source supplying a specific geographic area during that period. Requirements for special monitoring during construction and preparation of emergency plans are included in the construction contract.

CONCLUSION
Addressing the various imposed constraints by in depth consultation with all interests and by incorporating appropriate mitigation measures, enabled all the difficulties inherent in obtaining permission to form a large impounding reservoir in a very sensitive environment to be successfully overcome.

Chingford Pond Restoration

T R WANNER, Atkins Limited
H COUTTS, Atkins Limited

SYNOPSIS Chingford Pond comprises an earthfill embankment constructed in the mid 18th Century as an ornamental lake within the Grade II listed Burton Park, Petworth, West Sussex. In 1986 a notch was excavated in the reservoir embankment to remove the reservoir from the requirements of the Reservoirs Act, 1975. This resulted in a lowering of the reservoir top water level by around two metres.

In recent years a decision was made by the site owner to bring the reservoir back to its former state by repairing the embankment notch and restoring the historic water level. This resulted in restoration of an historic 'cascade' spillway and grotto, incorporation of new main and auxiliary spillways, installation of a scour pipe and chamber and inundation of around 3.5Ha of established wet woodland.

This paper describes the engineering involved in relation to the ecological and historical constraints associated with raising the water level and inundation of established habitats. This includes the mitigation requirements for protected species, protection of the downstream SSSI and compliance with various pieces of legislation including the EU Water Framework Directive. This is described through various stages of the project from development of an environmental impact assessment and planning application to detailed design. Conclusions are drawn to demonstrate how reservoir raising engineering works can be carried out effectively whilst managing environmental constraints and providing environmental enhancements.

INTRODUCTION
Chingford Pond comprises an earthfill embankment constructed in the mid 18th Century as an ornamental lake within the Grade II listed Historic Park and Garden, Burton Park, adjacent to the Grade I listed Burton House, Petworth, West Sussex (as shown in Figure 1). In 1986 a notch was excavated in the reservoir embankment to remove the reservoir from the requirements of the Reservoirs Act, 1975. This resulted in a lowering of the reservoir top water level by around two metres.

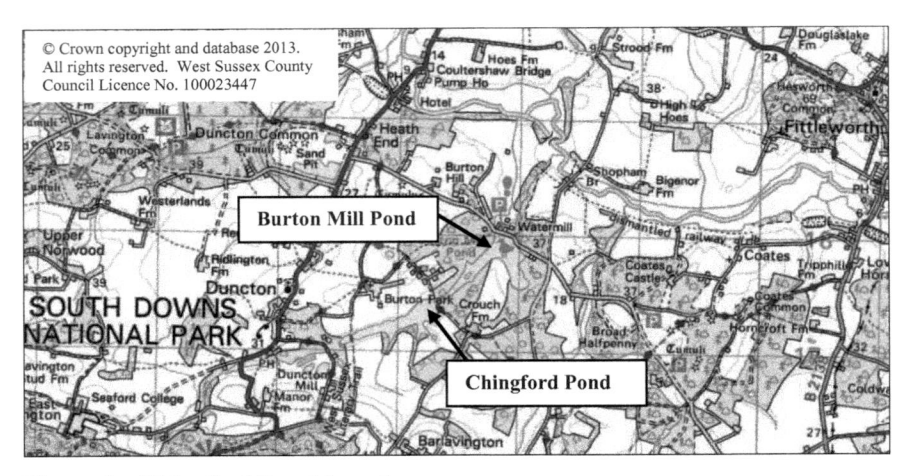

Figure 1. Chingford Pond location.

Burton Park, and the associated pond system incorporating Burton Mill Pond immediately downstream of Chingford Pond, is an environmentally important site. Burton Park lies within the South Downs National Park, and is itself a Grade II Registered Historic Park and Garden. Burton Mill Pond is a designated Site of Special Scientific Interest (SSSI) and Chingford Pond is a local Site of Nature Conservation Interest (SNCI), and together the two ponds form the Burton and Chingford Ponds Local Nature Reserve (LNR). The two ponds also support an important and diverse range of plants and animals, and some legally protected species, such as badgers and dormice, are known to inhabit the woodland areas that fringe the ponds. There are public footpaths that also provide public access to the site.

In 1996 planning permission was granted for limited residential development of Burton House. Conditions on the development under a Section 106 agreement included the restoration of the historic C18th parkland, the C19th and early C20th formal gardens, and the restoration of Chingford Pond to its original level and condition.

In recent years a decision was made to restore the reservoir to its former state by repairing the embankment notch and restoring the historic water level. The proposed restoration works at Chingford Pond seek to restore the pond to its historic water level of 18.3mAOD, to repair and reinstate the historic 'cascade' spillway and grotto and to restore the lost views to the pond through vegetation removal. This will give rise to an enhancement in the historic environment, both locally and to the wider historic landscape of Grade II listed Burton Park. The works also include incorporation of new main and auxiliary spillways, installation of a scour pipe and chamber and inundation of around 3.5Ha of land.

The restoration works were design by Atkins Limited in consultation with West Sussex County Council, which previously held a lease on the pond, and the freeholder Petworth Management Company, which is owned by, and acts on behalf of, the 49 properties in Burton Park.

Raising the water level of the reservoir will cause inundation of 3.5Ha of established wet woodland, home to bats, badgers, dormouse and a rare small snail, the Desmoulin's whorl snail. As a result the restoration design and works are being carried out sensitively with respect to the sites environmental importance and includes various environmental mitigations and enhancements to achieve this.

PROJECT REQUIREMENT
Burton Park is an historically important site, being a Registered Historic Park and Garden. Chingford Pond is an important feature within the Historic Park and Garden, and the lowering of the water levels and growth of trees and scrub in the areas that were once extensive open water has meant that the historic landscape has become degraded over time. The client, working in partnership with a number of other organisations, wishes to restore a number of historic features within the Park, including Chingford Pond. This is being implemented through the client's Historic Park Restoration Plan.

Chingford Pond forms part of a wider vision for the whole of Burton Park which seeks to restore the park to its *'previous grandeur of the early 1800s'*. The vision is to be achieved through the implementation of a Historic Park Restoration Plan. Chingford Pond forms part of this Restoration Plan and the proposed scheme seeks to restore the pond as a fundamental part within the historic landscape. The project is supported by Natural England and as such the client was awarded funding through the Countryside Stewardship Scheme, and more recently the Higher Level Scheme, to deliver its objectives.

Additionally, there is an existing planning condition (a Section 106 agreement) relating to the previous development of residential housing at Burton Park, which requires the client to undertake the restoration of Chingford Pond.

Although the reservoir is not currently within the Reservoirs Act 1975, its condition has deteriorated over time and so works are required to increase its safety. The reservoir currently holds just under 25,000m³ of water, and is in cascade with the downstream Burton Mill Pond, and so the reservoir could, in future, fall under the ambit of the Reservoirs Act as amended by the Flood and Water Management Act 2010. The raising of the water the level of reservoir will increase the storage to around 100,000m³, and so the reservoir will fall under the current ambit of the Reservoirs Act, 1975.

THE RESERVOIR
Chingford Pond is located on the southern boundary of Burton Park in the town of Duncton, West Sussex, and within the South Downs National Park. It is situated immediately upstream of the Category B Burton Mill Pond. Chingford Pond dam comprises a 6m high, 150m long earthfill embankment.

Prior to 1986
Chingford Pond is said to have been built around the mid 18th century as part of the landscaped gardens surrounding Burton House, as an ornamental lake to be viewed upon from the gardens and the upper levels of the house, and to be walked around, and boated upon.

Up until 1986 the pond held around 100,000m³ of water with a top water level (TWL) of 18.3mAOD. The TWL was set by one overflow spillway consisting of three oval brick culverts around 0.6m in diameter. The culverts passed under the embankment, onto a masonry chute and over a rock cascade. A grotto was located beneath the chute and cascade which could be entered from either side of the cascade. Once inside the grotto, visitors could look out from behind the cascading waterfall created by waters flowing over the spillway.

1986 to 2014
In 1986, and as a consequence of the enforcement of the Reservoirs Act 1975, and the owners at the time not being able to afford the costs, a notch was excavated in the embankment and the TWL of the reservoir was reduced to around 16.5mAOD. The storage volume was reduced to just below 25,000m³. During this period the reservoir was not maintained as a reservoir and some deterioration of the embankment and spillway occurred.

Present Restoration
The main purpose of the current restoration of Chingford Pond is to restore the reservoir to its pre 1986 state. This involves restoring the original TWL of 18.3mAOD and the 100,000m³ storage volume. The restoration will also ensure the safety of the reservoir is maintained through its registration under the Reservoirs Act 1975, including any amendments to that Act from the Flood and Water Management Act 2010.

DESIGN
The current design of the restoration of Chingford Pond is based on the pond being classified as a Category B reservoir, the same as the downstream Burton Mill Pond, therefore the reservoir must be able to safely pass the design 10,000 year flood event. However, with imminent changes to the Reservoirs Act from the implementation of the Flood and Water Management Act 2010, the restoration works have also to be designed to

safely pass the Probable Maximum Flood (PMF), with only an additional minor embankment crest raise, to be carried out at a later date if required.

The PMF was recalculated during the design phase and a peak flow of 41m^3/s was estimated. The existing cascade spillway was calculated to only be able to pass a maximum flow of around 3m^3/s and so a significant amount of additional spillway capacity was required.

In order to safely pass the PMF event an additional 7m wide Amorflex reinforced 'main' spillway and an additional 25m wide Enkamat reinforced grass auxiliary spillway were designed, along with a stilling basin for both spillways. The locations of these spillways, along with other works, are shown in Figure 2.

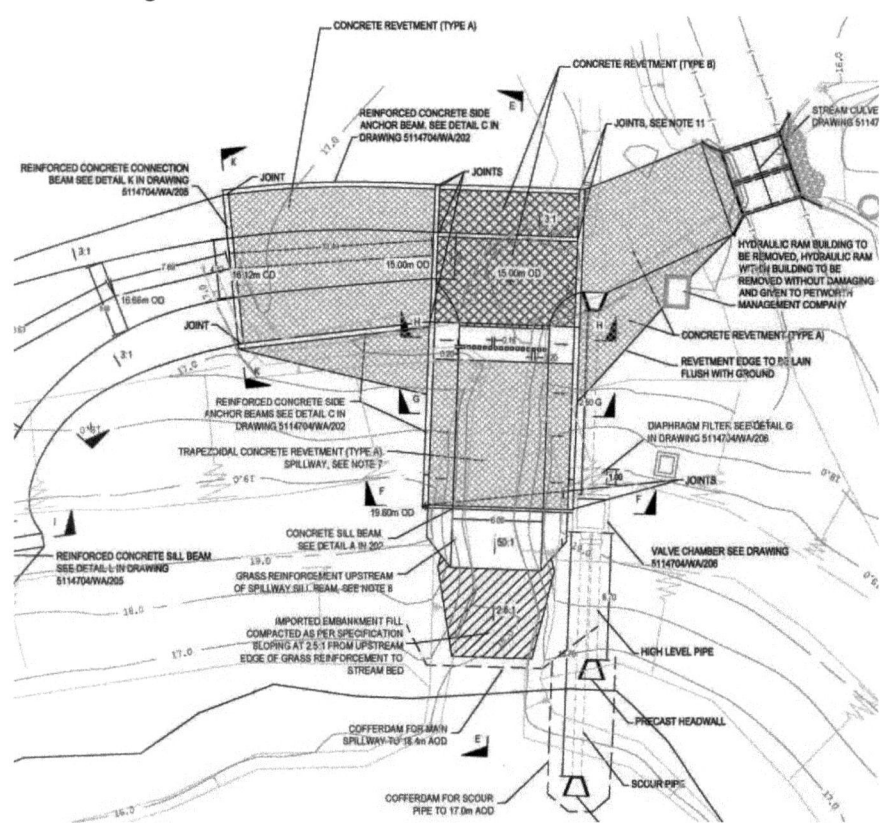

Figure 2. Location of site works

In addition to the new spillways, a new 500mm diameter scour pipe and chamber were designed to enable the reservoir to be lowered in an emergency. The scour contains two gate valves, a duty and standby. The valve chamber also contains a secondary outlet at a higher reservoir level to provide additional reservoir level control for environmental purposes during reservoir filling.

Other design works included a full restoration of the existing cascade spillway. This consisted of removal of the existing oval culverts and replacement with new culverts, removal of all vegetation growing on the spillway chute masonry and cascade rockery, and restoration / replacement of damaged / missing masonry and placed rocks. Where possible the restoration design was sympathetic to the original appearance of the cascade spillway. This was determined from the current state of the structure and historic photographs (an example is provided in Figure 3). For example, although the oval brick culverts are to be replaced as they were in a state of disrepair, the upstream and downstream masonry and brick 'façades' are being retained or rebuilt to match the existing.

Figure 3. Historic photograph of the cascade spillway

ENVIRONMENTAL REQUIREMENTS

Environmental Impact Assessment

In 2012 Chichester District Council confirmed that, given the size of the proposals and the environmental sensitivity of Chingford Pond and Burton Park, an Environmental Impact Assessment (EIA) should be carried out for the project. An EIA is a process which examines and reports on the potential implications of a development on the environment (both beneficial and adverse), and seeks to find ways to overcome any adverse effects that are identified. Chichester Council and South Downs National Park Authority also advised that the key issues that needed to be considered in the EIA were:

- The ecological effects of the proposals, including the effects on Burton Mill Pond SSSI, Chingford Pond SNCI and Burton and Chingford Ponds (LNR);

- Consideration of positive impacts on Burton Mill Pond SSSI due to reduced sediment inputs from Chingford Pond following the repair of the dam;

- Impacts on protected species in the area (bats, badgers, dormice, water voles and reptiles);

- Impacts on the flora and fauna that currently use Chingford Pond, especially the stands of aquatic plants that are established around the edge of the pond, and one particularly important snail, Desmoulin's whorl snail, which lives in this habitat;

- The effects on other aquatic species and habitats from a reduction in the area of shallow water and marginal habitats and an associated increase in deeper open water habitats in the pond;

- The implications of habitat losses due to water level raising from the Duncton Stream (the inflow to Chingford Pond) and the two areas of woodland that have developed to the south of Chingford Pond since it has dried out;

- Opportunities for ecological improvements for protected species or other habitats at the site; and

- Impacts of the scheme on the water quality of the pond system, both during the construction works and once the water levels are raised.

These, and a range of other environmental issues such as impacts on local residents, impacts on landscape and the potential for noise generation during the construction works, were also considered in the EIA.

Some of the main considerations of the EIA that impacted on the restoration works design and construction are discussed below.

Terrestrial Ecology
The terrestrial ecology of the area immediately surrounding Chingford Pond was investigated in some detail during the development of the design. The habitats and species found within Burton Park (including Burton Mill Pond) are considered to be nationally important, and some of Burton Park has therefore been designated as a SSSI. There are also several other sites nearby (including Chingford Pond) which are recognised as being of local importance for nature conservation, and are therefore designated as SNCIs.

Chingford Pond is bordered by areas of woodland. This includes important ancient and wet woodland habitat types. Within these woodlands the following legally protected species were recorded during surveys of the site, and required special consideration and management before, during and after the works:

- Dormice (in the woodlands to the south of the pond);

- Badgers (three badger setts in the woodland areas surrounding the pond); and

- Bats (recorded feeding over the pond area, hibernating in a tunnel in front of the cascade, and may also be using mature woodland trees as roost sites).

The habitats around Chingford Pond also support a diverse range of terrestrial invertebrate species.

Water Environment
The designation of Burton Mill Pond as part of the wider Burton Park SSSI, and the designation of Chingford Pond as a local SNCI, in recognition of the nature conservation value of the existing pond habitats, are key elements of the water environment within the site. In addition the numbers and diversity of aquatic invertebrates found living in the stream and ponds during ecological surveys indicated that the water quality is high.

Chingford Pond supports some important aquatic habitats, including the open water of the pond itself, but of particular importance are the extensive stands of reeds and other water plants that grow at the edges of the pond where the water is shallower. The Duncton Stream which flows into and out of Chingford Pond also contains a good variety of stream habitat types.

Chingford Pond, Burton Mill Pond and the Duncton Stream were found to contain a rich and diverse range of aquatic invertebrates, including some locally and nationally rare species. One species of particular importance is Desmoulin's whorl snail, a very small snail that depends on the stands of reeds and other water plants in the shallow margins of the pond. Populations of this snail have been declining across Europe and nationally in the UK. Chingford and Burton Mill Ponds are particularly important because they have some of the highest recorded densities of Desmoulin's whorl snail in the country.

Planning Application

The EIA was submitted as part of a full planning application to Chichester District Council. The planning application also consisted of the following accompanying assessments and studies:

- Arboricultural Impact Assessment (including tree survey)
- Heritage Statement
- Flood Risk Assessment
- Water Framework Directive Assessment
- Design Drawings

In addition to the above, extensive consultation was carried out to ensure the local community, stakeholders and statutory consultees had been engaged on the proposed restoration, and that their particular requirements would be addressed.

The main outcome of the EIA and planning process was that provided the identified good practice measures are followed in relation to the existing ecology and habitats, that the mitigation and compensation measures are secured and in place, and the recommended monitoring and ongoing management of the water level raising is implemented, the restoration of Chingford Pond could be constructed and operated without having a detrimental effect on the wider environment of Chingford Pond.

In addition, it was considered that if the works were not undertaken and the pond was maintained under its current regime and siltation of the pond continued, the area would eventually succeed to scrub and dry woodland habitat potentially threatening the more ecological significant aquatic and terrestrial invertebrate species the site supports.

ENVIRONMENTAL MITIGATION / ENHANCEMENTS

The outcomes of the EIA and planning process resulted in a series of environmental mitigation or enhancement requirements to be undertaken to ensure the restoration of Chingford Pond would not cause any detrimental effects to the ecology or habitats of the pond, or the surrounding area.

The design of the restoration works and associated methodology for raising of the water level of the pond were carried out in a manner to ensure a successful project that combines reservoir safety with restoration of an historical and environmentally important site, as described below.

Slow Filling Rate

The main mitigation requirement to emerge from the EIA and planning process was that the water level within Chingford Pond was to be increased at a controlled rate to reduce the impacts on the ecology and habitats. To overcome this the whole project was designed around the concept of raising the water level from 16.5m to 18.3m over the course of approximately 10 years at a rate of around 0.20m per year. Insofar as ecology is concerned, the impacts of this slow steady increase in water level were considered to have far less ramifications than if the water level was increased over a shorter period, of say 12 months for example. It was assessed that undertaking this slow increase in water level would allow marginal habitats, in particular the important fen and swamp habitats, to migrate and establish in other parts of the site over time, providing habitat for the diverse range of terrestrial invertebrates these areas support, including the Desmoulin's whorl snail population.

The new scour valve and pipe was designed with two valves to allow fine adjustments to the outflow for manual control of the reservoir water level. The slow filling rate will be achieved through keeping the new scour valve open at a percentage that will allow the outflow to match the inflow into the reservoir. This will be monitored at times of low and high flows and manual adjustments to the scour valve opening will be carried out to ensure the correct level is being maintained for the majority of the time. This will be carried out by Burton Park full time maintenance staff as part of their daily and weekly maintenance checks. There will be times when the water level will fluctuate but this would occur naturally.

In conjunction with the slow reservoir filling rate the vegetation clearance works within the reservoir inundation area will be undertaken in two phases. The first phase will be a clearance of trees up to an equivalent land level of 17.0mAOD. The second phase will clear trees up to the final reservoir top water level of 18.3mAOD. This phased approach is required to ensure that the existing fen/swamp and wet woodland habitats remain *in situ*, whilst associated on and off site environmental mitigation measures, as described below, become successfully established. If the required mitigation takes more time than anticipated to establish then water levels will be held at 17.0mAOD until these mitigation measures are satisfactorily in place. A secondary outlet pipe is incorporated into the outlet chamber at a level of just below 17.0mAOD, to allow for holding the level at this level, if required, with minimal manual control.

Wet Woodland

To compensate for the inundation of 3.5Ha of wet woodland a compensatory area of wet woodland habitat will be establish over time. An area to the north of the estate was considered to be suitable for this and has been set aside for this purpose.

Water Framework Directive
The EU Water Framework Directive (WFD) 2001 requires all natural water bodies to achieve both good chemical status and good ecological status (GES).

New activities and schemes that affect the water environment may adversely impact biological, supporting conditions (hydromorphological), supporting elements (physico-chemical) and/or chemical quality elements (WFD quality elements), leading to deterioration in water body status. They may also render proposed improvement measures ineffective, leading to the water body failing to meet its WFD objectives for GES/good ecologic potential (GEP). Under the WFD, activities must not cause deterioration in water body status or prevent a water body from meeting GES/GEP by invalidating improvement measures.

The overall ecological status of a water body is primarily based on consideration of its biological quality elements and determined by the lowest scoring of these. These biological elements are, however, in turn supported by the physico-chemical (supporting elements) and hydromorphological (supporting conditions) quality elements.

One WFD-classified waterbody was considered in relation to the scheme; the Burton Mill Pond lake waterbody. As set out within the South East River Basin Management Plan, this comprises Trout Pond, Burton Mill Pond and Chingford Pond itself, including the Duncton stream upstream of Chingford Pond. As the works at Chingford Pond have the potential to affect the water environment, an assessment of their compliance with the WFD was required.

The assessment was undertaken on the understanding that the Burton Mill Pond is classified as a lake waterbody under the WFD, and that an appropriate reference of the condition was the most recent condition assessment for Burton Mill Pond SSSI. This described the SSSI as being in 'unfavourable condition' because of the substantial loss of mesotrophic open water habitat to swamp, fen and wet woodland. The definition of 'lake' was taken to include the emergent and marginal vegetation of the water body.

The assessment concluded that the construction of the new spillways is not considered to fundamentally change the hydromorphological function of the pond which is a linear storage within the river network. Furthermore, the increase in capacity to store sediment in Chingford Pond may halt the reported decline in the condition of Burton Mill Pond which is thought to be caused by sedimentation and a consequent reduction in the area of open water habitats. The proposed works were therefore considered unlikely to cause deterioration or prevent the WFD status objectives from being reached for the hydromorphology element.

The biological assessment considered the Burton Mill Pond as a lake water body focussing on the open water habitat types (open water and marginal / emergent habitats). The assessment considered the proposed works not to be detrimental because the increase in water level generates a larger area of mesotrophic open water habitat and the proposed mitigation measures will ensure that the gross area of marginal and emergent vegetation is retained.

The overall WFD assessment conclusion was that the proposed works would be unlikely to cause deterioration or prevent the WFD status objectives from being reached.

SSSI
Chingford Pond is situated upstream of Burton Mill Pond, a designated SSSI since 1986. The SSSI is located immediately adjacent to the eastern boundary of Chingford Pond and covers an area of 57Ha. The SSSI is designated for its extensive areas of open water and aquatic and emergent vegetation and surrounding carr woodland. In addition, bog areas of wet heath and marshy grassland are present and the site includes several nationally rare invertebrate species and an important breeding water bird community.

Historically there has been anecdotal evidence that Chingford is having a negative impact on the status of the SSSI. It has been suggested that the level of sediment within the pond prevented it from functioning as a silt trap, and allowed nutrient enriched sediment laden water to flow out of the breach in the embankment and into Burton Mill Pond, reducing water quality. The area of open water in Burton Mill Pond is reducing every year and this may also be having adverse impacts on aquatic plant diversity.

Increasing the water levels of Chingford Pond will create additional depth of water, and a slight attenuation of the flows that enter the pond, allowing for increased sedimentation and a reduction of sediment reaching Burton Mill Pond and the SSSI area. In addition future sediment trap works proposed for the area upstream of Chingford Pond are aimed at providing a longer term benefit to the SSSI from reduced siltation.

The slow filling of Chingford Pond will cause a slight retention of some water entering the pond. This will have a corresponding slight reduction in flows into Burton Mill Pond downstream, which may result in impacts on water quality or flows through the Burton Mill Pond system. In order to mitigate this potential impact the raising of the water level will be undertaken in the wetter winter months (between November and February). This will reduce the risk of impacts on the downstream SSSI as rainfall inputs to the pond catchment are higher in the winter months, and there will be more 'available' water to fill Chingford Pond.

When the levels of Chingford pond are not to be increased (summer months) the scour opening will be set so that all flows into the reservoir will be the same as the flow out of the scour, and so there would be no reduction in flows to Burton Mill Pond.

General Requirements

The majority of the other identified environmental effects were addressed through standard good construction practice, as described in the accompanying Environmental Management Plan (EMP). Additional mitigation and compensation measures have been identified as being necessary during the construction and operational phases of the scheme. These have been implemented either by a planning agreement tied to the planning permission or conditions attached to the permission, and / or by the EMP, which is being used to ensure that environmental issues are addressed and mitigation measures identified in the EIA are put in place.

CONCLUSION

Restoration of the Chingford Pond reservoir will require significant engineering works to ensure the safety of the reservoir and compliance with reservoir legislation. The restoration works will also result in the inundation of 3.5Ha of wet woodland, home to important aquatic habitats and terrestrial ecology.

However, through thorough consultation, EIA and planning processes, clear environmental mitigation and enhancement requirements have been identified early within the design process of the project. This has allowed the reservoir upgrade works design to be tailored to suit these requirements, resulting in a successful project that combines reservoir safety with restoration of an historical and environmentally important site. It is envisaged that this project may be used as a case study for similar future projects on environmentally / historically important reservoirs, or for design and construction of new reservoirs where inundation of habitats is likely.

SECTION 2:
GEOTECHNICAL ISSUES OF DAM CONSTRUCTION AND MAINTENANCE

Maintaining the Safety of our Dams and Reservoirs
ISBN 978-0-7277-6034-0

ICE Publishing: All rights reserved
doi: 10.1680/mdam.60340.075

International Levee Handbook - new guidance and a UK approach to the use of Eurocodes

J SIMM, H R Wallingford
L CLARKE, CIRIA
R SANDHAM, Ove Arup and Partners Limited
A PICKLES, Ove Arup and Partners Limited

SYNOPSIS   Flood embankments, or levees, are a vital part of flood risk management. In the light of serious levee failures during major storms such as Hurricane Katrina in the US (2005), Tempête Xynthia in France (2010), torrential rain in Pakistan (2010) and torrential rain and snowmelt in the US (2011), organisations from six different countries (France, Germany, Ireland, the Netherlands, the UK and the USA) decided to come together to provide a guidance document on the design, construction and management of levees – the International Levee Handbook (ILH, 2013). The paper will describe the vision for the handbook, give an overview of its contents and provide a short summary of one of the key technical issues which emerged in relation to the UK and Ireland context.

As smaller reservoirs (down to 10,000m^3) come under regulation in at least some parts of the UK the distinction between flood embankments and dams will become less significant. A nascent community of practice in the UK and Ireland, which has developed around the ILH, supported the development of new national guidance on the design of flood embankments to Eurocode 7 (expected 2014). This will be of interest to dam engineers. The further development of this community is currently being discussed, with the possibility of it being developed in association with the BDS, to ensure that relevant knowledge is shared with the dam community.

INTRODUCTION –LEVEES AND THEIR ROLE
Levees, otherwise known as flood embankments or dikes, are a vital part of modern flood risk management, with several hundreds of thousands of kilometres of levees in Europe and the US alone. Raised, predominately earthen structures, the primary objective of levees is to protect against fluvial and coastal flood events along coasts, rivers and artificial waterways. They achieve this either by temporarily retaining water, and keeping it out of the leveed area, by channelling floodwater downstream to avoid

inundation of the leveed area or by permitting a controlled release of water into a designated location that will minimise inundation downstream. They come in a variety of forms, are not reshaped under normal conditions by the action of waves and currents, and can incorporate flood walls, pumping stations, gate closure structures, natural features and other associated structures. Typically long linear structures of modest height, levees are part of an overall flood defence system: a system that should be considered as a chain which is only as strong as its weakest link.

Despite their apparent simplicity, levees can be surprisingly complex structures. Historically, they have generally been constructed by placing a variety of local fill materials onto alluvial flood plains, with all their inherent natural variability. Older levees often comprise multiple stages of construction and varying materials and often have associated problems of:

- Inaccurate historical documentation regarding construction methods and material constituents used to build the levee.
- Embankment core and foundation materials lacking, due to improper placement or permeable characteristics, resulting in under-seepage and piping
- Inadequate surface erosion protection
- Poorly designed and/or constructed transitions in both transverse and longitudinal directions.
- Complexities of the structures arising from multiple stages of construction.

Unlike engineered dams, levees can be irregular in the standard and nature of their construction and will deteriorate if not well maintained.

In many cases, records of levee construction and historical performance do not exist. This means that, whilst conventional dam procedures might be appropriate (at least in some situations) for levees, there is often much poorer information available than is typically available for reservoir dams.

The effects of climate change on levees can also not be ignored. Mean sea level rise and increased fluvial peak flows are meaning the existing levee crest heights and profiles are proving inadequate. Climate variability may also affect the frequency and intensity of hydraulic loading, as well as soil erosion caused by significant precipitation, drought or high winds. Excessive vegetation and surficial cracking as a result of more prolonged dry seasons may also become problematic. All of these can affect the structural integrity of the levee.

Despite their critical importance in mitigating flood risk, interest and investment in levees has tended to be lower than in other critical water retaining infrastructure such as dams. Many levees can stand for much of their lives without being loaded to their design capacity, and can create a false sense of security about the level of protection they provide. This can also give the incorrect assumption that they are of lower risk and do not need as much money to maintain them. Indeed, in many countries, levees have lacked the legal and technical framework necessary to promote an appropriate level of performance.

IMPROVING PRACTICE IN LEVEE DESIGN AND MANAGEMENT
The role of levees in flood defence has received increased attention in recent years, following a series of events that caused risk to human health, economic damage and have had long term effects. These include the hurricanes Katrina and Rita in the US in 2005, the Xynthia storm in France in 2010, the flooding in southern Pakistan in 2010 that affected 18M people, and the worst flooding in a century along the Mississippi River in the US in 2011.

In the aftermath of these events the governments of France, Germany, Ireland, the Netherlands, the US and UK realised that there was the need for the production of a single reference source on good practice in the management and design of levees. This international collaboration led to the production of the International Levee Handbook (ILH) in 2013. Written by a core team of experts and practitioners from the full range of relevant disciplines drawn from all of the above partner countries, the handbook is based on more than five years work by an international team of experts supported by an international peer review process. It is more than just a revision or combination of existing documents from the participating countries.

The development of the handbook followed an agreed set of procedures that was managed by a technical editorial team, and supported by an executive steering board drawn from national groups of the partner countries. Management was provided by Construction Industry Research and Information Association (CIRIA) in the UK.

The handbook takes a performance, risk, and systems based approach. Performance is evaluated by means of evaluation of the all potential failure mechanisms which are grouped for the purposes of the handbook into three groups: external erosion, internal erosion and mass instability. The systems-based approach is vital because if any one segment of a flood defence system deteriorates, the probability of failure of that element increases and so the performance of the whole system can be compromised. The contribution of any one segment to the overall flood risk (probability times consequence) can be determined by an analysis of the entire system (Gouldby et al, 2008) taking account of:

1. The consideration of the source of the initiating event (e.g. an extreme rainfall event, storm surge).
2. The performance or response of the levee when exposed to that event.
3. The nature of the area protected by the levee.
4. The consequences incurred given that the flood occurs, reflecting the vulnerability of the receptors (e.g. people, property, environment).

The handbook also follows a tiered approach to all aspects of managing and maintaining a levee or levee system, such that concepts are applicable to both urban and rural settings. Useful information is also provided for both existing and newly designed levees.

The ILH is written to help a technically competent practitioner with a broad (but not necessarily expert) knowledge of the field to arrive at the best approach for a particular levee or levee system. It aims to provide information to support decision making rather than to direct it, and will also seek to provide a client who has a technical background, but no particular specialist knowledge, with sufficient information to understand the main issues and general procedures likely to be followed by an experienced practitioner. Useful information is also provided for constructors (or other organisations) that may be advising the manager or designer, carrying out maintenance, or carrying out new construction work.

Ten chapters cover four main areas of fundamentals, managing levees, the toolbox and making changes. These include dealing with three main parts of the levee life cycle (Figure 1):

- The operation, maintenance and assessment of existing riverine, coastal and estuarine flood protection levees (possibly for new or changed performance requirements).
- Emergency management.
- The adaptation or replacement of levees (including new levees), with management interventions ranging from major construction projects through to routine maintenance by the involved authorities' own work force.

The team behind the handbook believes that it has put together an extensive document on the safety, assessment, management, design and construction of levees; incorporating all elements of good practice. Although the ILH is not considered to be prescriptive it states that "the appropriate application of the guidance in this handbook will help to underpin long-term improvements in the management and design of levees, and will help to promote conservation of natural systems in balance with the proper protection of human life and property."

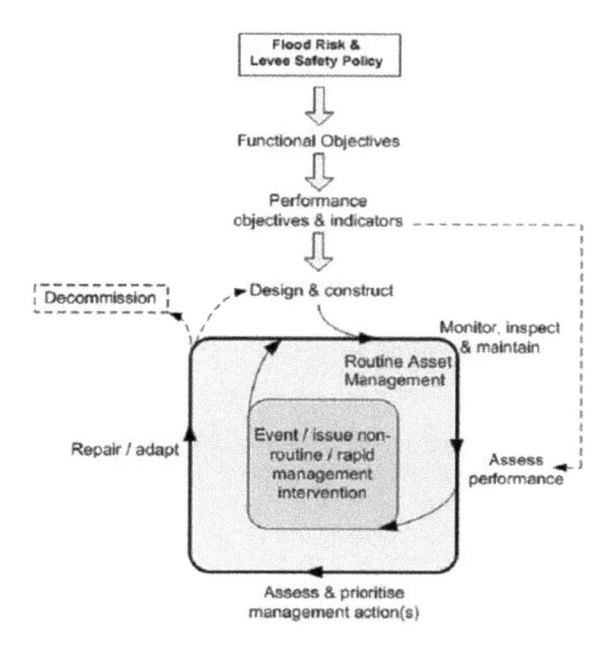

Figure 1. Levee life cycle

INTERNATIONAL LEVEE HANDBOOK IN MORE DETAIL

Details of each of the chapters are outlined in Figure 2 and explained below.

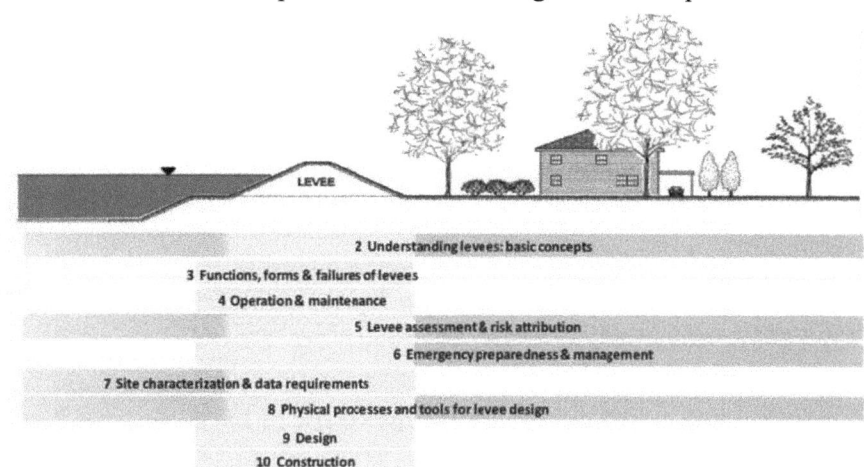

Figure 2. Illustration of the chapter contents relative to the levee.

The first chapter defines the structure, aims and scope of the handbook and explains how the document builds on and complements existing manuals and guidance.

Maintaining the Safety of our Dams and Reservoirs

Fundamentals
Chapter 2 Levees in flood risk management sets out the context for the use of levees, explaining roles and responsibilities and life cycle management.

Chapter 3 Function, forms and failure of levees overviews levee functions, illustrates the main forms of levees and associated structures and discusses processes of levee failure related to their forms and functions.

Managing levees
Chapter 4 Operation and maintenance addresses the management of existing levees, including the management of encroachments and vegetation. Maintenance requirements are related to the various types of defects arising from prevailing deterioration and damaging mechanisms.

Chapter 5 Levee inspection, assessment and risk attribution provides a tiered approach to the assessment of levee systems including risk analysis, assessment and inspection. Data collection and management is related to assessment activities. inspections, investigations and monitoring

Chapter 6 Emergency management and operations discusses preparedness and response and the various emergency intervention techniques for levees including equipment and activities for minimising overtopping and damage and for subsequent repair and closure of breaches.

Toolbox
Chapter 7 Site characterisation and data requirements describes the basic principles and investigation and analysis techniques to establish the morphological, hydraulic and geotechnical boundary conditions at levees and also the condition of existing levees, providing relevant equations and describing desk study procedures, techniques for field investigation and laboratory testing, approaches to data interpretation and methods and procedures for determining appropriate parameters for design.

Chapter 8 Physical processes and tools for levee assessment and design: provides equations and analytical tools for geotechnical and hydraulic analysis and design. It covers external and internal hydraulic, geotechnical and seismic actions on and physical processes in levees and associated floodwalls and methods of assessing levee breach and inundation.

Making changes
Chapter 9 Design sets out principles of levee design, roles and responsibilities and explains how to determine levee alignment and geometry. Specifics are given for the design and detailing for mass instability for seepage and internal erosion, surface protection and for limiting serviceability changes, selecting and compacting earthworks, designing spillways and any earthworks around embedded/associated structures, including crest walls and pipes.

Chapter 10 Construction describes organisational aspects, programming and the management of construction risk, focussing on earthworks including the suitability of the soils, their treatment and handling. The stages of construction for earthworks and the incorporation of structures are discussed.

DELIVERING BETTER PRACTICE
Uptake and implementation of the ILH is now underway following its launch in Arles in France in October and a UK launch in London in December 2013. The challenges of delivering better practice are not being underestimated and training slides are being prepared by the ILH core team for use in training sessions and webinars. In parallel with this, setting up of a levee community of practice is being explored, making appropriate links nationally with CIRIA, BDS and CIWEM and internationally with ICOLD.

The ILH focuses on identifying underlying principles and issues that need to be resolved in practice along with illustrative methods and examples from various nations in Europe and North America. However, it does not fulfil the role of providing nationally specific approaches to implementation

Through working with the different European nations a significant gap was identified in UK practice relating to the implementation of Eurocodes in regard to levees. A number of countries such as France, Germany and Netherlands have developed nationally specific non-contradictory complementary information (NCCI) to assist in this process. In particular such guidance has been prompted by the fact that water retaining structures are not covered well by the current version of Eurocode 7 (EC7), which deals with Geotechnical Design. These countries offer guidance for both flood embankments and small dams, recognising the similar challenges faced by designers and maintainers of these structures. However, such guidance is not available in the UK, despite the fact that from March 2010, all European public-sector clients have been legally required to commission Eurocode compliant designs.

The UK community recognised that there was a gap and identified the need to develop a guidance note for the design of flood embankments to EC7. This guidance note is currently being developed with the support of the UK community of practice, including flood risk management authorities in the UK, and is discussed in more detail below. Consideration is being given to the application of this guidance to the design of dams.

GUIDE TO THE DESIGN OF FLOOD EMBANKMENTS TO EUROCODE 7
The 'Application of Eurocode 7 to the design of flood embankments' is an interim guide for the UK that will eventually be superseded by the next revision of EC7 (due to be published in 2020). The guide takes into

consideration thinking from relevant EC7 Evolution Groups at the time of writing and aims to improve clarity on key issues, including:

- distinction between different design situations;
- application of Ultimate Limit States;
- risk categories for flood embankments; and
- design water pressures in marine or fluvial environments.

The guide refers primarily to the embankment cross section and the materials used to form the embankment, with other aspects of design being discussed in the ILH. It is intended to be used hand-in-hand with the ILH and relevant Eurocodes.

The embankment design process is considered in the first instance, with signposting to relevant sections of the ILH. This includes design considerations such as embankment alignment, ground conditions, materials, durability and serviceability. The concept of design situations within the context of Eurocodes is explained and four different situations are identified that commonly occur in flood embankment design:

- Construction
- Normal operating conditions
- Design flood event
- Rapid drawdown following design flood

All of these are *transient* situations except normal operating conditions, which are classed as *persistent*. Water levels and pressures are a critical aspect of these situations and sometimes it may also be necessary to consider *accidental* situations that could occur, such as extreme water levels as a result of a failure elsewhere in the system.

Limit states

The Eurocodes require that the assumptions that go into creating design situations should be varied and sufficiently severe to address circumstances that can reasonably be expected during the lifetime of the structure. For each design situation, the designer must verify that neither the ultimate limit states (ULSs) nor the serviceability limit states (SLSs) will be exceeded. That is, for the ULS the structure should not fail and for the SLS it should continue to perform.

In order to verify that limit states are not exceeded, calculations are usually undertaken to demonstrate that design values of the effects of actions do not exceed design resistances or design values of serviceability criteria. Actions may include the mass of the ground, earth pressures, external water pressures, pore water pressures, imposed traffic or impact loading. For flood embankments and embankment dams, resistance is typically provided by the strength of the material forming the embankment and the foundation soils.

Risk categorisation
Consequence classes are identified in Eurocode's basis of structural design, EN 1990, which classify a structure on the basis of the consequence of failure (loss of life, social and environmental considerations). These are broadly equivalent to risk categories proposed in the ILH and influence the levels of reliability in design. Differing reliability levels can be achieved through modification of partial factors and varying degrees of supervision of the design and construction process. The aim is that the level of reliability provided is proportionate to the consequence class. In addition, EC7 uses *geotechnical categories* (GC1 to 3). These are used for a similar purpose but consider the complexity of ground conditions as well as, to some extent, consequence of failure. The geotechnical category influences the level of ground investigation and analysis that is undertaken. It is unlikely that any flood embankment would fall into GC1, relating to small and relatively simple structures with negligible risk. Some may fall into GC3, where there are particularly challenging ground conditions, but many would be expected to fall into GC2.

Characteristic and design values
Within the Eurocode framework *design* values of actions, material properties or resistances have a specific definition which means that the appropriate margin of safety or partial factor has been applied. These margins or factors are normally applied to *representative* values of actions, *characteristic* values of material properties, and *nominal* values of dimensions. In EC7 these partial factors are on: material strength (γ_M); actions (or effect of actions) (γ_F); and resistance (γ_R). The new guide proposes a scheme to modify partial factors on the basis of consequence class. However, it is recommended that further work is undertaken to establish the most appropriate relationships between consequence class and modified partial factors for flood embankments.

Characteristic and design water levels
One of the challenges of developing guidance for the design of flood embankments to EC7 has been finding the best way to approach design water levels and pressures. For both flood embankments and embankment dams, water levels will control the hydraulic forces applied to the embankment by external water, which is impounded. They will also act as boundary conditions to the assessment of pore water pressure profiles within

the embankment and foundation soils. Changes in water levels define many of the design situations.

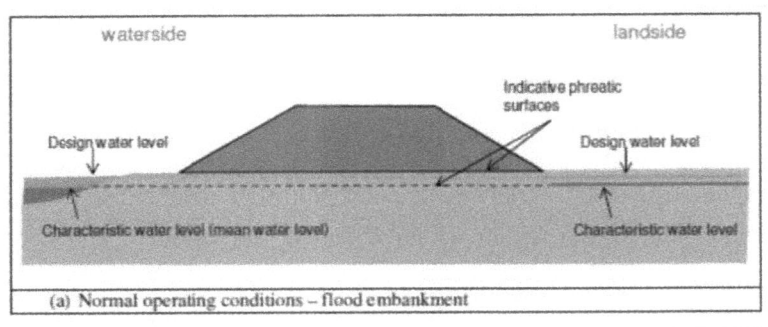

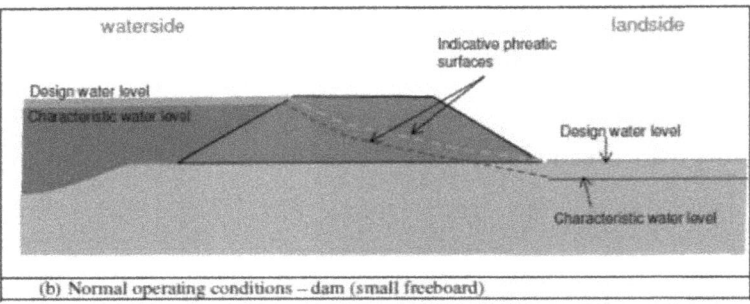

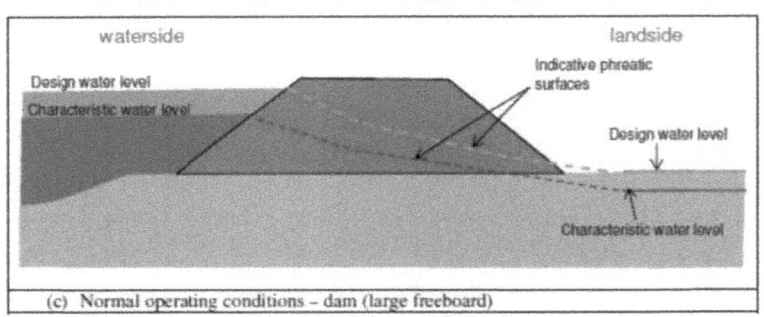

Figure 3. Characteristic and design water levels

At the time of writing, the guidance from Eurocode Evolution Group 9 (EG9) is to determine design water pressures directly from design water levels, rather than by applying a partial factor to characteristic water pressures. However, where the designer considers that this approach would not incorporate a sufficient margin of safety, such as in a low standard of protection embankment, a partial factor may be applied. Changes in permeability and seepage assumptions are likely to cause a much greater magnitude of change in the calculated pore pressures than applying the factor on actions.

There has been some debate on what constitutes a characteristic water level. In terms of flood embankments, characteristic water levels are proposed to be the most onerous which are likely to occur for a given design situation. Based on recommendations from EG9 this is taken to mean that the characteristic water level has a recurrence period of at least equal to the duration of the design situation of the structure. Design water levels are proposed to be the most onerous that could occur for a given design situation and EG9 proposes this to be defined as a water level that has a 1% probability of being exceeded in the design life of the structure.

APPLICATION OF NEW GUIDANCE FOR DAM ENGINEERS

The dam engineering community is accustomed to using the global factors of safety indicated in the table below, rather than applying margins of safety on water levels or partial factors. There is some work to do to establish an approach based on EC7 that gives equivalent reliability.

Table 1. Factors of safety commonly adopted for slope stability design of new embankment dams (Johnston et al)

Loading Condition	Typical minimum acceptable FoS
End of construction	1.3 to 1.5
Steady seepage with reservoir full	1.5
Rapid drawdown	1.2

The loading conditions above can be related to different design situations. It is not, however, straightforward to implement different factors for each of these within the current UK framework of partial factors.

Unlike the majority of flood embankments, the 'persistent' situation for dams tends to be a high water level. Steady state seepage is likely to have established and it may be difficult to verify that adding a margin on the water level provides the required level of reliability. While it is possible to consider factoring of water pressures, this is not desirable and instead it may be better to increase the factors on actions and material factors, which also act to provide overall levels of reliability.

Using the principles in the new guidance for flood embankments, consequence class could also be taken into consideration in deciding appropriate partial factors for design situations for dams. This could work for example in the rapid drawdown case. For this condition a lower factor of safety has been traditionally accepted as the associated failures tend to be shallow and therefore less likely to lead to a breach of the embankment than other deeper seated slips, which may affect the crest. As high pore pressures are normal in the upstream slope, rapid drawdown is usually more onerous than normal conditions. A lower consequence class could be applied, based on the mechanism of failure. This would allow lower partial factors to be used and avoid embankment design being governed by the rapid drawdown case. Any such modification to partial factors would currently need to be determined on a case by case basis, with other factors such as crest width and reparation of damage being taken into consideration. With further work, guidelines could be developed on specific partial factors for different design situations for dams and for different potential failure mechanisms.

The next revision of EC7 is due in 2020. There is an opportunity for the dam community to influence the future shape of EC7 and to ensure that the design approaches and partial factors are appropriate for use in dam design.

CONCLUSIONS

1. Flood embankments, or levees, are predominantly earthen structures, which form a vital part of flood risk management. They can be complex structures which will deteriorate if not well maintained.

2. Following a number of serious failures, an International Levee Handbook has been produced to provide guidance on the design, construction and management of these structures.

3. The ILH takes a performance, risk and systems based approach comprising evaluation of potential failure mechanisms, the nature of the area protected and consequences of failure.

4. The ILH covers the main parts of the levee lifecycle: operation, maintenance and assessment of existing levees; emergency management; and adaptation/ replacement/ construction of levees.

5. One key technical issue has emerged in relation to the UK and Ireland context; the design of flood embankments to Eurocodes.

6. As smaller reservoirs come under regulation in at least some parts of the UK (down to 10,000m^3) the distinction between flood embankments and dams will become less significant.

7. For both flood embankments and dams, one of the challenges of designing to Eurocodes lies in the choice of design water levels and pressures. Another challenge is factoring actions and material factors to achieve an overall level of reliability equivalent to the global factors of safety engineers have been used to.

8. There is an opportunity for the BDS to engage with a community of practice as it strives to influence revisions to the next issue of Eurocodes (due 2020). Further research could be undertaken to support designers of both levees and embankment dams.

REFERENCES

Gouldby, B, Sayers, P. Mulet-Marti, J. Hassan, M. and Benwell, D. (2008) A methodology for regional-scale flood risk assessment. *Proc. Institution of Civil Engineers – Water Management*, **161(3)**, 169-182.

Johnston, T.A., Millmore, J.P., Charles, J.A. and Teed, P. (1999) *An engineering guide to the safety of embankment dams in the United Kingdom.* Building Research Establishment, Watford, UK

CIRIA (2014 in print) Pickles, A. and Sandham, R. *Application of Eurocode 7 to the design of flood embankments.* C714, CIRIA, London

CIRIA, FRENCH MINISTRY OF ECOLOGY, USACE (2013) *The International Levee Handbook*, C731, CIRIA, London (ISBN: 978-0-86017-734-0). Go to: www.ciria.org/ILH

The nearly perfect nineteenth century embankment dam

P J RIGBY, United Utilities PLC
A N THOMPSON, United Utilities PLC
D E JONES, United Utilities PLC

SYNOPSIS   United Utilities (UU) is owner of over 170 embankment dams; most constructed during the late nineteenth century and include homogeneous and Pennine type dams. In support of its Portfolio Risk Assessment (PRA) approach prioritisation of dam work, UU has undertaken a number of seepage "Toolbox" workshops to quantitatively identify the main risks to its dams from internal erosion.

The paper provides a timeline of the development of embankment dam design and notable incidents and these are related to the construction of dams included within the Toolbox study

The Toolbox studies include both homogenous and clay core dams and the paper discusses common mechanisms of potential failure identified the embankment, associated with conduits and through the foundation. In addition some potential shortfalls in design, common defects and inconsistency in the construction practice are discussed. Following review of the homogenous and clay core dams, an idealised "C19th embankment dam" design is promoted that would address the majority of the potential failure mechanisms and associated risks for internal erosion.

HISTORY OF EMBANKMENT DAM DEVELOPMENT
United Utilities owns and operates some of the oldest embankment dams within the UK. Records indicate that the earliest was built around the turn of the 18th century to meet the demands of the rapidly expanding canal network around Rochdale. These early dams were constructed as homogenous dams using locally derived materials typically comprising Glacial Till and peat. Construction records for these early dams generally indicate no specific selection of materials, with little zoning or layering in the placement of the fill. However there is evidence from historical drawings to suggest that where impervious material was available it was used as a protection to the upstream slope or placed in thin layers to form a central core.

Design and as-built records indicate that the slopes adopted for the original construction of the dams were typically overly steep given the nature of the embankment fill, with downstream slopes of 1:1.5 not uncommon. Not surprisingly the records from this time detail remediation and reconstruction being required due to slope instability, with further works often required post first filling.

By the middle of the 19th century a fairly typical design of embankment dam had been adopted comprising an upstream slope of 1:3 with a downstream slope of 1:2.5 with a central puddle clay core, often with a cutoff trench extending to a lower impermeable strata beneath the dam. This often required deep narrow trenches to form the cut off with vertical sides; Delph reservoir, near Bolton is a good example of such techniques. These Pennine style dams represent the largest proportion of the UU dam ownership with the last of its type, Jumbles, completed in 1974.

Audenshaw No 2, completed in 2000, was the last dam to be constructed in the area. This was designed based on a more developed understanding of internal erosion and to suit modern plant. It has a wide central clay core with shallow core trench and incorporates a downstream transition filter adjacent to the core and a basal filter blanket between embankment and foundation. Internally, 300mm diameter drainage blankets were installed both upstream and downstream with a toe drain. Upstream slopes of 1:5 and variable downstream slope, typically 1:4.

Table 1 details the timeline from construction of the first to the latest of UU embankment dams along with others referred to in this paper. It references these against some key incidents and developments that have driven the ongoing development and understanding of these significant complex earth structures.

DAM FAILURES - INTERNAL EROSION

A review of historical internal erosion incidents by Charles (2002) and DEFRA (2011) show a similar distribution of internal erosion incidents in Europe and the UK, attributing to failures with conduits, embankments and foundations. In contrast data from the USA (USSD 2010) indicates that 70% of all its failures are associated with erosion within the foundations. This has been attributed to the lack of a fully penetrating cut-off across the length of the embankment and the permeable and erodible nature of the soils in the foundations. Figure 1 presents collated historical data for (category 1 and 2) incidents associated with internal erosion.

These internal erosion incidents have been separated between dams with a central core and those described as homogeneous for both the European and UK embankment dams (shown in Figure 2).

Table 1. Timeline showing the construction of embankment dams considered in Toolbox Workshops and key incidents and developments driving dam design and construction

Significant incidents in British dam construction practice, legislation and guidance (adapted from DEFRA, 2011) and UU Embankment Dams considered in this paper	Date	Type of dam
First use of clay core, Serpentine (Hyde Park)	1730	Homogeneous Dam Construction
First foundation cut-off	1766	
Construction of **Blackstone Edge (homogeneous)** 1st for Rochdale Canal Company	1794	
Failure of Blackbrook due to poor construction	1799	
Construction of **Hollingworth Lake (homogeneous)**	1800	
Construction of **Upper Chelburn (homogeneous)**	1801	
Construction of **Whiteholme (homogeneous)**	1816	
Construction of **Springs (clay core with cut off)**	1830	
Failure of Bilberry, 81 dead - Overtopping failure Construction of **Upper Rivington (clay core with cut off)**	1852	Pennine Type Dam Construction
Specifications for zoned fill construction become common practice.	1854	
Construction of **Warland (homogeneous)**	1857	
Failure of Dale Dyke, 244 dead - conduit failure	1864	
Construction of **Clowbridge (clay core with cut off trench)**	1865	
Introduction of Waterworks Bill	1866	
Construction of **Laneshaw (clay core with cut off)**	1889	
Construction of **Chapel House (homogeneous)**	1902	
Construction of **Delph (clay core with cut off)**	1921	
Overtopping Failure of Skelmorlie, 5 dead and Dolgarrog failure, 16 dead	1925	
Heaton Park open (clay core with upstream clay blanket)	1927	
End of puddle clay core construction and start of rolled clay core construction in UK	1960	Modern Dam Construction
Construction of **Jumbles (clay core - narrow core)** - last of its type in NW England	1974	
Reservoirs Act 1975; publication of the 'Flood Studies Report'	1975	
ICOLD Bulletin 70 - Dispersive soils in embankment dams: review	1990	
ICOLD bulletin 95 - Embankment Dams: granular filters and drains	1994	
Construction of **Audenshaw (using prevailing dam construction techniques)**	2000	
Internal Erosion incident within conduit at Upper Rivington	2002	
DEFRA funded research into internal erosion	2002 2004	
ICOLD Draft Bulletin - Internal erosion of existing dams, levee and dikes and their foundations	2013	

It is interesting to note that the incidence of failure associated with European central clay core dams is similar to the UK for these sample sets; other failures types are also broadly similar with one exception. The main difference with the UK dams is the significant number of incidents associated with structures within the dam, particularly conduits. It is also of note that no foundation incidents have been recorded associated with homogeneous dams.

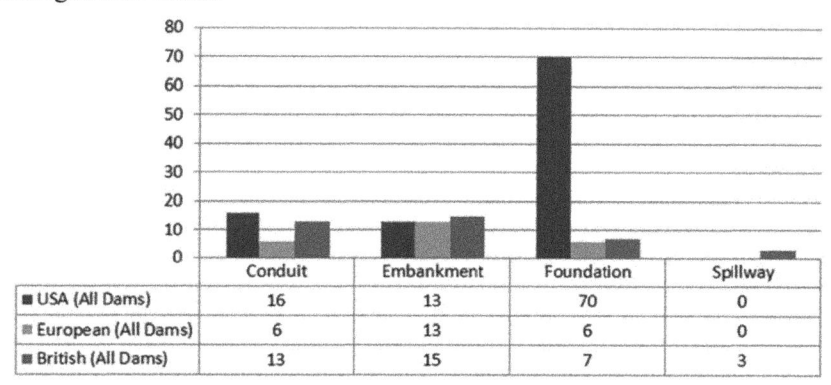

	Conduit	Embankment	Foundation	Spillway
USA (All Dams)	16	13	70	0
European (All Dams)	6	13	6	0
British (All Dams)	13	15	7	3

Figure 1. Historical Internal Erosion Incidents

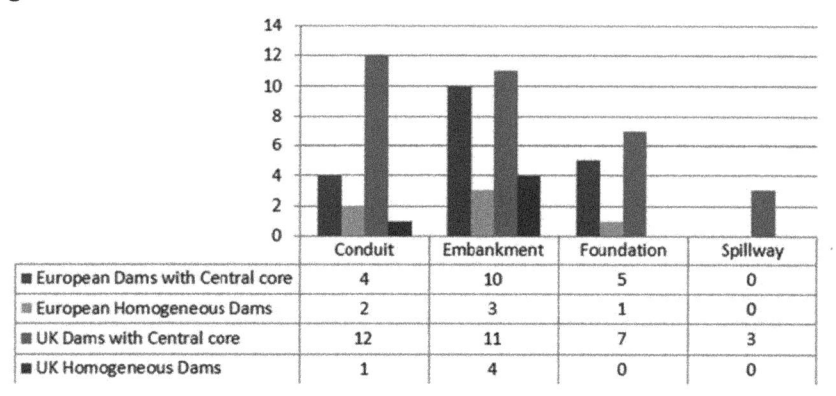

	Conduit	Embankment	Foundation	Spillway
European Dams with Central core	4	10	5	0
European Homogeneous Dams	2	3	1	0
UK Dams with Central core	12	11	7	3
UK Homogeneous Dams	1	4	0	0

Figure 2. Internal erosion incidents (Category 2 and above) for European Embankment Dams (based on Charles, 2002) and British Dams (based on DEFRA, 2011)

INTERNAL EROSION RISK ASSESSMENT
UU undertakes a Portfolio Risk Assessment (PRA) to provide for aspects regarding internal erosion. The PRA employs the University of New South Wales (UNSW) method described by Foster et al (1998). Based on the UNSW approach, a homogeneous dam over its life is considered to be over five times more likely than a central clay core dam to fail by piping failure. This has contributed to the 19th century homogenous dams at the top of UU

PRA ranking highly in the PRA and requiring further assessments based on piping modes of failure. In addition a number of clay core dams are considered to be at elevated risk of internal erosion due to the presence of conduits through the dam and increasing observable seepages. Figure 2 also suggests such an approach is appropriate.

UU has undertaken a number of "Toolbox" workshops to identify the key risks to 12 of its dams considered at PRA screening level to require more detailed assessment of the risk from internal erosion. The "Toolbox" uses an event tree approach to review the process of internal erosion from initiation, through continuation and progression and finally breach. The opportunities to intervene are also considered (USBR 2008).

INTERNAL EROSION THROUGH THE FOUNDATION

Ground conditions within the NW of England typically comprise a variable thickness of peat (typical of the homogenous dam) underlain by Glacial Deposits, primarily Glacial Clay. Rock comprises Coal Measures Strata, principally interbedded Mudstones and Siltstone and thick beds of Sandstone.

Amongst other things the Toolbox requires consideration of the potential for erosion of a soil foundations layer beneath the embankment by either backward erosion and/or suffusion in non-cohesive soils or through a crack in cohesive soils. Internal erosion through a rock foundation is considered by initiation of potential defects within the rock discontinuities.

The potential for initiation to occur depends on the seepage gradient driving erosion and the nature of the material through which seepage takes place. The range of seepage gradients acting on the foundation calculated during the workshops ranged between 0.1 and 0.2 for the homogeneous dams and in between 2 and >6 for the clay core dams. It can be seen that the gradients are understandably significantly greater if a cutoff trench is present giving seepage gradients in excess of 4. Whilst the clay core cutoff trench will restrict the backward progression of a foundation erosion mechanism the deep narrow cutoff trenches, present at Delph and Clowbridge, are susceptible to hydraulic fracture. Hydraulic fracturing is generally a key cause that could lead to leakage of a dam during first filling. The dams being considered in the paper are in excess of a 100 years old and the reservoirs are typically operated at top water level. It was considered that the initiation mechanism could be effectively excluded at screening level for Pennine type dams where the cutoff trench penetrates Glacial Clay.

SOIL FOUNDATIONS

A number of the dams included in the study are constructed on Glacial Till or peat materials without a cutoff trench (including both homogeneous and central clay core dams). With a combination of a low seepage gradient across the foundation and Plasticity Index values in excess of 13% the likelihood of backward erosion or suffusion within the soil foundation is considered to be negligible.

The potential for continuous or interconnected cracks in cohesive soils was considered during the early workshops. These cracks could be a result of differential settlement in the foundation or desiccation cracks where foundations had not been stripped. The $P_{failure}$ results (Figure 3 and 4) indicate a range of probabilities of failure from ALARP to Broadly Acceptable for this mechanism (HSE 2001).

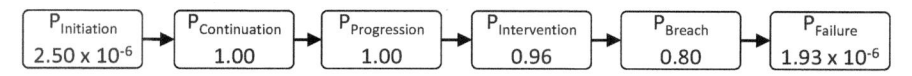

$P_{Initiation}$	$P_{Continuation}$	$P_{Progression}$	$P_{Intervention}$	P_{Breach}	$P_{Failure}$
2.50×10^{-6}	1.00	1.00	0.96	0.80	1.93×10^{-6}

Figure 3. Hollingworth Lake - differential settlement or desiccation cracks

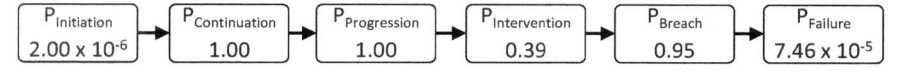

$P_{Initiation}$	$P_{Continuation}$	$P_{Progression}$	$P_{Intervention}$	P_{Breach}	$P_{Failure}$
2.00×10^{-6}	1.00	1.00	0.39	0.95	7.46×10^{-5}

Figure 4. Springs - differential settlement or desiccation cracks

In the case of the homogenous dams there is evidence to suggest that the foundations had not been stripped prior to dam construction. However the presence of a surface layer of peat overlying the Glacial Clay, whether natural (or placed as at Whiteholme) would effectively fill any cracks at this level. Records available for the clay core dams indicate that foundation preparation was carried out local to the core trench. In both cases foundation treatment would effectively prevent a continuous pathway beneath the embankment. On that basis the mechanism was excluded at the initiation stage of the event tree assessment.

ROCK FOUNDATIONS

During the Toolbox workshop sessions assessment of failure mechanism within rock foundations associated with discontinuities within the rock proved a challenging assessment. Geological knowledge of the rock types and local outcrop mapping was combined with ground investigation data and dam performance information gathered to inform the workshops.

Early Toolbox workshops assessed each potential defect in the rock at specific size ranges (from <5mm to >100mm), estimating the continuous nature of each defect (and each size) and whether the defects are open or infilled. The method resulted in seemingly very conservative assessment of the presence of continuous cracks and this generated high initiation probabilities. For a homogeneous dam with no central clay core cut off trench to limit flows this was a significant issue. Final probabilities in the Intolerable range were initially determined at Whiteholme (Figure 5). Additional ground investigation was undertaken to provide a better understanding of the nature of the rock discontinuities. The results of rock coring, groundwater data and insitu permeability testing, supported by inspection and performance data indicated that the results were unrepresentative of the actual situation.

As part of the reassessment process of this mechanism the toolbox was refined through the adoption of a somewhat simplified event tree using best practice given in USACE 2010.

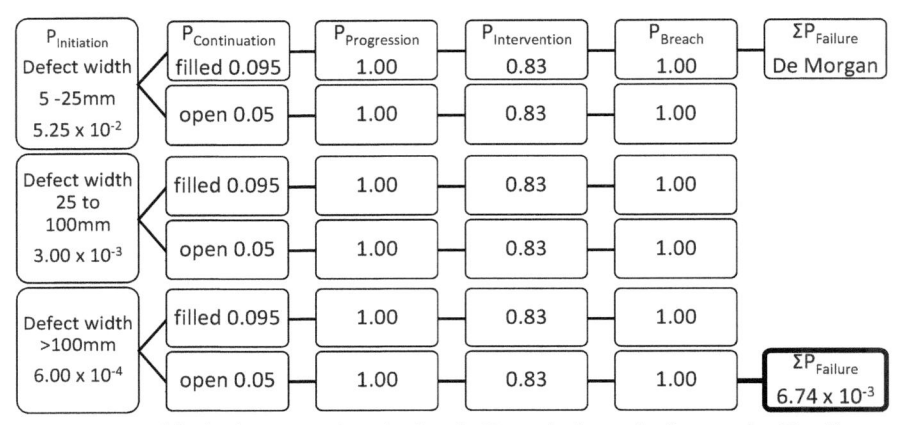

$P_{Initiation}$	$P_{Continuation}$	$P_{Progression}$	$P_{Intervention}$	P_{Breach}	$\Sigma P_{Failure}$
Defect width 5-25mm 5.25×10^{-2}	filled 0.095	1.00	0.83	1.00	De Morgan
	open 0.05	1.00	0.83	1.00	
Defect width 25 to 100mm 3.00×10^{-3}	filled 0.095	1.00	0.83	1.00	
	open 0.05	1.00	0.83	1.00	
Defect width >100mm 6.00×10^{-4}	filled 0.095	1.00	0.83	1.00	
	open 0.05	1.00	0.83	1.00	$\Sigma P_{Failure}$ 6.74×10^{-3}

Figure 5. Whiteholme Erosion in Rock Foundations (using early Toolbox methodology)

At Whiteholme the reassessment of continuous rock discontinuities resulted in a probability of failure of 6.56×10^{-7} (tolerable); there was more confidence in the result as it reflects the long dam performance of the dam (Figure 6).

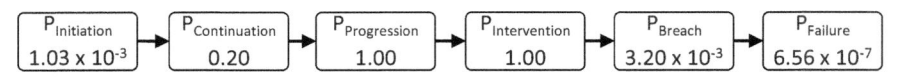

$P_{Initiation}$	$P_{Continuation}$	$P_{Progression}$	$P_{Intervention}$	P_{Breach}	$P_{Failure}$
1.03×10^{-3}	0.20	1.00	1.00	3.20×10^{-3}	6.56×10^{-7}

Figure 6. Whiteholme - Erosion in Rock Foundations (revised using USACE 2010 methodology)

CONDUIT

As may be seen from Figure 2 internal erosion is often associated with structures within or adjacent to the embankment dam and is a major cause of dam failure/incidents. If the interface of such structures with the soil forms a crack this can lead to development of pathways for concentrated leakage within the plastic embankment soils. The potential for this failure mechanism was evident at Upper Rivington in 2002 with water jetting through the culvert removing fines from the core (Gardiner *et al* 2004).

Typical causes of failure associated with conduits include (Fell *et al* 2005):
- Inadequate compaction due to the presence of cut off collars
- Inadequate compaction under and around the pipe
- Differential behaviour of soils in proximity of conduit or extremely weathered rock in the sides of a trench

Eight of the twelve embankment dams assessed using the toolbox had conduits, the majority of which were located within the embankment. Figures 7 and 8 detail the toolbox probabilities outcomes for a selection of homogeneous and Pennine type dams. The results for both the dam types place them close to Intolerable final probabilities (values $>10^{-4}$).

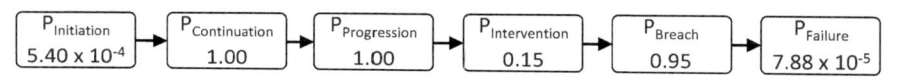

$P_{Initiation}$	$P_{Continuation}$	$P_{Progression}$	$P_{Intervention}$	P_{Breach}	$P_{Failure}$
5.40×10^{-4}	1.00	1.00	0.15	0.95	7.88×10^{-5}

Figure 7. Warland - Development of a crack around a conduit

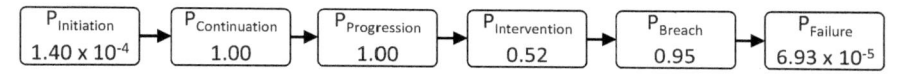

$P_{Initiation}$	$P_{Continuation}$	$P_{Progression}$	$P_{Intervention}$	P_{Breach}	$P_{Failure}$
1.40×10^{-4}	1.00	1.00	0.52	0.95	6.93×10^{-5}

Figure 8. Springs - Development of a crack around a conduit

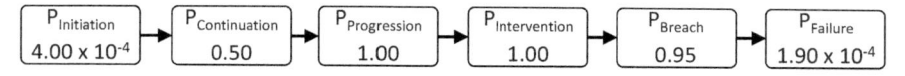

$P_{Initiation}$	$P_{Continuation}$	$P_{Progression}$	$P_{Intervention}$	P_{Breach}	$P_{Failure}$
4.00×10^{-4}	0.50	1.00	1.00	0.95	1.90×10^{-4}

Figure 9. Delph - Development of a crack around a conduit

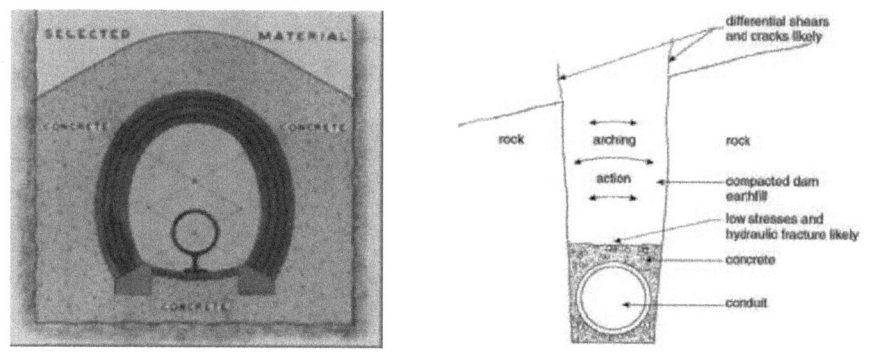

Figure 10. Conduit detail at Delph (left); example of bad practice of conduit embedment in rock (right, taken from Fell 2005)

At Delph efforts were made to prevent this particular mechanism (Figure 10). The conduit was constructed within a trench in the rock foundation and then encased in concrete with a puddle clay surround. However as the concrete did not extend to the top of the trench (and base of the embankment) it was considered there was potential for a crack to develop above the conduit within the backfill material above and along the trench walls resulting in a probability within the Intolerable Range (Figure 9).

EMBANKMENT
During the Toolbox study a failure mode has been identified common to both homogeneous and Pennine type dams. The mechanism is for the initiation of internal erosion by a poorly compacted or high permeability layer in the embankment.

The event tree allows for consideration of initiation of erosion due to a flaw in the core. The potential for Glacial Clay to provide some "filtering" function is justified given the potential beneficial affect and that this material is a common fill for UU dams. If this is a reasonable assumption then it would have the effect of reducing the probability of piping continuation that in the early Toolbox workshops was taken as a probability of 1.

The filtering properties of Glacial Clay were discussed by Bridle (2008) when assessing the vulnerability of British dams to internal erosion. He remarked that these dams have not shown vulnerability to long term internal erosion, which may be attributable in part to the low permeability and fine grained nature of the fill.

Figure 11 summarises the results of corrected average grading curves for core and embankment materials at Upper Rivington and Whiteholme. With reference to the US Army Corp of Engineers criteria for filter design (USACE 2000) the transition filter envelope has been predicted and compared to the filter erosion boundaries used in the toolbox (USACE 2008). The results suggest that the embankment material would offer filtering properties to the core and therefore give a significantly reduced probability for continuation, of the order of 0.0001.

Peat has been encountered in the homogeneous dams both within the embankment itself and at the embankment/foundation contact. Binnie (1987) makes the following comments on the use of peat in embankments such as Blackstone Edge "peat containing up to 50% partially decomposed organic material ...and squelches underfoot is a most unsuitable material to leave in any part of the foundations of an embankment". Such perceptions of the unsuitability of peat as a construction material, insufficient heed regarding the age of the dams and the limited guidance and developing nature over the application of the documentation in early Toolboxes influenced the early assessments. It was considered that the resultant probabilities (Figure 12) were, upon consideration of the probabilities of failure, very conservative.

Work by Xiao (2010) on subsurface erosion of peats concluded that internal erosion in peat was unlikely where it was consolidated under an embankment load, even at low loads. The high compressibility of peat under load would tend to close or heal piping channels that may have formed. The presence of a small percentage of organic matter (about 5%) results in a very significant increase in a soils ability to resist piping erosion and progression (Adams *et al*).

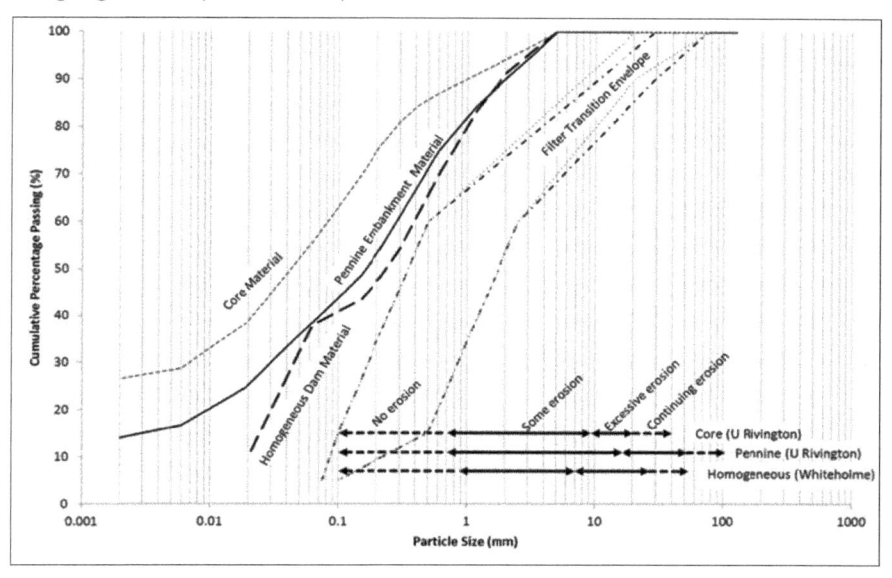

Figure 11. Averaged grading plots for the embankment and core materials at Rivington and Whiteholme showing the transition grading compared to the filter erosion boundaries

The results of the initial assessments for Blackstone Edge are shown in Figure 12 and the revised assessment (Figure 13), based on a refined understanding of the properties of the peat.

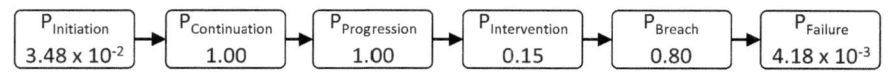

$P_{Initiation}$	$P_{Continuation}$	$P_{Progression}$	$P_{Intervention}$	P_{Breach}	$P_{Failure}$
3.48×10^{-2}	1.00	1.00	0.15	0.80	4.18×10^{-3}

Figure 12. Blackstone Edge Original toolbox assessment based on pessimistic properties of peat

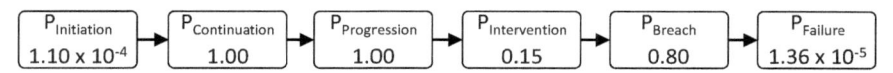

$P_{Initiation}$	$P_{Continuation}$	$P_{Progression}$	$P_{Intervention}$	P_{Breach}	$P_{Failure}$
1.10×10^{-4}	1.00	1.00	0.15	0.80	1.36×10^{-5}

Figure 13. Blackstone Edge Revision based on review of long term performance of the dam and contemporary assessment of peats resistance to erosion/piping

DISCUSSION

Figures 14 show the results of the 12 dams assessed as part of the Toolbox workshops. Comparison of the Intolerable probabilities of failure is very similar for the homogeneous and central clay core dams. The presence of structures within the embankment, particularly conduits close to the base of the embankments, has been identified as the main potential threat. Of the dams with conduits, all eight were assessed as having Intolerable probabilities of failure when the probabilities for erosion around and into the conduits were combined. Spillway problems lie within the ALARP range with the lower probabilities resulting from typically lower driving seepage gradients.

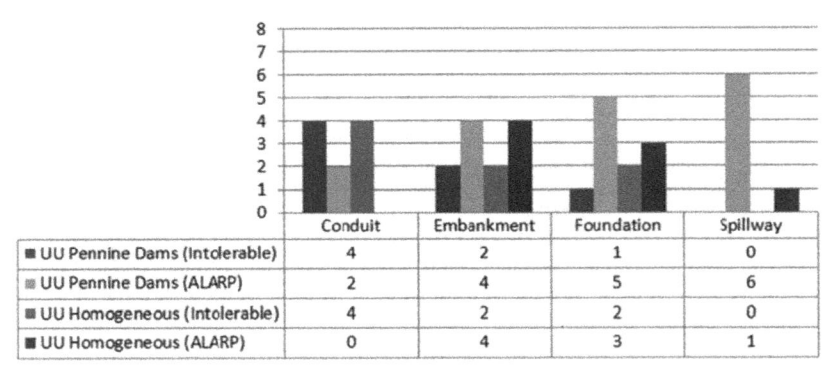

	Conduit	Embankment	Foundation	Spillway
UU Pennine Dams (Intolerable)	4	2	1	0
UU Pennine Dams (ALARP)	2	4	5	6
UU Homogeneous (Intolerable)	4	2	2	0
UU Homogeneous (ALARP)	0	4	3	1

Figure 14. Summary of Toolbox workshops undertaken on 12 dams

Foundation issues are relatively consistent for the two types of embankment. The presence of deep cutoffs and a relatively high seepage gradient on a number of the clay core dams is offset by the cohesive nature and selected grading of the Glacial Clay used in the core, reducing risk. The clay resists erosion and is considered not to be susceptible to suffusion; there is little evidence of backward erosion or suffusion in plastic soils. This may not be the case for dams constructed using more widely graded Glacial Clays in the core, such as the moraine clays used in Sweden.

Issues within the embankment are also consistent between the dam types. The presence of peat has been assessed and found to be beneficial in terms of resisting initiation of erosion both through the embankment and from the embankment into the foundation. The conclusions of the Toolbox workshop, when higher risk is likened to number of failures, support the findings of the draft ICOLD Bulletin (2013). Based on global data for dams built between 1800 and 1986 this concluded that the greatest proportion of failures are associated with the embankment. Of these, over half were associated with conduits through or walls adjacent to the embankment.

CONCLUSIONS

So did the 19th century dam builders get it right? It is fair to say that the homogeneous dams did suffer from slope stability issues both during and post construction. Overtopping issues were also a potential issue with the ongoing settlement of the peat resulting in loss of freeboard. However, after more than 100 years of operation with consolidation effectively complete the dams appear to be operating consistently to those of the later clay core dams. So whilst the homogeneous considered may not be nearly perfect, should they still be considered five times more likely to fail by internal erosion than the central clay core dams?

REFERENCES

Adams B, and Xiao M, (2010), *Piping potential of a fibrous peat*, 5th International Conference on Scour and Erosion proceedings, pp. 202 – 211.

Binnie G M, (1987) *Early Dam Builders in Britain*. London. Thomas Telford Limited 1987

Bridle R C, (2008), *Assessing the vulnerability of a typical British embankment dam to internal erosion*, Ensuring reservoir safety into the future, Thomas Telford, London

Charles, J.A. (2002) *A historical perspective on reservoir safety legislation in the United Kingdom. In Reservoirs in a Changing World*, Trinity College Dublin, 4-8 September 2002. London: Thomas Telford. 12th BDS Symposium Proceedings, pp. 494-509.

DEFRA and EA. (2011) *Lessons from historical dam incidents*. Bristol: Environment Agency, reference SCHO0811BUBA-E-E

Fell R, MacGregor P, Stapledon D, Bell G, (2005) *Geotechnical Engineering of Dams*, Taylor and Francis Group plc, London, UK, ISBN 041536 440x

Foster M, Fell R, Spanngle M, (1998), *Risk Assessment – Estimating the probability of failure of embankment dams by piping*. Proc ANCOLD/NZSOLD conference on dams. 1998 Sydney NSW 11p

Gardiner K, Hughes A, Brown AJ (2004), *Lessons from an incident at Upper Rivington reservoir*, Dams and Reservoirs, Volume 2, pp 35-37

ICOLD bulletin (2013) *Internal erosion of existing dams, levees and dikes and their foundation*, Volume 1, (draft)

USBR, USACE, The University of New South Wales and URS. (2008) *Unified Method for Estimating Probabilities of Failure of Embankment Dams by Internal Erosion and Piping Guidance Document*. (available at www.usace.army.mil/inet/usace-docs)

Health and Safety Executive (2001). *Reducing Risks: Protecting People*.

USACE, (2000), *Design and Construction of Levees*, Engineer Manual, CECW-EG Washington, DC 20314-1000, No. 1110-2-1913 (available at www.usace.army.mil/inet/usace-docs)

USACE, (2010), *A method for estimating probabilities of failure of embankment dams due to internal erosion*. Best Practices Guidance (draft available at www.usace.army.mil/inet/usace-docs)

USSD, (2009), *Managing our Water Retention systems* 29th annual USSD Conference, Nashville

Xiao M, Gomez J, Adams B, (2010), *Experimental study on subsurface erosion of peats*, Proceedings of GeoFlorida, Annual Geo-Congress of the Geo-Institute of ASCE

The successful rehabilitation of Heapey Embankment, Anglezarke Reservoir

C D PARKS, United Utilities, UK
M EDMONDSON, Edmondson Geotechnical Limited, UK
G MULREID, MWH, UK
D THOMSON, MWH, UK

SYNOPSIS   Parks and Walthall (2002) presented a paper that described how seepage through Heapey embankment, discovered in 1997, was seen to significantly increase when the reservoir level rose to within 800mm of top water level (TWL). The seepage emanating from the embankment was successfully sealed during 1999 by a phased grouting operation and the reservoir was returned to full operation.

Within three years the embankment was once again found to be exhibiting signs of seepage at the location of the earlier event. As on the previous occasion, the reservoir level was drawn down and was held 1m below TWL to curtail the leak. In 2010 the regular Section 10 Inspection included an "In The Interests Of Safety" (ITIOS) recommendation to "Treat the leak or reduce it to an acceptable level". A project commenced in 2011 to determine the available options to satisfy the ITIOS recommendation. This resulted in the award of a construction contract for a Weighted Filter to rehabilitate the embankment and bring the reservoir once more to full capacity operation.

This paper describes the background, the process of determining the rehabilitation option, the design and the construction of that option, completed in 2013, along with a summary of the post-rehabilitation performance.

CONSTRUCTION AND HISTORY OF HEAPEY EMBANKMENT
Anglezarke reservoir lies to the east of Chorley in Lancashire. With a capacity of about 4000Ml it is impounded by three embankment dams; Heapey embankment forms the northern limit of the reservoir. It was constructed between 1850 and 1857 for Liverpool Corporation and is one of the Rivington group of reservoirs. It has been under the ownership of United Utilities (UU) since 1995.

The embankment is approximately 85m in length and up to 10m in height. It was constructed from locally won Glacial Clay containing a significant proportion of weathered Coal Measures Mudstone and has a puddle-clay core. The crest is approximately 8m in width and carries a minor asphalt public road and a substantial masonry wave wall. Stone pitching covers the upstream face.

There are no ancillary structures at the embankment; a cast iron compensation flow draw off pipe of 18" (approximately 450mm) diameter is located under the eastern mitre of the embankment. The pipe formerly discharged into White Brook downstream and it was grouted in 1972 from the upstream valve. No records of the success of the grouting are available and the exact line of the pipe was assumed before the current project.

SEEPAGE AND REMEDIAL MEASURES - 1997 TO 1999
A seepage of water from within the embankment was discovered close to the eastern mitre upon refilling of the reservoir following a particularly dry summer in 1997. The seepage was sufficiently great to require an emergency draw-down. Investigations on the embankment led to a phased grouting operation which took place in 1998 and 1999. The first phase grouted the embankment along the 18" compensation flow pipe at depth. The second phase comprised two lines of grout holes at 1m centres along a 30m length aligned along the core and into the eastern valley side. Both phases utilised the "Tube-a-Manchette" technique with multiple pass grouting. Recording and plotting the grout injection volumes revealed zones of significant grout take. Further detail relating to these historical grouting works are presented by Parks and Walthall (2002).

The reservoir was allowed to refill to full operating condition in 1999 and the seepage had been reduced to a barely recordable level. This situation prevailed until 2002 when a leak was once again detected emanating from the same location as before. The reservoir was drawn down by 1m and remained drawn down in order to minimise the leakage.

Observed flows were collected and channelled through a dedicated headwall (referred to as Drain A) to allow regular monitoring of leakage flow rates against reservoir levels. This monitoring recorded flows of up to 0.83l/s during storm conditions as reservoir levels intermittently rose above the 1m drawdown below TWL.

2010 INSPECTION AND RECOMMENDATIONS
Within the 2010 Inspection Report, Reilly, (2011), made two ITIOS recommendations: 1) to treat the leak to seal it or to reduce it to an acceptable level and 2) to install piezometers on the downstream embankment face to determine the phreatic surface.

To meet the requirements of the second ITIOS recommendation, piezometers were installed in December 2012. As a result, the first Section 10(6) certification was granted on the 14th December 2012. In order to meet the requirements of first ITIOS recommendation a significant study and assessment of the options was required.

GEOTECHNICAL INVESTIGATIONS, REVIEW & INTERPRETATION

Ground Model

Ground investigation data were obtained from boreholes constructed to allow installation of piezometers on the embankment downstream face to satisfy the requirements of one the 2010 ITIOS requirements. This information was supplemented with historical data available from previous phases of investigation undertaken to inform earlier remedial works.

Borehole data confirmed the presence of a clay core to the embankment. The embankment shoulders were confirmed to comprise largely re-worked weathered Coal Measures Mudstone, Siltstone and Sandstone with possible Glacial Clay. The embankment was observed to be founded directly on Glacial Till and weathered Coal Measures strata. Figure 1 illustrates a typical cross section through the centre of the embankment.

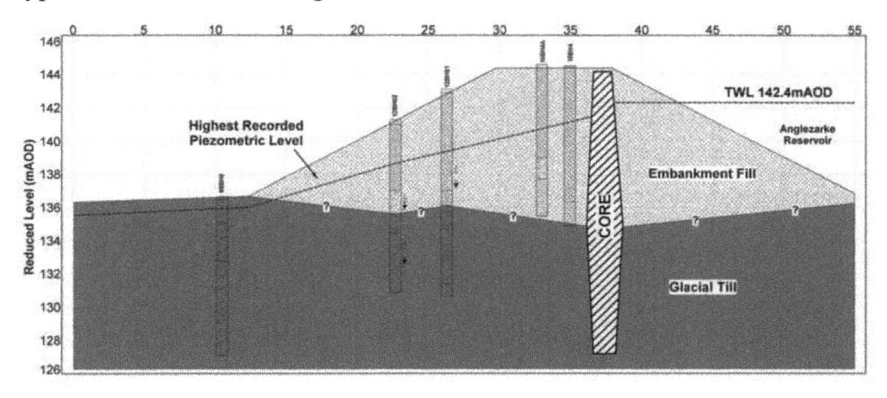

Figure 1. Geological cross section and Piezometric level through Heapey embankment

Ground water monitoring within the recently installed piezometers supplemented by existing instrumentation to the toe of the embankment indicated a phreatic surface falling gradually from reservoir water level to the embankment toe (Figure 1) with a small drop across the core. The data indicate that the core is not as efficient as might be expected from 'puddle clay'. A slight artesian head was however observed in the existing instrumentation to the toe of the embankment. This was not considered to be linked to reservoir levels and merely a function of the background local hydrogeology being linked to groundwater within the valley sides.

Ground Temperature Measurement

The ground temperature measurement data was overlain onto historical borehole information (Figure 2). This information was reviewed and appeared to indicate the potential location of the leak just above the bedrock / embankment interface to the eastern embankment mitre at an approximate level of 140m AOD (4.5mbgl). The results showed two distinct areas where seepage through the embankment may have been occurring, both of which were close to the known seepage.

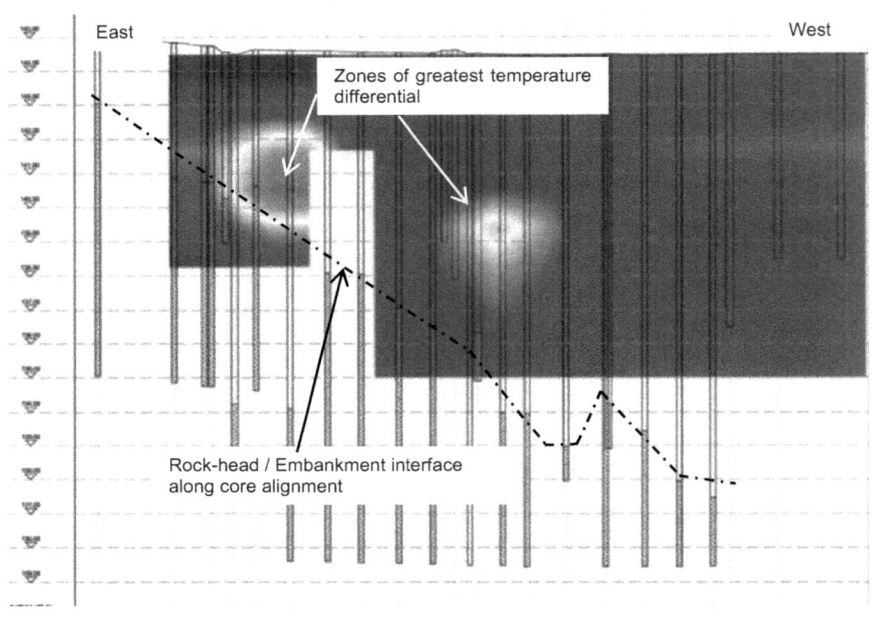

Figure 2. Borehole information overlain with ground temperature differential information

Willowstick Survey

A Willowstick survey was undertaken in 2006 (Willowstick, 2006) to assist interpretation of the possible leakage flow paths through the embankment. This survey also confirmed a direct flow path through the eastern mitre and added additional evidence to support the ground temperature monitoring previously undertaken.

Geophysical Surveys

Non-intrusive electrical resistivity and magnetometry surveys were undertaken beyond the eastern upstream and downstream toes. These works were undertaken to investigate the location of the 18" cast iron compensation flow pipe.

Interpretation of geophysical survey data to the eastern abutment indicated strong evidence of a linear feature aligned with the position of the abandoned valve control in the reservoir basin (Figure 3). This was considered to represent the alignment of the abandoned compensation flow pipe and was noted to pass directly beneath the 'Drain A' monitoring headwall.

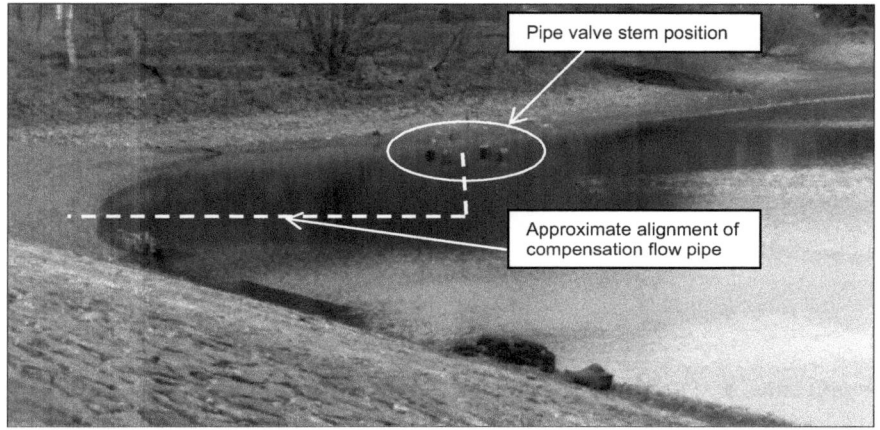

Figure 3. Photograph of location of the 18" compensation flow pipe inlet indicated by the four posts above the water.

DETERMINATION OF REMEDIAL WORKS SOLUTION
In order to meet the first ITIOS recommendation, which is 'Treat the leak in Heapey Dam to seal it or reduce it to an acceptable level' an initial desk study and report was produced. This report detailed a number of options and recommended a cutoff wall solution. An optioneering assessment process was instigated to give full consideration to each option and take the preferred solutions to a Project Team meeting with the Qualified Civil Engineer (QCE) to determine the final solution. The project team included project engineering, other reservoir experts, the United Utilities operational personnel and the Reservoir Safety Manager.

Before the options were evaluated the existing historical information was reviewed and considered in the optioneering process. One such issue was the uncertainty concerning the alignment through the embankment of the 18" compensation flow pipe. Since no records relating to grouting this pipe could be found the success of remedial works in the vicinity of this pipe, such as grouting or a cutoff wall, would be at risk. A diving team was contracted to ascertain if the pipeline was sealed and to locate the pipe through the embankment. Due to the mass of reservoir basin silt this exercise proved unsuccessful which led to the geophysical investigations detailed above to determine its position.

Remedial Works Options

Initially the following options were considered:

- Option 1A – Grouting along the alignment of the core
- Option 1B – Resin Injection along the alignment of the core
- Option 2A – Secant Pile Wall through the core
- Option 2B – Slurry Trench on the alignment of the core
- Option 2C – Sheet Pile Cut Off on the alignment of the core
- Option 3A – Weighted Filter on the downstream face in the area of the existing seepage pipe
- Option 3B – Weighted Filter on the downstream face with Targeted Grouting along the core alignment
- Option 4A – Permanently Lower Reservoir TWL
- Option 4B – Permanently Lower Reservoir TWL excluding the calculated cost to United Utilities for the loss of yield

Following internal review it was concluded that some of the options were not viable and they were discounted from further consideration. Initial analysis of the remaining options (Table 1) provided the following estimates and a basis for the next level of review along with the QCE.

Table 1. Remedial Works Options Estimated for QCE Review

Option	Detail	Cost
1A	Grouting	£1,400,000
2A	Secant Pile Wall	£604,000
4A	Permanently Lower TWL	£5,889,000
4B	Permanently Lower TWL (exclude loss of yield)	£244,000

During review with the QCE it transpired that the weighted filter option, despite it not reducing or substantially stemming the leakage (the ITIOS recommendation), would also be acceptable as a means of leakage control. As a result the appointed Panel Engineer confirmed that a weighted filter would suitable to meet the requirements of the original ITIOS recommendation. Re-puddling the core was also proposed since the likely area for treatment was small, however, it was recognised that construction of this option was not without risk.

The following options (Table 2) were progressed for further consideration in advance of deciding on the preferred solution.

Table 2. Remedial Works Options Progressed Following Initial QCE Review

Option	Detail	Cost
1C	Targeted Grouting	£405,000
3A	Weighted Filter	£340,000
5A	Re-puddled Core	£409,000

Preferred Remedial Works Solution
At the final review meeting these options were evaluated with the project team and QCE. It was highlighted that grouting had been previously attempted as a means of treating the leak and although initially successful, the leakage returned. Concerns were raised regarding the re-puddling of the clay core in respect of the method, including the integrity of the clay/rock interface and also the envisaged difficulties and safety considerations in constructing this option.

The most cost-effective solution was the weighted filter that had already been confirmed by the QCE as acceptable in respect of satisfying the ITIOS recommendation. This solution also avoided contact with the 18" compensation flow pipe from the reservoir, which was considered a beneficial outcome.

WEIGHTED FILTER DESIGN
The weighted filter was designed by UU Engineering as a Critical Filter in accordance with the procedures of Sherard & Dunnigan, (1989). The filter was designed to manage the observed seepage (maximum observed flow rate of 0.83l/s) and prevent potential future loss of fines from within the reservoir embankment fill and core.

Available relevant and representative ground investigation particle size distribution (PSD) laboratory test data from the embankment fill materials was collated for the filter design. Of the available data, the range of fines (<75μm) was observed to be generally between 15% and 39%, thus initially classifying the material as a Category 3 base soil.

Initially it was assessed whether the base soil was gap-graded or potentially subject to internal instability in order to determine a filter design able to protect against the potential movement of finer particles. This was achieved by comparison with a theoretical PSD curve based on research by Sherard (1985). Base soil grading curves were adjusted at the point of inflection where the base soil curve became flatter than the theoretical curve. The corrected base soil grading curves were reassessed to determine a revised design base soil Category 2 (between 40% and 85% <75 μm). The procedure outlined by Sherard & Dunnigan (1989), allowed the design of a grading envelope for the primary filter designated material 6T (Figure 4).

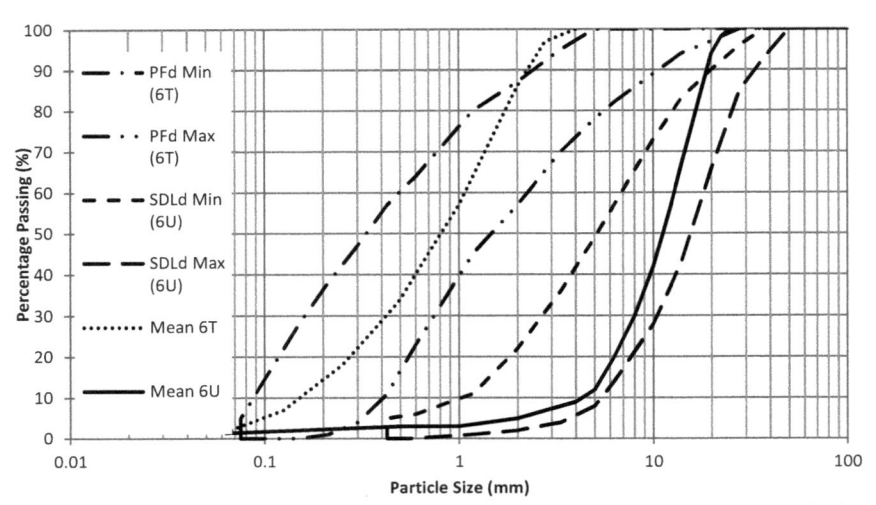

Figure 4. Primary Filter (PF) & Secondary Drainage Layer (SDL) design envelopes (d=design) and actual Mean 6T and 6U grading

A secondary drainage layer, designated material 6U, was designed to act as a filter to the primary filter. The same principles of design were applied to design of the secondary drainage layer, the design grading envelope for which is also illustrated in Figure 4. Each grading was further checked to ensure that it was a minimum of five times greater permeability than the material it filters. That is, the primary filter being five times more permeable than the base soil (embankment fill) and the secondary drainage layer five times more permeable than the primary filter. Due to constructability constraints a minimum construction thickness of 300mm was specified for both the primary and secondary filter layers. Weighting material placed over the filter was specified as Class 1A material (Department for Transport, 2009). The thickness of the weighting material was designed to withstand the theoretical maximum pressure exerted by a seepage flow entering the filter from the embankment face.

WEIGHTED FILTER CONSTRUCTION

The contract to construct the weighted filter was awarded in August 2013 to the A E Yates Ltd. with a programme extending to mid-December 2013. The specification required the Contractor to confirm the filter design and procure the appropriate graded filter materials. The filter system comprised 300mm 6T filter material, below a 300mm 6U drainage material, below a permeable geotextile separator layer, below a Class 1A 'weighting' material and topsoil and turf (Figure 5). The bespoke 6T filter material and 6U drainage layer were sourced from Glensanda Quarry in Scotland and delivered to Liverpool docks. Due to the restricted site access, materials were delivered in 10T loads from the stockpile at the docks.

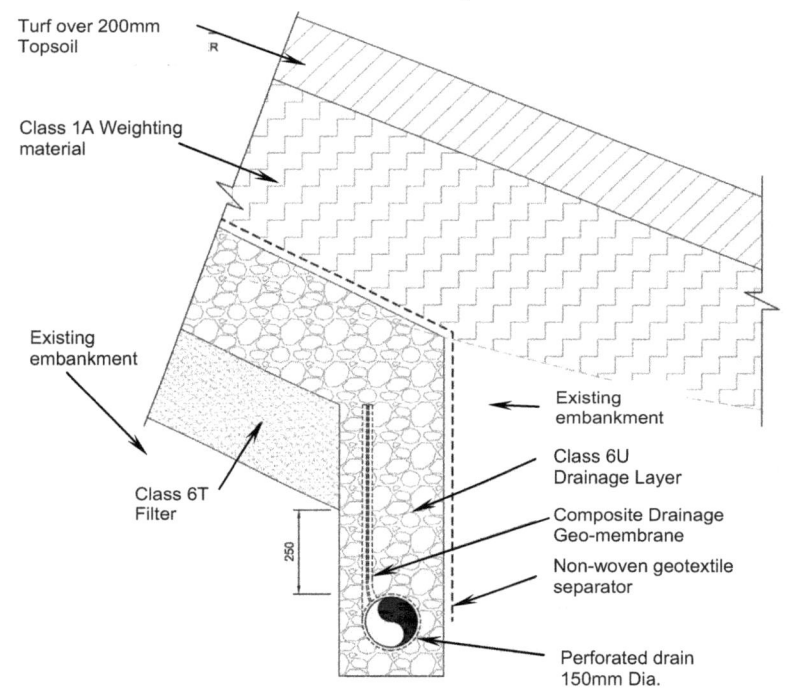

Figure 5. Detail of filter and drain construction

The sourced 6T and 6U materials were not fully compliant with design envelopes, however, the materials were acceptable to the QCE and UU on recommendation from UU Engineering as the PSD curves indicated approximately 95% compliance with design envelopes.

Turf and topsoil were removed in 1m lifts to limit the possibility of weather adversely affecting the formation. The filter was constructed in controlled lifts by placing the class 6T filter on the slope in a 300mm thickness and compacting using a vibratory plate compactor. The 6U was then directly placed on top of the 6T using the same compaction method (Figure 6). The geotextile was placed above the 6U and then the Class 1A was compacted in layers to the required thickness using the same method as the filter and drainage materials.

Routine measurements of the density of the 6T, 6U and 1A material on the slope were determined by a nuclear density gauge; the in-situ density and moisture content of the material within the body of each layer being determined to confirm the level of compaction achieved. At the toe of the filter construction a 500mm deep trench was excavated below the level of the 6T to ensure that all flows from the filter passed into the drainage layer. A geocomposite drainage membrane attached to a 150mm diameter perforated pipe was installed to carry the flows to a solid discharge pipe which discharges at a new reinforced concrete headwall structure beyond the toe of the embankment.

Figure 6. Weighted Filter under construction

During excavation for the toe drain the abandoned 18" cast iron compensation flow pipe was exposed in the position indicated by the geophysical survey. It was noted during construction that the base flow emerged up through the filter construction from the area of the existing seepage and was directly above the draw off pipe. A 300mm wide clay stank on the downstream side of the drainage trench was constructed to force the flows through the filter to the discharge pipe.

SEEPAGE FLOW MONITORING

Monitoring improvements
So that a clear understanding could be gained of the flows being monitored at the headwall, the toe and mitre drains were each constructed separately. Perforated pipes within a gravel filled trench and wrapped in a non-woven geotextile separator were constructed along appropriate routes, each discharging to a separate point in the new headwall, alongside the filter drain discharge pipe.

Post Construction Monitoring

Monitoring of flows passing through the filter has been undertaken since completion of the weighted filter construction and rehabilitation of Heapey Embankment. Flows ranging between zero and 0.58l/s have been recorded.

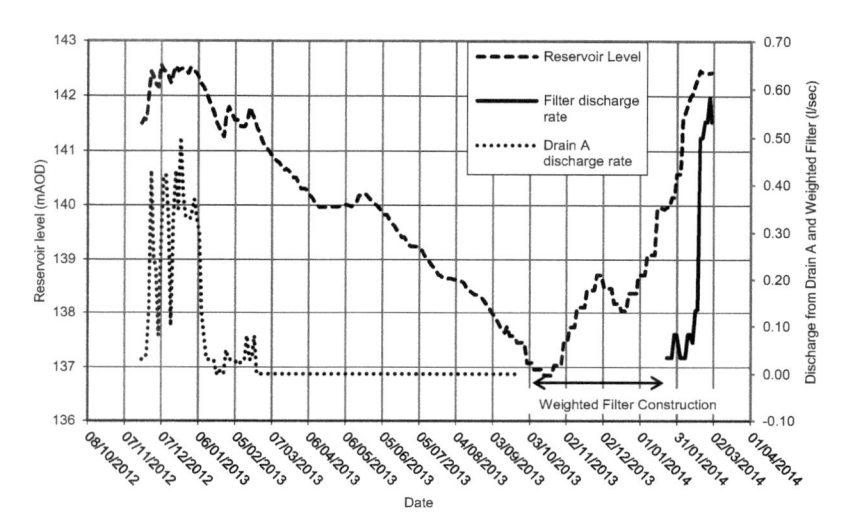

Figure 7. Reservoir Level vs discharge from Drain A (pre –construction) and Filter (post-construction)

Figure 7 shows the flow rates recorded pre-construction from Drain A and for the relatively short post-construction period from the weighted filter. The peak flows are clearly linked to reservoir level with high reading being recorded only when reservoir level is above approximately 141.6mOD. The sudden spikes in discharge rate and coincident rise in reservoir level are associated with rainfall episodes (rainfall rates are not shown on Figure 7).

The flow rate from the filter is seen to be slightly greater than the previous Drain A rate for a similar reservoir level. This can be explained by considering that the weighted filter captures all of the flow, post-construction, whereas pre-construction some seepage water may have been lost through flow to ground before reaching the old headwall discharge pipe. As a measure of the success of the filter, suspended solids testing was specified. It has been confirmed that suspended solids within the seepage water are within expected levels of turbidity and this indicates that the filter is not passing eroded material.

CONCLUSIONS

The early performance data indicates the weighted filter is operating as expected, passing water with minimal solids. The filter is apparently capturing a slightly higher flow, which is likely to be a result of intercepting a great proportion of the seepage, compared with pre-construction flows. The new drainage network now provides flow separation making the measurement of flows more straightforward and ensuring enhanced monitoring accuracy in terms of operational safety.

Obtaining filter materials that complied with the specified grading envelopes was not easy to achieve. However, the contractor was able to procure materials that were sufficiently close to the specified gradings to offer no detriment to the successful operation of the filter.

The construction of the weighted filter has enabled the reservoir to be returned to full operational level once more, after a gap of more than ten years. The collaborative working relationship with the contractor also ensured that the construction phase was completed within budget and programme. Close liaison with the QCE at all stages of the design and construction process has ensured that the project was efficiently executed.

ACKNOWLEDGEMENTS
The authors wish to thank Mr N Reilly for his guidance and the staff of A E Yates Ltd for their commitment and collaboration during the construction of the project.

REFERENCES
Department for Transport (2009), *Manual of Contract Documents for Highway Works, Volume 1 - Specification for Highway Works, Series 600 Earthworks.* HMSO, London, UK

Parks, C. D. and Walthall, S. (2002) The successful grouting of Heapey embankment, Anglezarke Reservoir. *Proceedings of the 12th Conference of the British Dam Society, Reservoirs in a Changing World,* Thomas Telford, London, UK

Reilly, N. (2011) *Report on the result of an Inspection and Inspecting Engineer's Certificate under Section 10(5) for United Utilities (Water) Ltd,* Pell Frischmann (Unpublished)

Sherard J L. (1985) Hydraulic Fracturing in Embankment Dams. *Proceedings ASCE Symposium on Seepage and Leakage from Dams and Impoundments,* Denver, May 1985. (Ed R L Volpe and W E Kelly.) New York, ASCE. pp 115-141.

Sherard, J.L, and Dunnigan, L.P. (1989), "Critical Filters for Impervious Soils", *Journal of Geotechnical Engineering,* ASCE, 115(7), 927-947

Willowstick Technologies LLC, (2006). *Heapey Dam (Anglezarke Reservoir)* Aquatrack Survey Report. (Unpublished)

Ground Engineering aspects of the Grane Valley Reservoir Safety Improvements

D THOMSON, MWH Global, Warrington, UK
H TAYLOR, United Utilities, Warrington, UK
M EDDLESTON, MWH Global, Warrington, UK

SYNOPSIS A recently completed improvement project on two United Utilities owned dams in the Grane Valley Cascade in Haslingden, Lancashire had to overcome many design and construction challenges due to geotechnical constraints. The works included;

Ogden Reservoir: Improvements to the overflow and spillway facilities by the construction of a 32m wide wedge block central auxiliary overflow and appurtenant structures, improvements to the northern spillway and southern by-wash channel, and a new combined transition channel to connect all of the structures.

Holden Wood: Improvements to the stability of the embankment and its susceptibility to internal erosion by slope slackening with an integral weighted filter, improvements to the overflow by the construction of a 15m wide reinforced concrete (RC) central overflow and improvements to the scour facilities.

This paper describes the background of the scheme, and how the key ground engineering challenges were overcome throughout each stage of the project to improve the overall safety of the dams.

HISTORY AND BACKGROUND

The Grane Valley Reservoir System is a cascade of three impounding reservoirs (IR); Calf Hey, Ogden and Holden Wood, constructed in 1905, 1912 and 1841 respectively. Located to the north of Haslingden, Lancashire, they supply Haslingden Grane Water Treatment Works. The main features of the cascade are shown in Figure 1.

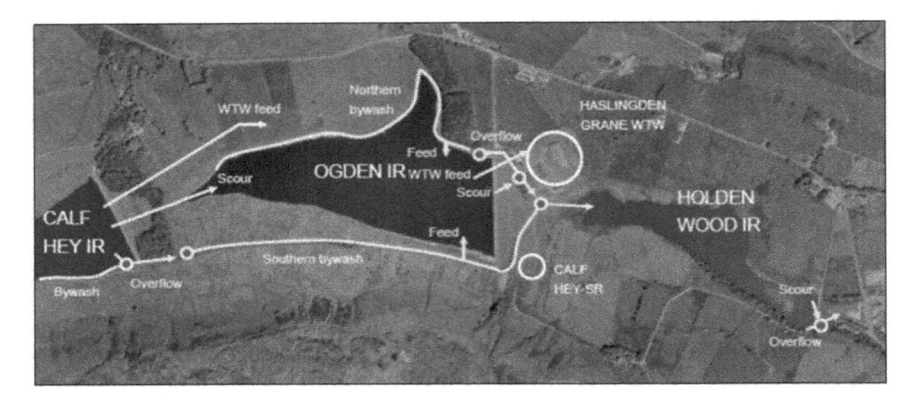

Figure 1. Aerial photograph showing the main features of Grane Valley Reservoir System and proposed improvement works.

Ogden IR
Ogden IR was designed by Joshua Cartwright and commissioned for construction by Bury and District Joint Water Board in 1902. As construction neared completion in 1909, the ground near the southern by-wash and the toe of the dam started to move. The solution was to weight the dam with 25,000T of soil and rock whilst the spillway channels at the toe of the dam were converted to tunnels. The dam became operational in 1912.

Ogden IR has a capacity at top water level (TWL) of 1,501,000m^3. The dam is a Pennine earth fill embankment, 400m long and 15m high. It has a crest width of 4.5m, a downstream slope at an incline of 1 in 2.5 with a berm half way down and upstream slope of 1 in 3. By-wash channels extend along both sides of the reservoir which discharge down cascades at each mitre.

Holden Wood IR
Holden Wood IR lies at the bottom of the cascade and has a capacity of 367,000m^3 at TWL. Construction was completed in 1841 to supply local mills. In 1896 the site was redeveloped with a bleach works sited at the toe of the dam. The works closed in 1998 when it was demolished; the resultant rubble was spread against the toe of the dam and downstream areas, whilst extensive foundation structures remained intact. It now supplies compensation water to the Ogden Brook to ensure water levels are suitable to sustain fish habitats. The reservoir is retained by an earth fill embankment with a clay core; it is likely to be constructed from locally sourced materials, however there were no design or as-built records available. The dam is 100m long and 18m high, it has an upstream slope of 1 in 3 and had a downstream slope of 1 in 2. The dam crest is 3.5m wide and carries an access road to a local property and farm land.

DESCRIPTION OF PROJECT
Inspections under Section 10 of the Reservoirs Act 1975 and subsequent flood studies and physical model tests identified that the existing overflow facilities at Ogden and Holden Wood reservoirs were inadequate to pass the design Probable Maximum Flood (PMF) safely. Subsequently recommendations were made in the interests of safety (ITIOS) to investigate stability, seepage and the requirement to improve overflow facilities in the reservoir cascade.

In 2009 UU Engineering with MWH Global commenced the concept design of the scheme. By the summer of 2012 Eric Wright Civil Engineering (EWCE) was appointed as the main contractor and P&S Consulting Engineers were appointed as the detailed designers.

Geological setting
Geological maps indicated that glacial drift deposits cover the valley floor and lower flanks, whilst peat is present over the upland plateaus. The solid geology is shown to comprise sedimentary rocks from the Millstone Grit Series dating from the Carboniferous Period. The rocks are shown to generally comprise sandstone known as the Brooks Bottom Grit.

Geotechnical constraints at Ogden IR
British Geological Survey mapping and detailed geomorphological mapping of the local area show that the valley in which the Ogden embankment dam was constructed, is within an area of potential ancient post glacial landslips.

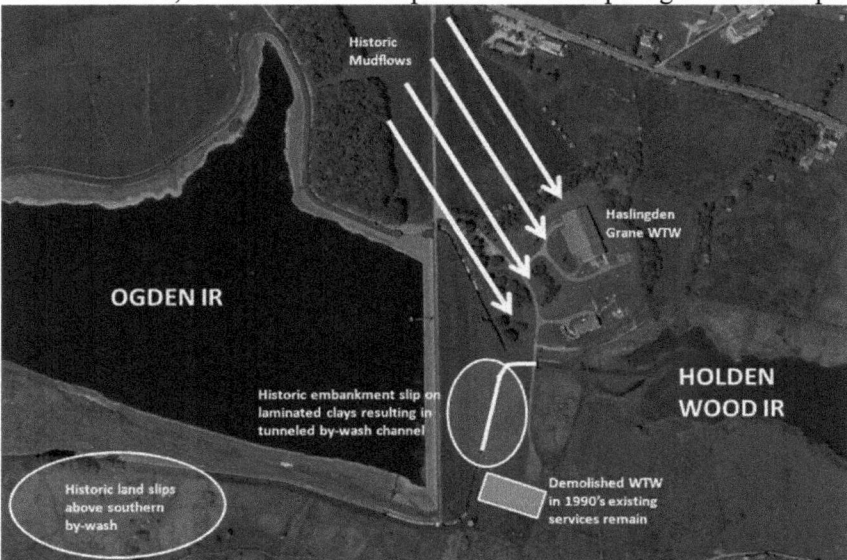

Figure 2. Ogden Embankment before the works showing areas of potential instability

On the northern valley these landslips occurred on steepened valley sides produced by ice and melt water. Whilst it appeared likely that many of these slips had not moved for thousands of years, it was considered that the forming of excavations to enable construction of a deep overflow channel on the northern abutment slope of the dam was unsafe. It would require management of the risk of re-activated movement along pre-existing shallow and/or deep seated failure surfaces and costs associated with that.

In addition, there was evidence of deep seated laminated clays and a history of mudslides on the southern valley sides adjacent to the existing spillway, making positioning of a new spillway on this bank problematic at best. This confirmed that the optimum solution was to upgrade the existing spillway and design an auxiliary crest overflow structure over the centre of the dam, Figure 2.

Geotechnical constraints at Holden Wood IR
The existing spillway at Holden Wood ran along the right mitre in a series of large steps, where it turned 90° at the toe of the dam into a gabion lined channel. As a result the spillway was hydraulically inefficient and it could only convey approximately 10% of the PMF. In addition the very steep southern abutment slope adjacent to the existing spillway also showed signs of instability, which made the option of an on-line modification impracticable. Historic mudflows present in the valley at Ogden IR were also present on the northern abutment at Holden Wood IR.

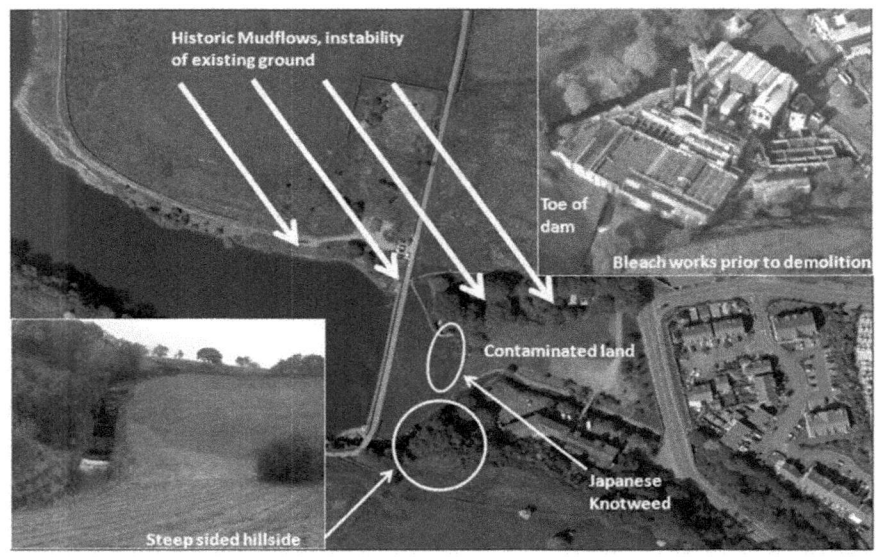

Figure 3. Holden Wood Embankment before the works showing areas of historic and potential instability, contamination and former bleach works

Preliminary assessments indicated the following potential contaminants on site: Polycyclic Aromatic Hydrocarbons which may pose a long term risk to human health if left at surface, Asbestos Containing Materials and asbestos fibres in made ground and concentrations of leachable inorganic contaminants. Japanese knotweed was also identified on site during previous phases of works and required excavation and removal from site during the construction phase. The described constraints are depicted in figure 3.

Ground Investigations and Studies
Ground investigations were undertaken as part of the slope stability and seepage studies, and for the new overflow arrangements at both dams. Extensive data gathering at both sites was required in order to design a filter to resist internal erosion at Holden Wood and to facilitate the temporary and permanent works design of the new overflow arrangements.

A slope stability analysis of the existing embankments was undertaken which confirmed adequate stability of the embankment at Ogden and highlighted that the downstream slope of Holden Wood required re-grading to meet current safety standards.

Internal erosion assessments concluded that a granular filter was not required at Ogden. However, a weighted filter would need to be incorporated into the profiling works at Holden Wood.

OGDEN DESIGN AND CONSTRUCTION
The abutment slope stability considerations at Ogden led to the decision to renovate the existing spillway and the construction of the auxiliary wedge block spillway over the crest of the dam to take a proportion of the PMF, once the existing spillway had reached design capacity.

Wedge Block Spillway
Wedge block spillways originated in the early 1970s, with full size operational spillways constructed in Russia. The first application of a wedge block spillway in the UK was at Brushes Clough (Baker, R, 1994 and Baker R and Gardiner, K., 1994) in 1993. The spillway concept relies on pre-cast concrete blocks which interlock and overlap to form a stepped chute that assists with energy dissipation (Hewlett, et al., 1997). Due to its stepped nature, a low pressure zone is formed on the downstream side and a series of drainage vents are provided on the downstream face of the step to connect the low pressure zone to a designed under drainage layer. This pathway allows seepage at low flows, but when flowing sufficiently it effectively 'sucks' the blocks onto the embankment. The stepped spillway itself dissipates energy, removing the requirement for deep excavations with extensive temporary works. Physical model tests were undertaken for the spillway, however, due to scaling factors it could only be modelled as a

stepped spillway and not using individual blocks. The only significant limitation with using this type of spillway was its hydraulic limitations, with testing only carried out to 2m^3/s/m run, resulting in such a wide spillway.

An inherent feature of the wedge block spillway is its flexibility, enabling it to move with the existing embankment, allowing the undertaker to see any movement in the dam. The wedge block spillway was chosen to alleviate concerns with the surrounding geotechnical history of the area, the slippage of the embankment during construction and more specifically that settlement is prevalent at the centre of the dam.

An important element of the wedge block design was the filter and erosion protection layer underlying the blocks. The wedge blocks were to be placed on a bespoke granular regulating layer, comprising an equal volume of single sized 10mm and 4mm to 7mm clean stone. A geotextile separator layer was placed between the regulating layer and the wedge blocks to prevent suction of the regulating layer into the vent holes when water is overflowing.

The wedge block spillway required a toe restraint designed to accommodate sliding forces from the wedge blocks. The solution comprised 52 No. 600mm Continuous Flight Auger (CFA) piles; these were installed in two rows, to 15m depth, through soft clays overlying stiff glacial clay. In-situ steps, matching the profile of the upper levels of the wedge block spillway, were cast integral with the pile cap to provide resistance to the turbulent flow conditions that could occur at this location due to the hydraulic jump.

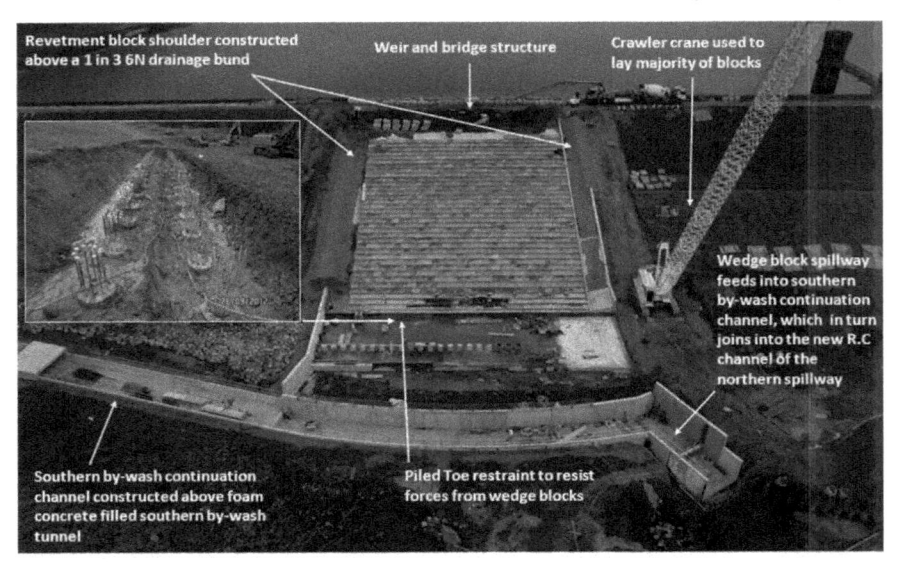

Figure 4. Construction of wedge block spillway

The 1300 wedge blocks units were pre-cast off site to fulfil the requirement of permissible deviation from line and level of 10mm along the entire 32m width and to ensure repeatable quality. Benching was stipulated to ensure that the granular material did not slide down slope.

In order to comply with the tight tolerances, a washed dust filter layer was incorporated into the design whilst on site to bed the blocks. Timber templates were used to screed the material under the geotextile, Figure 5.

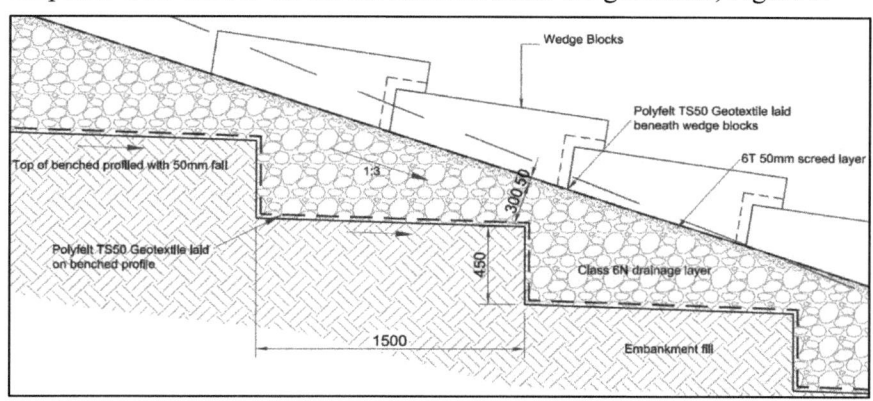

Figure 5. Section through wedge block spillway illustrating the design and construction

Crest overflow structure
The crest overflow structure on the wedge block spillway had several key geotechnical challenges. Determination of the permissible settlement over the 120 year design life of the structure utilised the historical settlement data over the past 20 years. It showed that the potential long term future settlement was approximately 80mm. Consequently four movement joints were designed into the crest structure to allow for this movement.

To construct the crest structure, a cut 40m wide and 2.5m deep was removed from the crest and a water tight connection was designed to key the existing clay core to the new RC structure. This cut was protected from rising levels in the reservoir by constructing the upstream section of the apron prior to removal of the watertight clay core. However, to ensure reservoir levels did not have a detrimental effect on the integrity of the dam and construction personnel, the reservoir was controlled by a flood control plan. This was reviewed weekly to ensure activities at each site were prioritised in relation to expected flows and predicted weather.

As part of the design it was decided to incorporate a 1 in 7 slope with an RC nib to the abutment of the structure and an RC downstand into the clay core. To minimise the risk of seepage around the structure, sheet piles with pile sealant were cast into the nib and downstand to provide connection between the existing and new clay cores.

The reinstatement of the clay core was specified as rolled clay (class 7R) which was compacted in 150mm layers using a mini sheepsfoot roller and a hand-held compactor around pile locations. Site trials were undertaken to ensure that the consistency of the clay was according to the specification by wetting and breaking up of the 7R clay and using a hand vane to ensure that the remoulded un-drained shear strength was within the acceptable limits. Small holes were dug between the layers during the site trials to ensure that there were no seepage paths visible and that the clay was homogenous.

To ensure that there were no leakage paths, the final barrier against seepage was installed on the downstream side of the clay core reinstatement. A leakage detection system was constructed using granular filter and drainage materials with a geotextile surrounding the combined filter. At the bottom of the drainage layer, some 2.5m below TWL, a perforated drainage pipe was installed which transitioned into a solid pipe as it exited the leakage detection system. This pipe discharges within the wedge block spillway and permits any flow experienced to be seen.

It is also noted that the details constructed on the Ogden IR crest were replicated at Holden Wood IR.

Southern by-wash tunnel
As part of the southern by-wash continuation channel and wedge block spillway construction, the existing southern by-wash tunnel had to be infilled as its alignment clashed with the formation of the new structures. A structural assessment of the tunnel ruled out subsequent man-entry and as such a no-man-entry solution was required. Works to stabilise the dam when it was built had covered the existing by-wash channel with material varying from 1.2m to 3.5m depth.

Open excavation was precluded primarily due to a concern that an open excavation through the stabilising material could reactivate the slip that occurred during first filling and also for reasons of safety during construction. Foam concrete, due to its lightweight filling properties, was used to infill the southern by-wash channel. This was to minimise the risk of further settlement of the by-wash tunnel and to minimise the settlement of the structures being constructed above the line of the tunnel.

HOLDEN WOOD DESIGN AND CONSTRUCTION
The abutment slope stability considerations and impracticalities of an on-line modification led to the decision to construct an RC spillway over the crest of the dam to take the full PMF. Width limitations of the toe of the dam made a wedge block solution unworkable as it was in excess of the 15m allowable toe width.

Weighted granular filter

At Holden Wood IR a weighted filter was to be incorporated into the downstream embankment re-profiling to improve stability and accommodate the new spillway profile. The purpose of the filter was to prevent piping failure through the core and the embankment fill by blocking movement of eroding soil particles. The filter and drainage system intercepts water emerging at the original surface of the existing embankment construction.

The fine filter was designed in accordance with current best practice (ICOLD, 1986, USDA, 1994) and USBR, 2011)). The granular filter was designed to be placed directly onto the embankment fill following a topsoil strip and stone picking. The filter serves a filtration function and a secondary drainage layer is placed over the filter, prior to the placement of general fill and topsoil. The design of the drainage layer followed the same procedure, using the designed fine filter as the base soil so that water emerging from the filter layer did not instigate erosion of the filter materials. The relatively high permeability of the drainage layer encourages flows across the fine filter rather than along it. The filter layers were specified as a minimum thickness of 300mm (normal to the slope). The general fill used to re-profile the downstream slope consisted of imported Class 1A (DTp, 2009). Specific requirements for the placement of this fill, which superseded the published specification, were that the Class 1A material was to be deposited in layers not exceeding 250mm uncompacted thickness and the maximum particle size should not exceed two thirds of the layer thickness.

Contaminated land

The former Bleach Works at the toe of the dam had contaminated the ground, resulting in elevated levels of metals, and the demolition of the works contributed chrysotile (white) asbestos to the site. A Land Quality Risk Assessment (LQRA) report and a Detailed Quantitative Risk Assessment (DQRA) report were prepared, to establish the suitability for re-use of excavated materials, in accordance with the CL:AIRE guidance (2011). In addition a Remediation Strategy was produced assessing how potential risks to human health and controlled waters could be avoided, allowing the safe re-use of the material and approvals to be gained from the local authority. The remediation solution required a 600mm clay capping over the entire 2,500m^2 area along with a 10mm geo-synthetic clay liner having a permeability of 1×10^{-9}m/s along the northern wall of the stilling basin and outlet channel.

Having identified asbestos fibres in soil samples, air monitoring was undertaken during construction; no asbestos was detected.

Approximately 120m³ of Japanese knotweed was removed from the footprint of the new overflow and spillway facilities. It was not possible to remove all of the knotweed at the toe of the dam for dam stability reasons. However a root barrier, to prevent spread, was installed and covered with 100mm of concrete to prevent burrowing animals damaging it. The removal of the Japanese Knotweed resulted in an uneven formation level underneath the new spillway and foam concrete was used to reduce the potential effects of differential settlement, Figure 6. Movement joints were required in the foam concrete to provide the required stepped profile for the new spillway.

Construction of the 4.0m deep stilling basin base required excavation in rock and needed 200mm diameter anti flotation tension piles. These piles range from 6m-8m in depth and were constructed using centralised 57.5mm diameter Dywidag GEWI+ bars cast in cement grout with a strength of 40N/mm². The piles were connected to the stilling basin with a galvanised header plate, GEWI sleeve with the annulus to the bars filled with Denso void filler and a Visqueen bond breaker between the formation and stilling basin base. Water flows within the rock were controlled using filter media prior to blinding. The works were phased to coincide with low reservoir levels and low rainfall to mitigate the risk of flooding the works. The congested construction processes are shown in Figure 6.

Figure 6. Top left: aerial photo mid construction, Top right: Temporary works to stilling basin, Bottom left: Foam concrete, Bottom right: Installation of anti-flotation piles.

CONCLUSIONS

October 2013 saw the successful completion of the largest wedge block spillway constructed on a constantly impounded reservoir (i.e. excluding flood storage reservoirs) in the UK. All three reservoirs have been returned to TWL with Ogden IR supplying Grane WTW at full capacity of 16Ml/day. The completed reservoir safety schemes are depicted in Figures 7 and 8.

Figure 7. Aerial photo of completed Holden Wood IR remedial works

Figure 8. Aerial photo of completed Ogden IR remedial works

The reservoir safety improvement measures implemented on the scheme have been completed to the latest prevailing geotechnical and hydraulic standards. The design provided a solution which fulfilled all the project and industry safety requirements and detailed attention to construction techniques and quality on site was maintained throughout the project to secure many decades of future operation.

ACKNOWLEDGMENTS

The authors acknowledge the permission of United Utilities to publish this paper, Gavin Hulme, EWCE, Paul Thurlwell P&S Consulting Engineers and Adam Wood UU for their valued contribution.

REFERENCES

CL:AIRE(2011) *Definition of Waste: Development Industry Code of Practice (the Code of Practice) Version 2*. Contaminated Land: Applications in Real Environments, London, UK

Baker, R. (1994*). Brushes Clough Wedge Block Spillway - Progress Report No. 3. SCEL Project Report No. SJ542-4*, University of Salford, UK, Nov., 47 pages.

Baker, R., and Gardiner, K. (1994). The Construction and Performance of a Wedge Block Spillway at Brushes Clough Reservoir. *Proc. 8th Conf. British Dam Society, Exeter University, Sept 1994*, Thomas Telford Publ., pp. 214-223.UK

Department of Transport (2009) *Manual Of Contract Documents For Highway Works. Volume 1 Specification For Highway Works. Series 600 Earthworks*. HMSO, London, UK

http://www.dft.gov.uk/ha/standards/mchw/vol1/pdfs/series_0600.pdf

Hewlett H., Baker R., May R.W.P and Pravdivets Y., 1997. *Design of stepped block spillways. CIRIA Special Publication 142*, CIRIA, London, UK.

ICOLD Bulletin No. 55 (1986) *Geotextiles as Filters and Transitions in Fill Dams*.

United States Department of Agriculture National Resources Conservation Service. *Part 633 National Engineering Handbook, Chapter 26 – Gradation Design of Sand and Gravel Filters*, October 1994.

USBR Design Standard No. 13 (2011): *Embankment Dams, Chapter 5: Protective Filters*, USBR, USA

Remedial Grouting at Shon Sheffrey Dam, Wales

D A BRUGGEMANN, Atkins
O J FRANCIS, BAM Ritchie

SYNOPSIS   Shon Sheffrey dam is a 15m high x 260m long embankment dam located about 3km north-northwest of Tredegar, Wales. The impervious element is provided by a puddle clay core. The dam is owned and operated by Dŵr Cymru Welsh Water. The dam was completed in 1896 and the spillway and embankment crests were raised 1.83m (6ft) and 2.44m (8ft) respectively between 1945 and 1948. Further remedial works were carried out in 1984 when a 34m long x 6m deep steel sheet pile wall was installed in the embankment adjacent to the spillway.

The dam had a history of leakage into the spillway chute and drawoff works culvert under the embankment. In recent years, the loss of embankment material into the culvert was observed and a need for remedial works was identified.

The paper will describe the results of a geophysical leakage detection survey carried using a Magneto-Metric Resistivity (MMR) technique (Willowstick), development of the grouting layout, the grouting plant, grout mixes and methods including the selection of the grouting pressure. The results of the grouting operations will be presented and conclusions drawn to demonstrate the advantages offered by targeted grouting.

INTRODUCTION

Description
Shon Sheffrey Dam is owned and operated by Dŵr Cymru Welsh Water and is located about 3km north-northwest of Tredegar, Wales. The dam was completed in 1896 and is about 15m high x 260m long with a free overflow spillway located on the right abutment. The impervious element is provided by a central puddle clay core. About 50m from the overflow weir, there is 60° upstream change in direction of the axis of the dam and a 30° upstream change in direction about 60m from the other end of the embankment. A wet well valve shaft is located within the upstream shoulder just upstream of the crest towards the right hand end of the embankment. The wet well discharges via pipework passing through a tunnel which is plugged in line with the core of the embankment.

The geology of the site is mainly orthoquartizites and marine mudstones of Millstone Grit which lie beneath the mudstones and siltstones of the Lower Coal Measures which outcrop on the hill slopes above and downstream of the reservoir. The abutments of the embankment and the shore of the reservoir are underlain by Glacial Till.

Figure 1. General View of Shon Sheffrey Dam (Courtesy of J C Ackers, Black and Veatch)

Modifications
Between 1945 and 1948 the Top Water Level was raised by 1.83m (6ft) and the embankment crest by 2.44m (8ft). Modifications were made to the upstream slope and a conventional concrete gravity wall was installed at the top of the upstream face to provide a crest width of 3.81m (12.5ft).

Investigations were carried out into leakage in 1983 when it was also suggested that the top of the original puddle clay core might have cracked due to desiccation. When the dam crest and core were raised, the desiccated zone might not have been replaced as part of the works. It was postulated that such cracks were the cause of the observed leakage.

In 1984 a 34m long x 6m deep steel sheet pile wall was driven into the core just to the left of the spillway to cut off leakage in this area. Pressure grouting was also carried from the top of the upstream wall and along the spillway crest. Primary hole spacing was 4m with secondary holes at 2m intervals and the depth 5m below the embankment crest level.

INVESTIGATION
Leakage into the spillway chute and tunnel continued over the years, and in recent years fines were detected in the seepage water emerging inside the

tunnel. An investigation to identify the locations of the seepage sources was commissioned in February 2012. The investigation comprised a Magneto-Metric Resistivity (MMR) survey using the proprietary system developed by Willowstick Technologies LLC, USA. The system maps changes in electrical conductivity of the ground when a potential difference is imposed between the reservoir and seepage water downstream of the dam. With application of Electric Current Flow and Electric Current Distribution models, sources of seepage can be identified. Processing of data is undertaken as the field work progresses so that any additional data can be collected as the work proceeds. All field work was completed in five days.

Figure 2 shows a typical plan output from the MMR survey which indicates the location of preferential seepage paths. Vertical slices are also produced which provide an estimated elevation at which seepage is taking place.

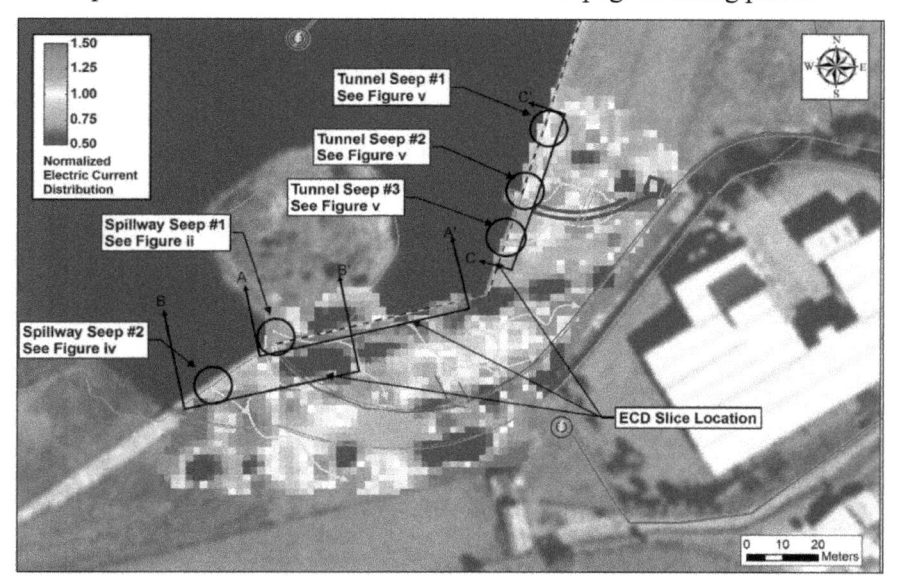

Figure 2. Typical Output showing preferential seepage paths (Courtesy of Willowstick Technologies, LLC)

The investigation identified five sources of leakage. Two sources were associated with the spillway, each adjacent to the ends of the spillway weir block, and three sources were associated with the tunnel. The vertical sections suggested that the seepage was associated with the interface of the raised and the original core.

Examination of historical drawings showed that there was a 2m wide "window" between the spillway side wall and the start of the sheet piles so the potential for seepage in this area was high. However, the geophysical survey showed that seepage was also taking place around the end of the sheet piles remote from the spillway and making its way to spillway chute.

There was also a suggestion that there could be preferential seepage paths under the spillway apron and chute. This source of seepage was not treated as part of the grouting exercise. It was deemed appropriate to deal with this area as part of the spillway works described in Terrell *et al* (2014).

Vertical slices in the vicinity of the tunnel suggested that seepage originated at the elevation of the interface between the raised and original cores. This seepage then moved downwards towards the tunnel.

GROUTING DESIGN

Using the MMR survey results it was possible to concentrate grouting efforts in the specific zones where preferential seepage was identified. The depths of the grout holes were also determined to limit treatment to the elevation of the identified preferential seepage zones.

Table 1. Summary of Areas Treated

Ref	Location	Plan Length (m)	Depth of grout holes(m)	Remarks
S1	Right hand side of spillway weir	13	5	2.5m below the interface of raised and original core.
S2	Left hand side of spillway	8	5	
S3	Left hand end of sheet piles	10	10	4m below toe of piles
T1	Right of tunnel	12	5	
T2	Over and left of the tunnel	30	18 max 9 min	To the left of the tunnel depth stepped from 18m to 9m. 11m deep over the top of the tunnel.

Table 1 shows that the grouting was limited to 73m plan length out of a total crest length of 260m. Moreover, as the elevation of the source of the leakage had been identified, the depths of the holes were limited to apply treatment at the required elevation rather than over the full depth from crest to foundation level.

To provide control over the point of injection of the grout Tube à Manchette grouting was selected. A primary hole spacing of 2m was selected with the provision for split spaced secondary holes once the grout take in the primary holes had been reviewed. This spacing was selected to ensure that all parts of any preferential seepage paths were filled with grout.

The grout volume limit for each port was set at 45 litres at the specified pressure. Alternate sleeves were grouted on the first pass. If the volume limit was reached before the required pressure, grouting was stopped and the port was re-grouted. Grouting was carried out on the ports above and below ports where the take was 45 litres, either on the first pass or subsequent passes.

The selection of the grouting pressures was based on grouting trials described below.

CONSTRUCTION

Plant

The method of grout injection was by Tube à Manchette (TàM) as this allows grout to be injected at discreet locations and also allows for multiple injections within a single borehole.

Figure 3. TàM pipes being installed on the crest of the dam with a lightweight rotary drilling rig (Courtesy of BAM Ritchies)

Lightweight rotary drilling rigs were used to install the TàM pipes as these could safely access the crest of the dam without overloading the structure. The diameter of the holes was 100mm and drilling was carried out using a segmented flight auger. This method of drilling was adopted as it offered the least potential for damage to the embankment fill compared with water or air flush.

During the installation of the TàM pipes they were grouted in place using very low gravity) pressure. Some of the installations required significant volumes of grout to ensure that the seepage pathways were filled with grout.

The TàM pipe selected was plastic due to the relatively low pressures required for injection. The 50mm diameter pipe has three injection ports per metre. By using a double packer, an individual port can be isolated from within the TàM pipe and the grout injected into the ground from that specific point.

Figure 4. Section of TàM pipe (Courtesy of BAM Ritchies)

The grout was mixed in a colloidal grout mixer and pumped with a progressive cavity pump. The grout pumps were fitted with data-loggers which recorded the flow rate and pressure of each injection.

Grout Mix

The grout was required to penetrate the existing seepage paths to restore the impervious core of the dam. The grout was not required to be high in strength, but a low viscosity was essential to ensure that the grout could penetrate easily in the narrow flow paths.

Following some initial mixing trials, the following grout mix was adopted for the injected grout:

Table 2. Grout Mix Components

Mix	Water (litres)	Cement (kg)	Bentonite (kg)
Sleeve (annulus) Grout	50	25	5
Injection Grout	50	50	1.5

The specific gravity of the fresh grout was 1.45 and the Marsh cone viscosity was 35 seconds.

The compressive strength of the grout was measured by crushing 100mm cube samples. The average results are shown in the table below:

Table 3. Average Compressive strength of Injection Grout

Age (days)	7	28	56
Average compressive strength (MPa)	4.9	7.4	7.6

Grouting Trials

The aim of the grout injections is to fill existing seepage pathways with grout, thus preventing water from seeping through the dam. If the grout injection pressure is too high, new pathways can be created by hydro-fracture of the existing clay. Conversely, if too low a pressure is used, the grout will not penetrate all of the available pathways and water will continue to seep through. It is essential, therefore, to establish the limiting pressure at which injections should be terminated. This pressure is linked to the fresh properties of the specific grout, the characteristics of the ground and will vary with depth (overburden pressure).

In addition to identifying areas of seepage the MMR survey results also provided information on key areas where there was a low probability of seepage occurring. Trial injections were carried out in these areas, where the pressure required to inject a small volume of grout into the clay was recorded over a range of depths.

**Average Injection Pressure Vs Depth
(Corrected for Line Loss)**

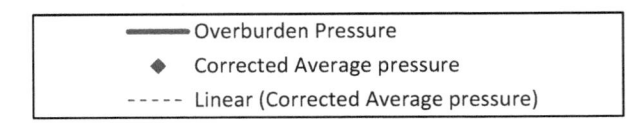

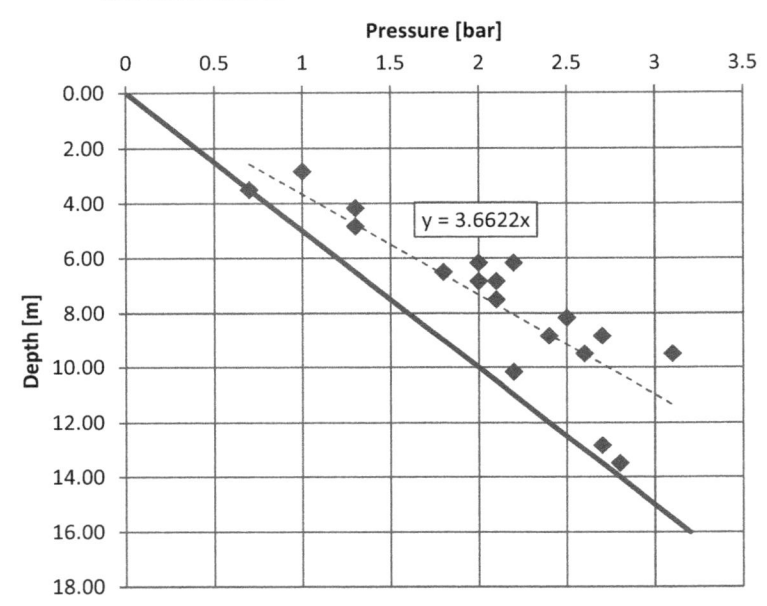

Figure 5. Plot of Injection Pressure Limits with Depth

By plotting the acceptance pressure against depth for a range of injections it was possible to establish a pressure limit that was then applied to all the working injections. This gave the best possible chance of filling the existing pathways whilst minimising the risk of damaging the ground.

On the basis of the grouting trials above, the works grouting pressure was set at 0.25bar/metre.

Programme
The general sequence of the works was to install all the TàM pipes, carry out the grouting trials then follow on with the working grout injections.

The site works were programmed to start in February, but due to a prolonged spell of cold weather, with temperatures frequently below 4°C, it was late March before the drilling and installation began in earnest. Low temperatures prevent grout from curing effectively, which can lead to durability issues in the long term.

The drilling and installation of the TàM pipes was completed within three weeks. Having de-mobilised the drilling plant from site, the working grout injections were then completed over the following 10 weeks. The project was completed in late July 2013.

GROUTING RESULTS

Where alternate ports took 45 litres on the first pass and the ports on either side also took 45 litres on the second pass, a secondary hole was installed on either side of hole. Secondary grouting was concentrated in the zones of the 45 litres takes in the primary holes.

A summary of the grouting work is given in Table 4 below.

Table 4 Summary of Grouting Results

Item	Value	Remarks
No. of Holes	49	34 Primary, 15 Secondary
Length Drilled	580m	
Volume of sleeve (annulus) grout	10,560 litres	Ratio of actual to theoretical volume 3.05
Volume of grout injected	8,146 litres	

The ratio of the actual to theoretical volume of the sleeve grout suggests that that some of the voids or seepage paths within the embankment were treated by the sleeve grout.

The success of the grouting was observed with the drying up of the seepage into the spillway chute on the side of the grouting. In the tunnel, fines were no longer detected in the seepage water collected in the V notches and there was a general reduction in seepage.

CONCLUSIONS

The execution of a Magneto-Metric Resistivity (MMR) survey enabled grouting to be targeted at four specific areas on the embankment covering 73m plan length of the 260m length of the embankment. The survey also enabled the depth of the grouting to be limited to depths where seepage had been detected. The survey also detected seepage around end of the steel sheet pile wall which was unlikely to be expected from conventional ground investigation.

The length of drilling was limited about 580m, which is significantly less than would have been required if the holes had been installed along the full length of the embankment.

The targeting of the specific zones minimised the length of drilling required and also the quantities of grout consumed. In the absence of the MMR survey, more TàM pipes would have been installed and more sleeve grout would have been consumed, with additional time being spent attempting to grout the additional ports.

Observations indicate that seepage into the spillway chute has been eliminated and migration of fines into the tunnel has been arrested as well as a general reduction in seepage.

The area of seepage identified on the far side of the spillway with a possible seepage path under the spillway apron is to be treated as part of the spillway remedial works described by Terrell *et al* (2014).

The targeting of grouting on the basis of the MMR survey resulted in significant savings in time and materials, and thus cost, as efforts were concentrated over 73m out of 260m of the embankment length.

ACKNOWLEDGEMENTS
The authors would like to thank Dŵr Cymru Welsh Water for permission to publish this paper.

REFERENCES
Terrell *et al* (2014). 'Shon Sheffrey Reservoir – Labyrinth overflow and replacement of masonry spillway', *Maintaining the Safety of Our Dams and Reservoirs: Proceedings of the 18th Biennial Conference of the British Dam Society, Belfast.* ICE Publishing, London

Maintaining the Safety of our Dams and Reservoirs
ISBN 978-0-7277-6034-0

ICE Publishing: All rights reserved
doi: 10.1680/mdam.60340.135

Bransholme Lagoon: Problems in design and implementation of a sheet pile solution to remedial works

A PETERS, Arup
S A PRYCE, Black and Veatch
A K HUGHES, Atkins
M WHEELER, Black and Veatch

SYNOPSIS   Bransholme Lagoon is located adjacent to the River Hull and is operated by Yorkshire Water Services Ltd. The lagoon had a pre-raised volume of 74,000m³ and is a Category A reservoir. The lagoon is kept at a low level, operating as emergency storage for the adjacent pumping station, providing flood storage for the surrounding area should there be a major storm event. Following an inspection under Section 10 of the Reservoirs Act, 1975 a recommendation was made to reconstruct the spillway. Additionally, to provide the increased capacity for the design event, the reservoir embankments needed raising.

Installing sheet piles through the existing embankments using a silent press technique enabled the operating water level to be raised by 2m. Following installation of the sheet piles, a water test of the reservoir noted seepage from the toe of the embankment. Investigation of the seepage through observations and monitoring with additional instrumentation produced an unexpected result from the changes in reservoir level.

This paper describes the original design solution, the issues encountered during testing, and the subsequent monitoring and remedial works. The paper summarises the lessons learnt in the application of sheet pile solutions to clay embankments for water retention and in understanding the behaviour of a composite solution.

INTRODUCTION
Bransholme lagoon was constructed in 1970 and is formed by earth embankments around all four sides with typical height of 2.2m. The upstream slopes are 1 on 2.5 and are protected by concrete slabs. The crest is approximately 2.25m wide and grassed, with the downstream slope 1 on 3 and grassed.

The lagoon has a surface water area of 3.67Ha and plan dimensions of 250m by 125m, with an overflow elevation of 3.505mOD. The capacity of the lagoon was approximately 72,000m³ and it was categorised as a Category A reservoir, as a breach of the dam could endanger lives in a community. A volume of water is retained in the lagoon at all times to keep the clay lining or naturally occurring alluvial clay in the base of the lagoon wet, thus the available operational storage volume at times of flood is approximately 50,000m³.

Bransholme lagoon is located next to the River Hull. There is a flood defence embankment along the river, which protects Bransholme from fluvial flooding and is maintained by the Environment Agency. The west side embankment of the lagoon and the flood defence embankment are effectively a common embankment.

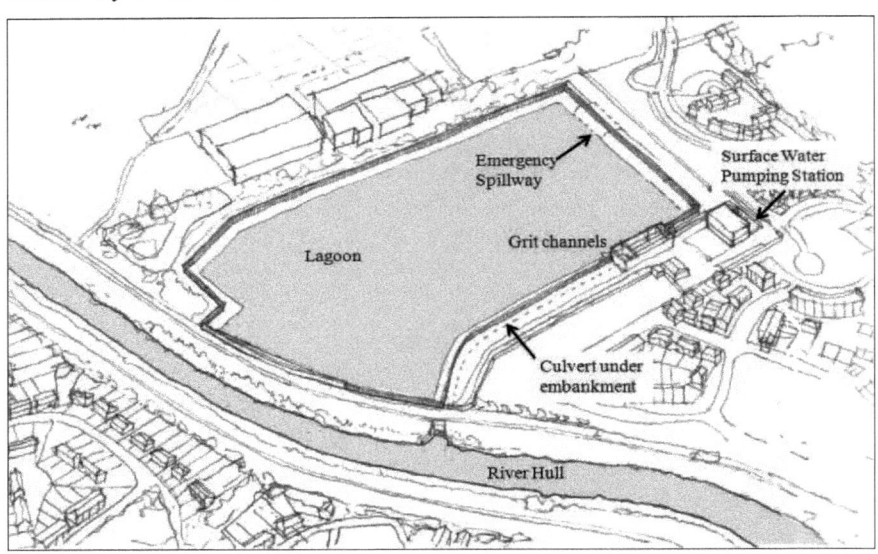

Figure 1. Site layout at Bransholme Lagoon

The lagoon is filled from a surface water pumping station (SWPS) through a grit channel structure. When the tide is low the SWPS discharges directly to the River Hull. In times of high tide any flows from the SWPS are diverted into the lagoon. When the tide levels in the river fall the stored volume in the reservoir is discharged by gravity into the River Hull. Other than its surface area, the lagoon does not have any direct catchment.

The original spillway was constructed as a reinforced concrete spillweir, discharging via the downstream slope to a stilling area, complete with energy dissipator blocks. The spillway was arranged to discharge inland, across the access road to the former Sewage Treatment Works (STW).

Between 1999 and 2006, the Sewage Treatment Works was decommissioned and the land was sold to a private housing developer.

A report in 2006 made under the Reservoirs Act 1975 included a number of recommendations. The Inspecting Engineer noted that the surrounding ground levels had been raised downstream of the spillway, housing development had taken place and that boundary fencing, comprising close boarded timber fencing, had been erected at the toe of the stilling basin. In an overtopping event this would prevent flows from freely discharging and would result in flows along the toe of the embankment. He therefore recommended that the existing spillway provision be improved to allow free discharge, or that a new spillway be constructed to discharge directly to the River Hull. The reservoir was also re-categorised from Category B to Category A due to the housing immediately downstream and the potential for a breach of the dam to endanger lives in a community.

The client decided a new overflow facility discharging directly to the River Hull was preferred. The design maximum top water level was determined by the head over the spillway weir when discharging future design flows from an upgraded SWPS and the PMF rainfall on the lagoon area. The final raised level was set to create an additional storage of 57,400m^3 in the lagoon, thereby meeting operational requirements for the facility up to 2020. A notional design for the new works was developed by Arup prior to the award of a design and build contract to Black & Veatch Ltd under the YWS large scheme framework arrangement.

Historical changes
A report under Section 10(2) of the Reservoirs Act, 1975, dated 1999, noted that in parts of the embankment it appeared that the crest had been previously raised. The S10 report dated 2007 stated that raising of the embankment was undertaken by September 2001 to provide a minimum crest level of 4.375m OD. No details of the raising works were available and no as-built drawings could be found, therefore construction details of any raising works were unknown.

Historically, the performance when the lagoon had been filled was generally satisfactory. Minor seepage had previously been noted between the eastern embankment and the grit channel structure. In both May and September 2002, seepage flows were noted flowing from the southern embankment at a number of locations and the toe was found to be very wet. Also water was found standing on the access road. The source of leakage was not known at the time although the outlet culvert, which runs through the southern embankment and parallel to the slope, was thought to be a likely source.

Geology and ground conditions

The geological maps show superficial deposits of tidal flat deposits (Estuarine alluvium), underlain by Quaternary Glacial deposits and bedrock of the Flamborough Chalk Formation. Site strata are shown in Table 1.

Table 1: Strata sequence from ground investigation

Horizon	Description
Embankment fill	Stiff sandy slightly gravelly clay
Estuarine Alluvium	Soft sandy slightly organic silty clay with occasional organic lenses
Glacial Till	Stiff slightly sandy slightly gravelly clay
Glacio-lacustrine	Firm laminated clay
Flamborough Chalk formation	Weathered chalk over very weak fractured medium density white chalk

Groundwater level is around 1m below original ground level.

OUTLINE DESIGN SOLUTION

Various discharge methods were considered and all involved raising the existing lagoon embankments by some degree to ensure that the lagoon could still discharge to the River Hull above top water levels in the river.

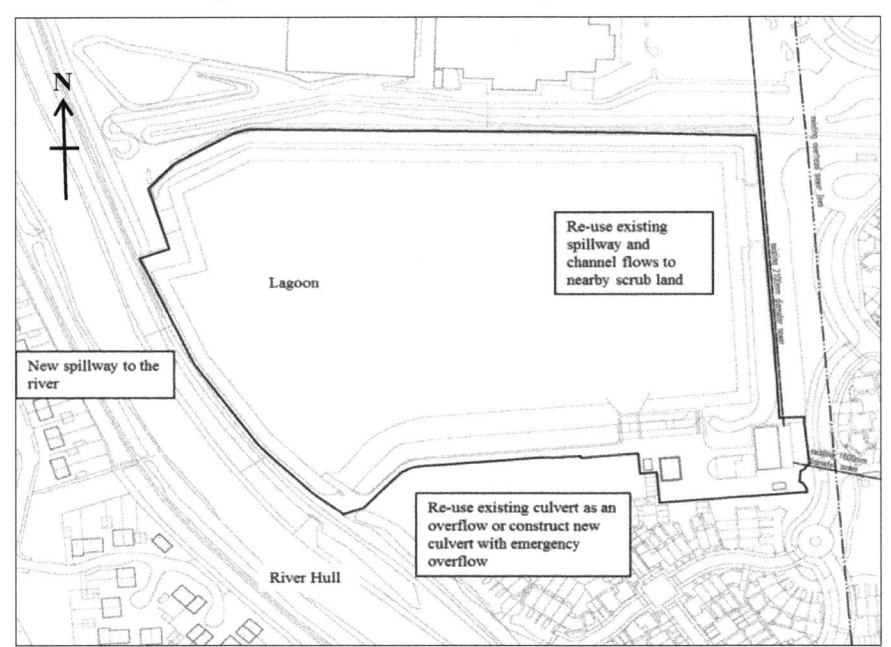

Figure 2: Options considered for the position of the spillway

The preferred solution was to construct a new spillway on the west side so that it discharged directly into the river, with an overflow at elevation 5.10mOD. The other three sides of the lagoon would need to be raised above the new spillway level to provide freeboard and head over the weir. This would raise the banks from 4.37mOD to approximately 6.00mOD.

Repairs to the culvert and other ancillary structures were also required.

Various options for embankment raising were considered, including using earth fill to minimise the visual impact of the scheme, but due to the site boundary constraints and the need to maintain existing storage, sheet piling was considered by the client as the most viable option. The sheet piles, while raising the existing embankments, would also create an additional cut-off through the existing embankment. The sheet pile wall was positioned along the inside edge of the existing embankment crest.

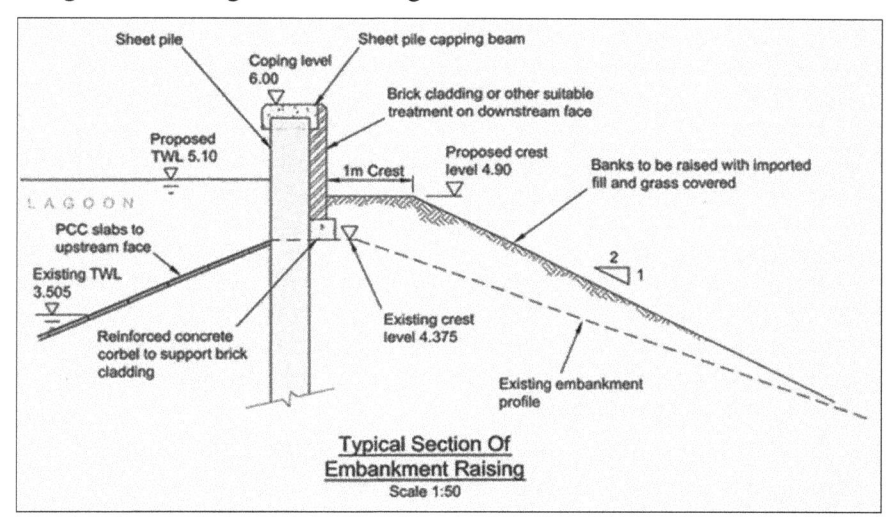

Figure 3. Notional design solution for embankment raising.

CONSTRUCTION

The notional design solution shown in Figure 3 was further developed and optioneered within the design and build contract. The design operational water level of the lagoon was increased by up to 2m using sheet piles with a 600mm high concrete capping beam. The notional design was for piles to toe into the glacial till, which would have given pile lengths of between 12m and 15m. Design development gave pile lengths of 8m in length. These were installed using a silent press technique through the embankment fill into the underlying alluvial soil with some longer piles (up to 12m) required to provide sufficient reaction for the piling contractor's chosen installation method.

A revised overflow arrangement for spillway discharge to the river was developed, incorporating precast culvert units.

Figure 4. Aerial view of Bransholme Reservoir Lagoon during raising works (August 2010)

On the inside (or upstream) side of the piles the client requested a footpath be added to allow the lagoon and internal faces of the sheet piles to be inspected. Part of the upstream slope of the embankment was cut down to provide this. To meet the requirements of no visible leakage through the sheet piles the clutches were welded from capping beam level down to embankment footpath level. The clay embankment present either side of the sheet piles was expected to restrict seepage through the embankment.

Filling test of the reservoir
The Contractor was required to undertake a water test on the lagoon on completion of the sheet piling works. This involved filling the lagoon to design operational top water level and holding at this level for an extended period of seven days to allow for inspection of the embankment and the sheet pile walls and other structures.

During the test filling seepage was observed at the toe of the southern embankment and at the north-west embankment corner. The test filling was abandoned before the top water level had been reached. The water level in the reservoir was reduced back to its normal operating level.

Investigation of groundwater response and seepage to reservoir filling
A ground investigation provided further information on the ground profile of the lagoon embankments and the underlying deposits, and piezometers were installed in the different soil horizons to observe the groundwater response to filling the lagoon reservoir.

The ground investigation showed signs of fissuring in the embankment clay which had not been observed or recorded in previous investigations. The fissures were due to the alluvium used in the previous raising of the embankments, the method of construction and severe desiccation. The ground investigation also encountered a layer of chalk gravel near the base of the embankment fill in a number of boreholes. This layer was up to

400mm thick in places and was observed on three sides of the lagoon. This layer was interpreted as being hardcore used as a running layer for plant movement most probably at the time of constructing the original lagoon embankments.

Thirty four Casagrande style piezometers were installed with response zones at varying depths in the embankment fill and underlying alluvium to investigate potential seepage pathways. The piezometer response zones were positioned to investigate potential seepage flow through the fissured embankment fill, through the chalk hardcore layer, through the alluvium, and flow under the toe of the piles.

Frequent monitoring of the piezometers and recording of seepage observations along the embankments was undertaken during a repeat lagoon test filling. The readings were used to validate groundwater modelling using industry standard software to help understand the problem.

Conclusions of the investigation and seepage observations
Analysis of the groundwater monitoring information and the site observations of seepage indicated that a significant groundwater response in the piezometers and the first signs of seepage both started to show when the water level in the lagoon rose to within 400mm of top of embankment crest level. Prior to reaching this level the response was negligible indicating that the internal embankment fill and the foundations were of low permeability.

The responses shown in the piezometers and from observations of seepage were even more pronounced when the lagoon water overtopped the footpath and the lagoon water had direct access to the sheet piles and the unwelded clutches at footpath level.

These observations along with analysis of the piezometer readings and mathematical modelling provided sufficient evidence that seepage was occurring through the sheet pile clutches when the rising water gained access to the clutches and possibly any space between the soil and the sheet pile to 'gain access to' an unwelded clutch. The water then appeared on the other side of the sheet pile wall by travelling along the most permeable layer available, *viz*:

- Initially this flow route seen as seepage through fissures or other defects in the clay embankment fill (Figure 5);
- The fissures appeared to be present as a connected network that allowed water to flow through the upper part of the embankment emerging on the face when the storage capacity of the fissures was exceeded;
- Flow also took place along the chalk gravel layer where it connected to the sheet piles (but is confined by overlying clay layer);

- Where fissures were not present in the clay of the embankment - such as at the area of the old spillway which was reinstated as part of the works - then seepage was not observed on the embankment face. However, water was still able to flow through the sheet piles clutches at crest level and discharge over the crest and downstream slope of the embankment.

Seepage between the eastern embankment and the grit channel was found to be from a pipe within the embankment connected to the grit channel.

Figure 5. Seepage emerging from fissures and a rabbit hole in the embankment fill

Sheet pile permeability
Steel sheet piles themselves can be considered impervious and therefore the only possible route for fluid to traverse the wall is via the joints (clutches). A design guide to Impervious Steel Sheet Pile Wall (Arbed Group (1998)) states that permeability values for empty sheet pile joints strongly depend on the soil properties, the variations being very large. Clays would normally be expected to impart very low permeability.

Sellmeijer (1993) and Arcelor (2008) give characteristic values of joint resistance for sheet pile clutches from the results of *in situ* tests. For sheet piles without sealant the value is stated as 1×10^{-7} m/s. The values given for sheet pile joint resistance are for horizontal seepage through the clutches; any component of vertical flow is ignored. Vertical permeability of a pile clutch could be much higher than horizontal permeability.

At Bransholme falling head tests carried out against the sheet pile clutches gave an indication of the permeability for the sheet piles and gave results of a similar order to published values.

The piles used in the construction works at Bransholme lagoon were Japanese 3W piles. During the investigations there was some discussion as to whether the clutch detail varied significantly from the original design solution using Arcelor PU18 piles. It was considered that a more open clutch detail could result in a greater horizontal permeability value, than that stated in Sellmeijer (1993). The clutch profiles of the Japanese 3W piles and the Arcelor PU18 piles were overlain and it was shown that the clutch details were comparable. Discrepancies in clutch design and therefore permeability were not considered further. Seepage differences between new and re-used piles were also investigated as both had been used at Bransholme, to determine if re-used piles had looser, more permeable, clutches. Leakage through the re-used piles was observed to be less.

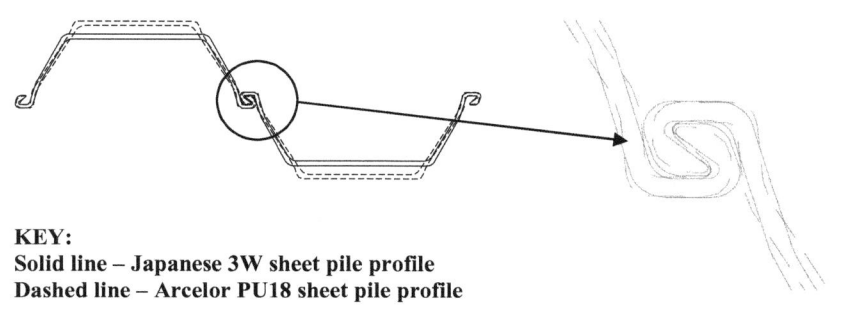

KEY:
Solid line – Japanese 3W sheet pile profile
Dashed line – Arcelor PU18 sheet pile profile

Figure 6. Comparison of sheet pile clutch tightness between Japanese 3W and Arcelor PU18 sections

Developing a solution
An embankment without defects in the form of cracks or fissures would be expected to provide an effective seal between the welded piles and the clay. Seepage would therefore not be expected. The investigations at Bransholme demonstrated that the combination of cracks or fissures in the clay and pile clutches that are open provides an opportunity for seepage to develop as the water level in the lagoon rises to the top of and above the internal clay fill. In order to control this, a remedial option had to provide an efficient sealing mechanism that would prevent water having access to and passing through the unsealed portion of any pile clutches. More than a dozen options for sealing and enhancing the performance of the embankment were examined and assessed in relation to practicability and effectiveness. With the mechanism of seepage described, remedial works applied to the inside of the lagoon were likely to be most effective.

The preferred approach was to combine extending the welding of each clutch to below the footpath level with backfilling using a cement:bentonite slurry. Observations from the water filling tests and mathematical modelling were used to support the decision on the additional depth needed for sealing of the sheet pile clutches.

The information from the tests and groundwater monitoring suggests a fissured zone of around 400mm deep but this does not account for further desiccation that might occur in long dry periods. Engineering judgement supported by visual observation in trial pits on the downstream side suggested extending the depth of the upstream fissured zone to 600mm. This is taken as a pessimistic assumption.

Mathematical modelling considered different weld lengths and the effect on seepage. The required depth of the weld is dependent on the depth to which the fissured layer extends in the model. To give the appropriate level of confidence in the remedial works to limit seepage to an acceptable level, welding over the full depth of the fissured layer was required. A weld depth of 600mm was adjudged to achieve this. This was based on assessment of the groundwater monitoring results and the assumption that the fissured layer extended to 600mm depth. However, an additional 200mm of welding was included to take the welding length to 800mm in order to provide an improved level of confidence in the works and an element of 'safety factor'.

Remedial works

In order to complete welding to a depth of 800mm below the footpath formation level, excavation by hand to 900mm depth was needed to permit access to the clutches for the welding.

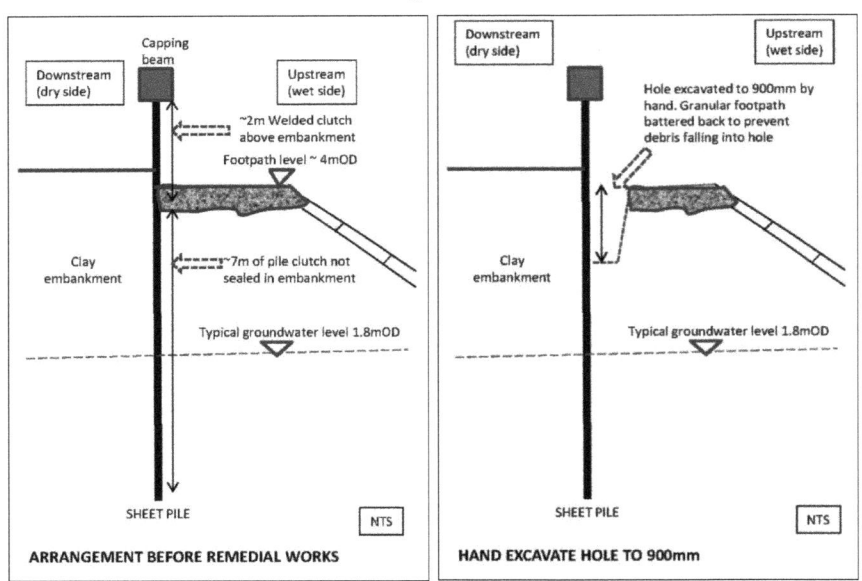

Figures 7a and 7b. Construction sequence of remedial works

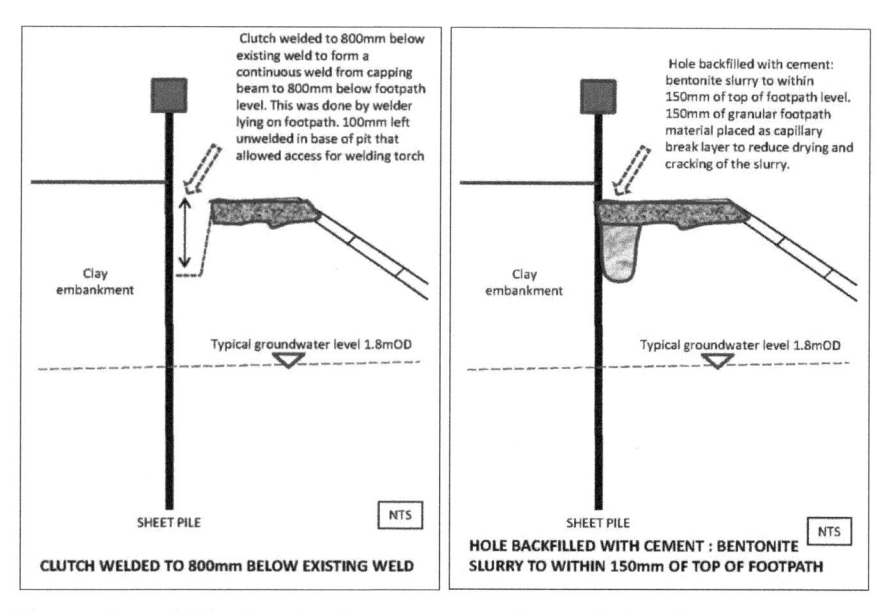

Figures 7c and 7d. Construction sequence of remedial works

After welding the hole was backfilled with a cement:bentonite slurry up to a level of 150mm below footpath formation level. In order to minimize evaporation from the ground surface and the slurry that might cause cracking to develop, a 150 mm thick granular capillary break layer was placed above the slurry. This layer ties into the formation level of the existing footpath.

Testing of remedial works

Figure 8: Final water test

A water test followed completion of the remedial works and seepage was not observed, indicating that the remedial sealing works had been successful.

CONCLUSIONS

Water can flow through untreated sheet pile clutches and this flow can be significant. Published values for the permeability of sheet piles with untreated clutches is 1×10^{-7} m/s in the horizontal plane, or greater. This should be considered in the detailed design and incorporated into any seepage modelling. Sensitivity analysis of a varying permeability of pile clutches should also be considered carefully, as the stated permeabilities are taken as a lower bound only and the actual permeability can be much greater.

Where sheet piles are installed into dams consideration should be given to extending welding or sealant to sufficient depth in clay embankments to provide an effective seal to account for defects such as cracks or fissures. An alternative approach would be to consider a seal for sheet pile clutches over their whole length. Future designs should not assume that driving sheet piles through a clay embankment will form a watertight structure. As experienced at Bransholme, the vertical flow of water along a pile clutch, connecting more permeable layers within an existing embankment, can cause significant seepage problems and lead to extensive remedial works.

Welding provides an effective means of sealing clutches, including second-hand piles. The interface between clutches that have had sealant installed and clutches to be welded presents a problem with on site welding potentially affecting the sealant. One approach is for a plate to be welded at the interface and for grout or sealant to be injected behind the plate.

Sealant can be factory applied for a relatively low cost (around £5/m), however, this assumes that new piles are being used. More work is required and it is potentially less effective applying sealant to second hand piles.

Keeping record drawings of works is important. Even small raising works can result in fill being placed that conceals defects that may affect future operation of a facility.

Final lesson – be careful when relying on piles.

REFERENCES

Arbed Group, 1998. *The impervious steel sheet pile wall, Part 1 Design.*

Arbed Group, 1998. *The impervious steel sheet pile wall, Part 2 Practical aspects.*

Arcelor, 2008. *Piling Handbook, 8th Edition.* ArcelorMittal, Solihull, UK

Institution of Civil Engineers, 1996. *Floods and Reservoir Safety, 3rd Edition*. Thomas Telford, London, UK.

HMSO 1975. *Reservoirs Act, 1975 (1975) Elizabeth II. Chapter 23*. Her Majesty's Stationery Office, London, UK

Sellmeijer, J., 1993. Steel sheet pile resistance seepage. *Proc. 4th International Landfill Symposium, Cagliari.*

Maintaining the Safety of our Dams and Reservoirs
ISBN 978-0-7277-6034-0

ICE Publishing: All rights reserved
doi: 10.1680/mdam.60340.148

Skhalta Dam – Design of a hardfill dam founded on deep alluvium and lacustrine deposits

J R PAWSON, Mott MacDonald
E RUSSELL, Mott MacDonald

SYNOPSIS The proposed Skhalta Reservoir is part of the Shuakhevi HPP scheme to develop the hydropower potential of the Adjaristsqali River and its tributaries. It is located in the Autonomous Republic of Adjara, in southwest Georgia. The reservoir will provide diurnal storage and will facilitate the transfer of flows between catchments. It will have a gross storage volume of 2.0Mm³ and will be impounded by a 27m high Faced Symmetrical Hardfill Dam (FSHD) approximately 160m long at crest level.

The geology at the site consists of deep alluvium deposits (typically sands and gravels) with a significant variation in depth across the valley. At each abutment the valley slopes are steep and rock is found relatively close to the surface, whereas in the centre of the valley bedrock is up to 50m below the surface. Within the upper alluvium deposits there is a layer (4m-5m) of low strength finer grained material. This is associated with a historical reservoir formed following damming of the river by a landslide downstream of the proposed dam site in 1989.

The foundation conditions require a structure appropriate for low strength foundations that can tolerate significant differential settlements during construction. A Faced Symmetrical Hardfill Dam was therefore selected.

The significant variations in thickness of the alluvial deposits and the presence of a low strength layer have been major considerations in development of the design. Ground improvement measures comprising interlocking soil mixed panels have been developed to provide the required foundation strength and a full cut-off to rock has been specified to reduce seepage to acceptable levels.

This paper describes design aspects of the dam with particular emphasis on the dam foundations.

PROJECT DESCRIPTION

Adjaristsqali Georgia LLC (AGL) owns the rights to develop the hydropower potential of the Adjaristsqali River and its tributaries in southwest Georgia.

Mott MacDonald (MM) was appointed by AGL in 2011 to produce a feasibility study and environmental and social impact assessment (ESIA) for a three-stage hydropower cascade. Subsequently MM was commissioned to undertake tender design, detailed design and site supervision for the uppermost scheme in the cascade, Shuakhevi HPP.

The Shuakhevi HPP scheme is a run-of-river hydropower scheme collecting flows from the Chirukhistsqali, Skhalta and Adjaristsqali rivers. Transfer tunnels will convey flows to a reservoir on the Adjaristsqali River, from where a headrace tunnel will lead to a powerhouse further downstream on the Adjaristsqali River. Diurnal storage will be provided by reservoirs on the Adjaristsqali and Skhalta rivers to ensure that power generation can be timed to coincide with periods of maximum demand. The power generated will be exported to both the Georgian and Turkish power systems.

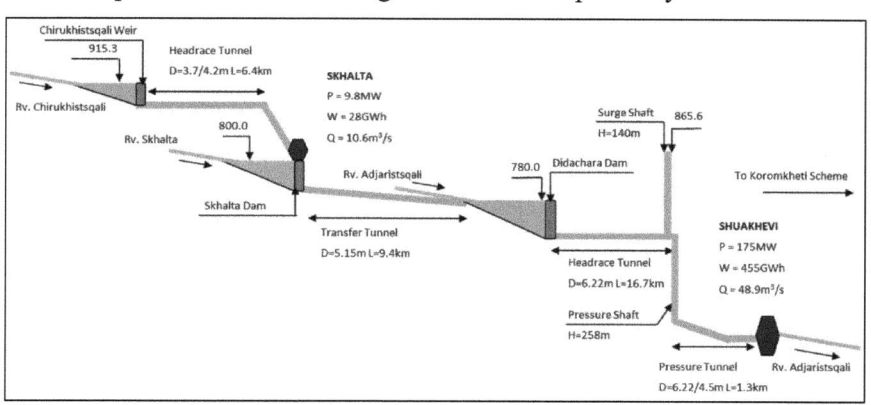

Figure 1. Shuakhevi HPP schematic

A summary of the main elements of Shuakhevi HPP is given below and shown in figure 1:

- Chirukhistsqali headworks comprising a weir intake structure and sedimentation basin on the Chirukhistsqali river;
- Transfer / headrace tunnel to Skhalta reservoir;
- Skhalta powerhouse;
- Skhalta headworks comprising a hardfill dam and intake structure on the Skhalta river;
- Skhalta to Didachara transfer tunnel;

- Didachara headworks comprising a concrete dam and intake structure on the Adjaristsqali river;
- Headrace tunnel;
- Shuakhevi powerhouse.

The layout of the scheme is dictated by:

- a suitable reservoir site on the Adjaristsqali river at a level of around 780m
- a suitable reservoir site on the Skhalta river at a marginally higher level with sufficient head difference to enable flow transfer
- a suitable site for an offtake weir, desilting basins and tunnel intake below the existing hydropower scheme on the Chirukhistsqali

There were no obvious viable alternatives to the selected reservoir sites. This was mainly due to the need to avoid potential landslide locations in all three river valleys coupled with the locations of settlements.

HYDROLOGY AND CATCHMENT
The Skhalta River is a tributary of the Adjaristsqali River and originates in the western part of the Arsiani mountain range, 2,435m above sea level. Its inflows are provided by snowmelt, rainfall runoff and ground water.

Melting of winter snows feeds a major period of higher flows in spring through to early summer. A second period of high flows occurs in late autumn driven by rain storms.

The river system is active with erosion, transport and deposition of sediment creating a wide range of channel morphology along the length of the river and tributaries. The catchment also has landslides which add large quantities of sediment to the river.

Sediment modelling indicates that approximately 111,000m^3 of sediment will be deposited in the reservoir annually. Therefore provision of low level outlets designed to allow periodic flushing is vital to ensure that adequate live storage can be maintained.

GEOLOGY AND GROUND INVESTIGATIONS
The bedrock within the scheme area comprises volcanic and volcaniclastic sediments. The Skhalta site is located on the southern limb of the regional syncline with flow-banding/bedding dipping shallowly to the north (out of the slope) on the left bank, but much more steeply on the right bank. A large thrust fault is postulated to be present trending NW-SE along the river and the steep dips on the right bank are considered to occur within an upthrust block bounded by this main thrust fault and a fault spur. The rock types present are predominantly medium strong to strong Andesite-Basalt

Breccias interbedded and dyked with strong to very strong Andesite-Basalt, and occasionally interbedded with medium strong to strong Tuff.

Figure 2. Skhalta dam site

The large volumes of sediment entering the catchment from landslides and erosion have resulted in a sediment-heavy river system in the reaches of the dam site and both upstream and downstream (Figure 2). The steep sides of the valley reach down to a broad river plain with braided channels. Significant thicknesses of alluvial sands and gravels are deposited within the steep sided valley and locally increased thicknesses and variations in materials occur due to the effects of previous landslide damming of the river.

Geomorphological and geological mapping was used to conduct landslide hazard assessments for the potential reservoir and was the deciding factor in limiting the reservoir extent. Mapped slope morphologies and rock exposures were used to delineate areas of stable, unstable and potentially unstable slopes allowing a qualitative risk assessment for the project to be undertaken. In general slopes surrounding the reservoir showed steep slopes (30°-45° right bank; 35°-45° left bank) indicative of near-surface rock, which was largely apparent and visible on the right bank, particularly at river level, while the denser vegetation on the left bank was indicative of a colluvial/talus wedge at the slope toes. Breaks in slope high above the reservoir on the left bank and the identification of *in situ* rock in accessible areas gave confidence in this interpretation. A deep seated failure, extending down to river level, identified on the left bank defined the upstream reservoir extent.

A large relict landslide located downstream of the dam site was of particular concern with regard to the reservoir in the early phase of the scheme. The Tsablana landslide occurred in 1989. The debris flow run-out from the failure blocked the river to a height in excess of 25m, filled a tributary on the opposite side of the river, and destroyed a village with significant loss of life. Mapping of this feature defined the source area in the upper reaches of a tributary valley and identified very poor quality rock in the back scarp. It appeared that similar failures may have occurred in tributary valley catchments downstream of Tsablana. Although the mechanism for failure would be independent of the presence of the reservoir, mapping in the head catchments to tributary valleys upstream and above the proposed reservoir was undertaken to identify any signs of developing instability. Active erosive processes and shallow creep movements were noted in these areas but there were no apparent indicators for larger scale instability.

Ground investigations for the Skhalta dam were undertaken in three phases; feasibility, tender and detailed design, and comprised a total of 12 boreholes, 2 trial pits, 11 seismic refraction profiles and rockmass mapping focused on understanding founding conditions for the structure and associated cut-off. The feasibility investigations initially targeted an area just upstream of the Tsablana landslide to maximise the size of the reservoir. However, following initial investigations, the location was proven unsuitable for the dam based on an apparent detached block in the valley side and an 11m to 13m thickness of very loose silty sand with silt and clay layers, encountered in two boreholes. The steep sided nature of the valley resulted in poor data quality from seismic refraction and it was not possible to determine the valley profile by this method. The loose material encountered was interpreted to be a lacustrine layer representing the deposits of the landslide-dammed lake, which had not been visible from the valley surface. It was anticipated that the thickness of this loose fine grained material would be greater at the location of the landslide and would reduce in thickness upstream to a point where it wedged out.

During the tender phase a location some 500m further upstream was investigated. Four boreholes indicated that the lacustrine layer was of limited extent and at this location comprised medium dense (typical SPT N 12-20) slightly gravelly sand from 9 to 12 mbgl, with local areas of loose (SPT N 3-5) silty sand with a little organic material as noted from 10.5 to 12mbgl in two to the boreholes. However, during detailed design investigations the layer was again found to be more variable and extensive. Four boreholes encountered the layer between approximately 7.5 and 13 mbgl as slightly silty sand interbedded with very soft sandy silts and occasional sandy clays, SPTN values were typically between 8 to 12 in the upper part of the layer but decreased to less than N=1 in some locations at the base of the layer.

The alluvial deposits in the centre of the valley at the Skhalta dam site have been proven to depths of 50m. The sequence of materials encountered typically comprises an upper alluvial layer of medium dense silty gravelly sand with cobbles up to 9m thick, underlain by a laterally and vertically variable lacustrine layer, between 3m to 6m thick, of very loose to medium dense silty sand variably interbedded with very soft sandy silts and occasional sandy clays with variable organic content. The lacustrine layer is underlain by medium dense to very dense silty sandy cobbley gravel over bedrock. Rockhead is anticipated to be deepest towards the centre of the valley based on investigation results. The typical properties of the materials are summarised in table 1 and figure 2.

Table 1. Skhalta dam ground profile

Layer	Thickness range (m)	Typical SPT/UCS	Typical Permeability
Upper alluvium	7 – 9	N=22	2.5×10^{-3} m/s
Lacustrine layer	3 – 6	N=7	5×10^{-5} m/s
Lower alluvium	7 – 36	N=45	5×10^{-3} m/s
Bedrock	-	UCS=40to90	150 to 50 Lugeons (decreasing with depth)

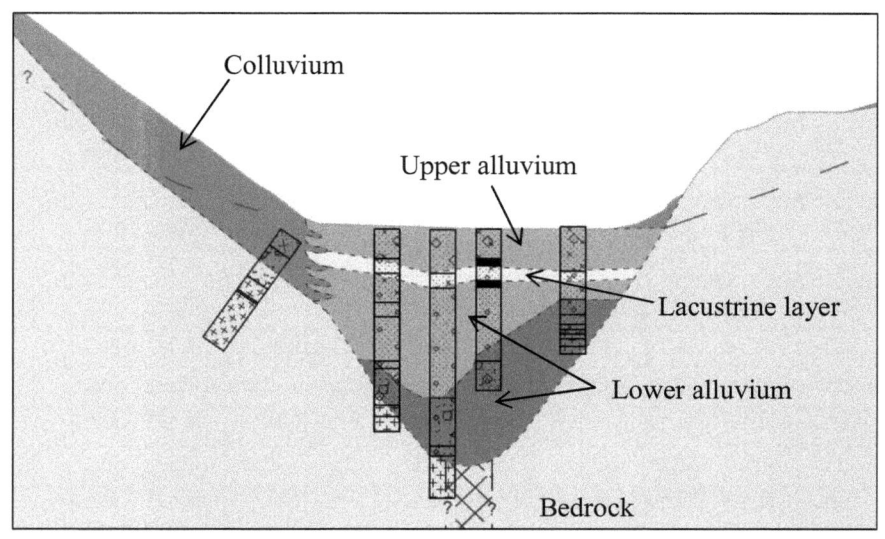

Figure 3. Skhalta dam ground profile

The density of the silty sands encountered within the lacustrine layer were such that they would be expected to liquefy during a seismic event, resulting in a temporary complete loss of strength and support provided by the material. The layer thus posed a significant issue that needed to be addressed in design of the dam foundations.

SELECTION OF DAM TYPE

Skhalta reservoir will have a gross storage volume of 2Mm³ and cover a surface area of 20Ha at its Maximum Operating Level (MOL) of 800mAD. The dam will be 22m high above existing bed level (27m above foundation level) with a crest length of approximately 160m.

The design floods adopted for the scheme reflect the additional risk posed by dam types that are susceptible to overtopping (Table 2). This had an impact on the dam type selected since the steep valley sides mean that there is limited space for a chute spillway and this is exacerbated by the higher design flood for embankment dams.

Following dam break analysis using 1D hydraulic modelling software it was established that the dam breach wave has a negligible impact on the PMF flood extents. Consequently the Safety Check Flood design criterion was reduced from PMF to the 1 in 10,000 year flood.

Table 2. Skhalta design floods

Item	Structure	Design criterion
Design Flood (DF)	Concrete barrages and weirs (<15 m and gross storage < 1 Mm³),	1 in 100 year flood (330m³/s)
	High concrete dams (>15 m) and/or storage > 1 Mm³	1 in 1000 year flood (510m³/s)
	Embankment dams	1 in 10,000 year flood (700m³/s)
Safety Check Flood*	Dams <15 m	1 in 10,000 year flood (700m³/s)
	Dams > 15 m and/or storage > 1 Mm³	PMF (1750m³/s)

A number of dam types appropriate for deep alluvium foundations were considered for the site. The key considerations in determining the dam type adopted were:

- Improvement/replacement of the lacustrine material beneath the dam foundation was required for all dam options due to the low strength and liquefaction potential. It was therefore important to minimise the dam footprint in order to minimise the costs of ground improvement.

- The steep valley slopes mean that there is little space for a chute spillway located at either abutment.
- The material available from the local borrow areas typically comprises gravels and cobbles. There is little cohesive material available locally.

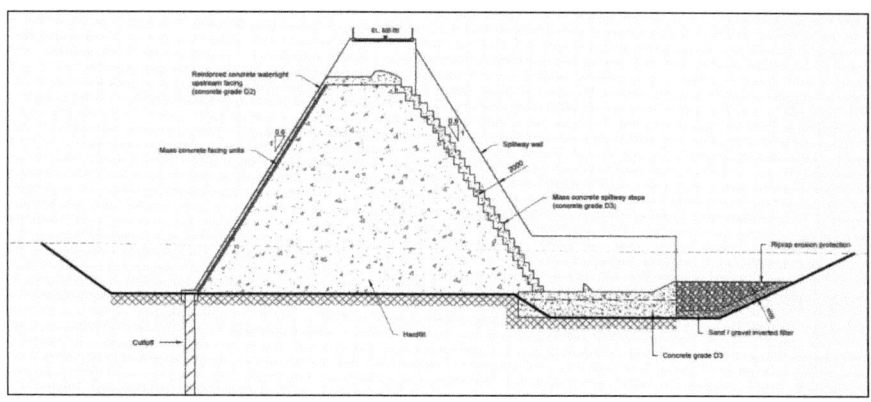

Figure 4. Skhalta dam typical section

From these criteria a Faced Symmetrical Hardfill Dam (FSHD) was selected as this can be tolerant to the predicted settlement, the spillway can be over the dam body and the dam footprint is relatively compact compared to a CFRD, providing a saving on the ground improvement required. In addition the fact that this dam type is overtoppable enabled the design flood to be reduced to the 1:1000 year flood.

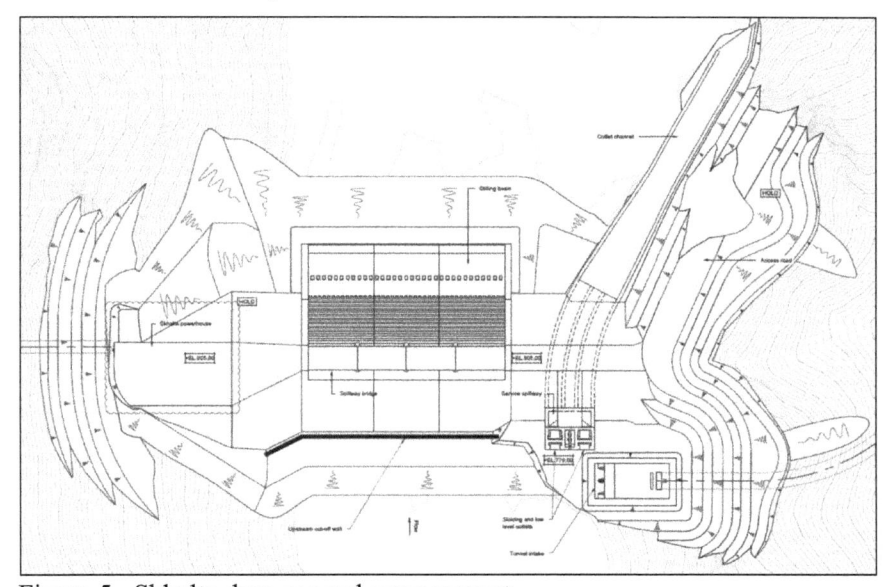

Figure 5. Skhalta dam general arrangement

The alternative dam types considered were:

Earthfill – rejected due to relatively large footprint, difficulties of constructing a chute spillway and lack of suitable cohesive material to form the central core zone.

CFRD - ruled out due to the relatively large footprint and consequent additional area of ground improvement required

Barrage - ruled out as the stilling arrangements required meant that the area of ground improvement was similar to that required for the FSHD option, and the complex reinforced concrete structure proved expensive.

DESIGN OF DAM FOUNDATION TREATMENT

Selection of treatment approach
At tender stage compaction grouting had been considered an appropriate method to improve the lacustrine layer. However, additional information obtained during detailed design indicated greater thickness and variability of the lacustrine materials, the potential for liquefaction of the layer during a seismic event and insufficient frictional properties of the upper alluvium during a seismic event. It was therefore necessary to reassess the ground improvement design.

Ground improvement was required to both the upper alluvial layer between 0m and 8mgbl and the lacustrine layer between 8 and 13mbgl. The difference in material type between these two layers (and also within the lacustrine layer), and the depth of treatment required careful consideration of the various techniques to select the most suitable method with due consideration to anticipated costs.

Taking into account the required improvement a number of methods were given consideration. Two methods were selected for further design development to allow a final decision to be taken; the other methods considered and reasons for their rejection are included in Table 3.

The ground improvement options selected for further development were:

Wet soil mixing using trench cutting equipment – gives the flexibility to deal with the dominantly granular materials of the upper alluvium as well as the mixed granular and cohesive materials of the lacustrine layer as the method disaggregates the soil and mixes in a fluid binder to form a homogeneous material. The binder is designed to achieve the required strength for the most critical material and verification can be undertaken though sampling of unset soil mixed material and later sampling of the set material to confirm compliance. Foundation material is treated in-situ, thus minimising the required excavation depth.

Excavate and replace within a dewatered excavation – unsuitable material is removed and replaced with selected backfill. This is a straightforward earthworks operation but requires the dewatered excavation to extend to the underside of the lacustrine layer (~13 mbgl). The high permeability of the upper and lower alluvium presents potentially significant difficulties in dewatering an excavation of this depth. Consequently the level of temporary works required to facilitate dewatering is critical to the economics of this solution.

Table 3. Ground improvement options considered

Method	Reasoning for rejection
Excavate and replace - underwater	Nature of lacustrine materials is such that they may run upon excavation. Risk of instability of excavation that could result in loss and delay.
	Anticipated requirement for subsequent improvement of fill placed underwater.
Compaction grouting	Method considered appropriate for upper alluvium.
	Improvement >300% required for the lacustrine layer, which may not be achievable, particularly in clay and silt layers.
	Uncertainty over a consistent level of improvement due to variability of lacustrine layer. Significant post treatment assessment required across area. Risk of excessive grout takes.
	Concerns over integrity of single grout columns in lacustrine layer during seismic event.
Vibrocompaction	Suitable for upper alluvium, but unsuitable for silty sands, silt and clay of the lacustrine layer.
Vibroreplacement (stone columns)	Suitable for upper alluvium.
	Concerns over stability of isolated columns within lacustrine layer during a seismic event.
	Requirement for very high replacement ratio.
Jet grouting	Method considered suitable.
	Concerns over requirements for highly skilled operators and quality assurance team.
	Cost of jet grouting not anticipated to be cost effective for this application.

Method	Reasoning for rejection
Mass wet mixing (Horizontal soil mixing)	Depth of treatment on the limit of available equipment. Discussions with manufactures and users indicated that equipment would not be suitable for mixing materials in the upper alluvium.

Design of soil mixed panels

The design of the soil mix panels addressed both external and internal stability of the foundation system under static and seismic loading. External cases considered included bearing capacity, sliding and slope stability and internal cases addressed shear failure of the soil mixed panels, acceptable stresses within the panels and deformation of the soil mixed panels due to seismic loading.

The use of trench cutting methods was selected following discussions with manufacturers over the ability of equipment to deal with the presence of cobbles and boulders within the upper alluvium and toeing into the lower alluvium. The ability to produce interlocked panels allows the element to have a greater bending stiffness than individual isolated panels/columns. This item was important considering the potential for negligible support to be provided by the unimproved lacustrine layer that would be left between panels.

The initial design concept was to construct continuous soil mixed panels perpendicular to the dam axis and extending to a distance 2/3 the depth of the improved layer beyond the load bearing parts of the structure and the cut-off. The panels were to be toed one metre into the underlying lower alluvium to ensure a good frictional contact. The end panels were to be closed off (with a row of panels parallel to the dam axis) to ensure no mechanism for materials of the lacustrine layer to flow out from between the panels if they were to liquefy.

In developing the design the original concept was amended to incorporate axis-parallel panels throughout the footprint of the structure. The primary function of these axis-parallel rows was to reduce the induced stresses in the axis-perpendicular panels to an acceptable level; the addition of axis-parallel panels was preferred over reducing the spacing of axis-perpendicular panels as it produces a more robust structure. Taking consideration of the imposed loadings and the inherent variability in strength of soil mixed materials the soil mixed panels required a strength of 1.7MPa to ensure an adequate factor of safety in both static and seismic load cases.

The final design arrangement was developed assuming an 800mm wide cutter-soil-mixing tool with a length of 2800mm. The arrangement comprised interlocked rows of 800mm wide panels (with 200mm overlap) parallel to the dam axis and spaced at 3200mm centres. The rows of axis-perpendicular panels were spaced at 1600mm centres along the dam axis and the spacing of axis-parallel rows of panels allowed for the installation of these panels between these rows with a 200mm overlap.

SETTLEMENT ANALYSIS

Due to the significant variation in depth of the alluvial foundation beneath the dam structure, long-term differential settlements were of concern for the facing which would have limited tolerance to such movements. The thickness of the alluvial strata was anticipated to range from 0m to approximately 36m below the 9m thick soil mixed zone upon which the dam was founded.

A three dimensional model was set up to consider the variation in settlement beneath the structure and three main load cases were assessed: end construction, reservoir full, and reservoir full and silted. The settlements relevant to design of the facing were those occurring post construction, after its placement, during reservoir filling and subsequent operation. These loadings result in higher loadings at the toe initially with additional loading towards the heel with sedimentation. Due to the absence of detailed data on the stiffness of the alluvial materials, non-linear stiffness was not considered for the materials in the model and a constant stiffness was adopted. Although not ideal the use of a constant stiffness was considered conservative as predicted settlements would likely be an overestimate of true settlements and thus safe for design of the facing.

In addition to settlements predicted by the model, it is known that significant creep of granular foundations can occur over time (Burland and Burbridge, 1985) and may be particularly large in cases where the loading fluctuates. An assessment of time dependant settlement was deemed necessary considering the sensitivity of the dam facing and this was achieved by using the time factors proposed by Burland and Burbidge (1985). The maximum long-term settlement, anticipated in the deepest part of the valley, was determined to be in the region of 70mm-80mm, with the potential for differential settlement equating to an angular distortion of 1:450.

ADAPTATIONS TO ACCOMMODATE SETTLEMENT

FSHDs are typically constructed on weak rock foundations and so are not required to tolerate significant foundation settlement. Therefore a number of adaptations to the design have been necessary in order to accommodate the predicted movement.

FSHDs are often constructed without joints; however, for the Skhalta site the maximum predicted post-construction differential settlement is approximately 70mm-80mm. Whilst the hardfill dam body could tolerate this movement the reinforced concrete upstream facing would be expected to crack and leak under this imposed settlement.

Consequently vertical movement joints have been provided in the facing at approximately 20m centres. However, due to the steep 1V:0.6H upstream slope it is necessary to anchor the facing to the dam body by means of dowel bars. It is therefore necessary to ensure that joints in the facing are continuous through the dam body since the dowels mean that the facing will move with the main body of the dam.

The details for movement joints in the facing, peripheral plinth and cut-off connection slab are taken from CFRD practice, with a triple barrier waterstop system (centrebulb waterstop, copper rearguard andsurface mastic bulb) employed at joints experiencing the highest potential differential movements.

SUMMARY
The design of this 27m high dam on deep alluvial foundations has required detailed consideration of dam type, foundation ground improvement and expected settlements. An FSHD was selected for the site due to the relatively small footprint, tolerance to settlement and ability to withstand overtopping. Ground improvement comprising either wet soil mixing or excavation and replacement of the upper alluvium and lacustrine material will be carried out beneath the dam footprint.

REFERENCES
Burland, J.B. & Burbridge, M.C. (1985). Settlement of foundations on sand and gravel. *ICE Proceedings, Vol 78, Issue 6* p1325-1381. Institution of Civil Engineers, London, UK

Didachara Dam – Site Constraints in Dam Hydraulic Design

P J HARVEY, Mott MacDonald

SYNOPSIS   This paper described the issues related to the design of the 40m high conventional concrete Didachara Dam, which will impound a reservoir with a gross storage of 2.3Mm³. It is the principal component of the headworks for the 175MW Shuakhevi HPP located in south-west Georgia near the border with Turkey. Construction of enabling works has already commenced and the dam should be built between 2014 and 2016.

The reservoir is required to provide both diurnal storage for the power generation and also temporary storage for sediment. This sediment will be largely flushed from the reservoir on an annual basis to maintain the diurnal storage requirement and also to provide ongoing storage for sediment management.

The principal issues with the design of Didachara dam are the constraints imposed by the geology of the area, which is dominated by landslips; the topography of the site; and the need to provide both spill and flushing facilities within a relatively tight site. These are discussed in the paper.

INTRODUCTION
Adjaristsqali Georgia LLC (AGL), owns the rights to develop the hydropower potential of the Adjaristsqali River and its tributaries in south-west Georgia.

Mott MacDonald was appointed by AGL in 2011 to produce a feasibility study and environmental and social impact assessment (ESIA) for a three-stage hydropower cascade.   Subsequently they were commissioned to undertake tender design, detailed design and site supervision for the uppermost scheme in the cascade, Shuakhevi HPP.

Shuakhevi is a 175MW hydropower project comprising an intake on the Chirukhistsqali river; tunnel transfer to the Skhalta river; a 20m high dam and reservoir at the transfer outfall; and a further tunnel transfer to the upstream end of Didachara Reservoir. A schematic of the project with key project dimensions is shown in Figure 1.

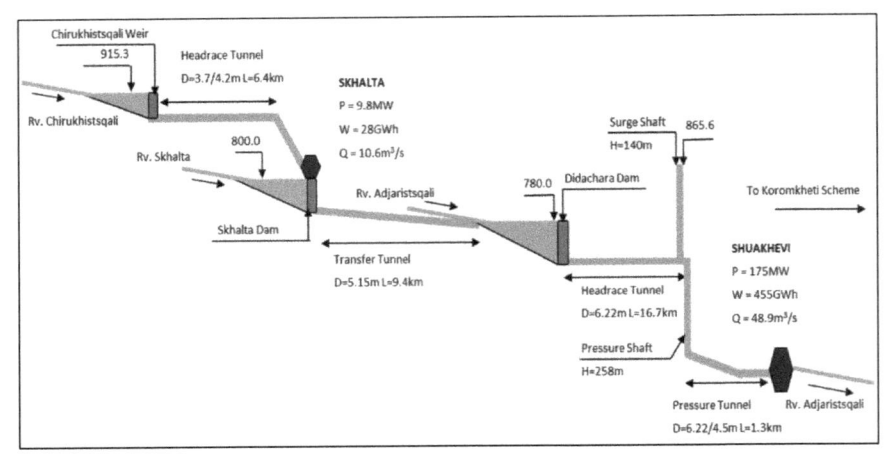

Figure 1. Shuakhevi HPP Schematic

The proposed Didachara Dam is located on the Adjaristsqali river in the south east of Geogia close to the border with Turkey approximately 65km upstream of its confluence with the Chorokhi river close to the Black Sea resort of Batumi.

The location of the dam and its height were constrained by the sites available for storage at the upstream extent of area covered by the licence agreement and the efficiency of the overall scheme layout. Consideration had to be given to the levels in the Skhalta reservoir and the need for storage on the Adjaristsqali. These two elements had to work in combination with the storage in Skhalta to provide both diurnal storage and temporary storage for sediment being carried by the Adjaristsqali river.

The site available on the Didachara river that provided the optimum storage at the required level was located in a narrow gorge just downstream of the Ghorjomi tributary as shown on Figure 2.

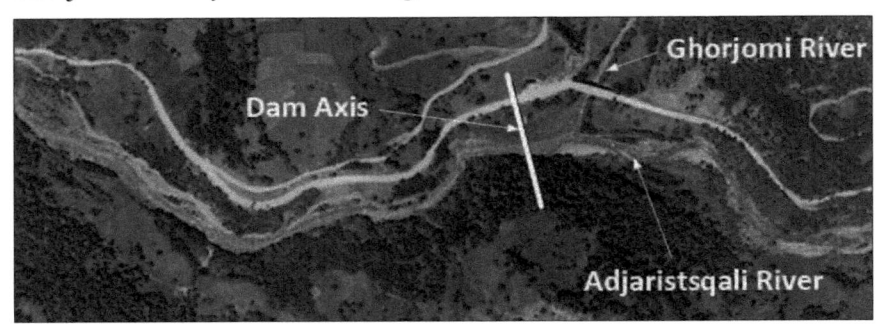

Figure 2. Site Layout

This site requires a dam of 40m height and the reservoir created has a gross storage of 2.3Mm^3.

The microlocation of the dam was constrained both by a landslip, which caused the dam to be moved during the design development as the site investigation results became available, and the topography which placed constraints on the layout on the dam.

In addition to the normal spill and low level outlet facilities the dam also needs to be able to pass sediment temporarily stored in the reservoir, as the reservoir volume available is too small to provide long term storage whilst providing the necessary diurnal storage for the more valuable peaking generation. The cost of providing sand traps on the headrace tunnels would have been both more expensive and also prohibitive for the scheme to develop.

The development of the design to address the constraints of the site and the requirements of the project whilst satisfying both the spill and sediment management are discussed below.

SITE CONDITIONS

Geological Conditions

Extensive site investigations had shown that the local geology to be made up of weak to strong Andesite Basalt Breccias, Tuffs and strong to very strong Andesite Basalt. However on the right bank of the dam site the rock was found to exhibit a highly variable weathering profile leading to a reduction in rock strength. It thus became necessary to consider how any proposed civil work could be minimised within this area, as the required slope protection would greatly increase the cost to the client.

Additionally, general site investigations had revealed a significant risk posed by landslips within the area. The results of the study found two areas of significant risk at the Didachara Dam site.

One area is associated with a ridge of bedrock located downstream of the final dam site on the left abutment. This was found to have channelled slipped material from a large landslide into the river valley. Thus the location of this landslip ultimately limited the downstream position of the dam.

A second area related to a large landslip complex located further upstream on the right abutment. Although relatively removed from the dam location, bedrock located in the upper reaches of the right abutment was found to be overlain by shallow colluvium. Concerns were raised that this colluvium could be related to the larger complex further above. As a result the maximum flood rise was limited to El.785m to limit any potential future problems.

Figure 3 shows the location of the landslip locations and how they relate to the overall dam location.

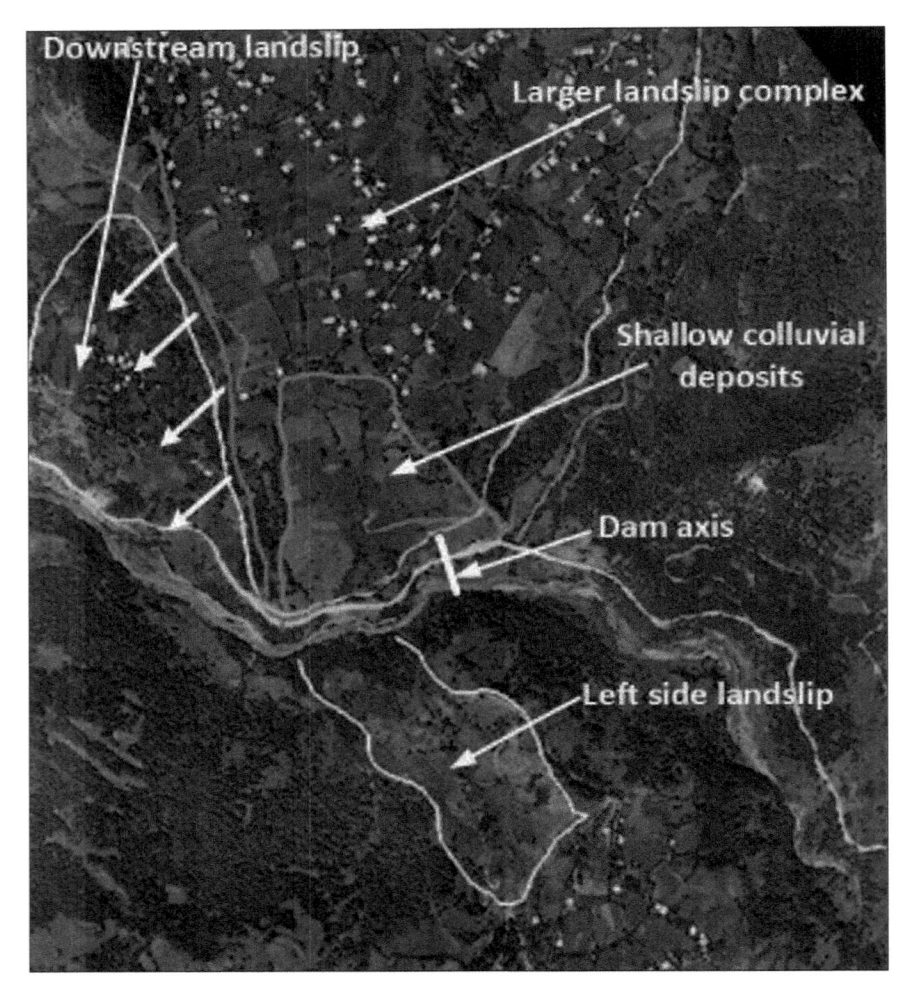

Figure 3. Location plan of landslip areas.

Topographical Constraints
In addition to the geological conditions there were various topographical constraints which influenced the overall arrangement of the dam. The location of the dam had been primarily fixed due to the location of landslips and the need to maximise the hydropower potential of the scheme, including the need to provide diurnal storage. Thus, these topographical difficulties required careful management. The principle difficulties were the available width within which flood discharges could be passed and general alignment of the Adjaristsqali river. Figure 4 shows the topographic layout of the site.

It can be seen that the limited widths within the river valley could lead to high flood rises within the reservoir. A maximum width of 55m was available within which all spillway flows needed to retained. Furthermore the river alignment can be seen to limit the space to provide a suitable downstream condition for any proposed stilling arrangements.

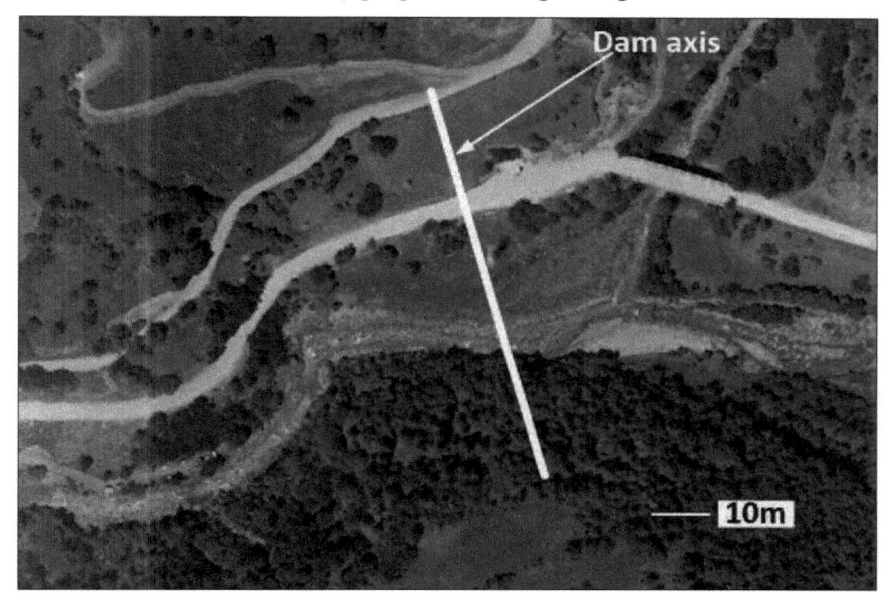

Figure 4. Aerial photograph of the dam site

Site Hydrology

Catchment Characteristics
The Adjaristsqali River originates from the western part of the Arsiani mountain range, 2,435m above sea level with a total length of 90km.

At the Didachara Dam Site the catchment area of the Adjaristsqali is 211km^2 with inflows provided by snowmelt, rainfall runoff and ground water, with the primary inflow supplied by rainfall.

The hydrological regime is dominated by two periods of high flow. In the first instance melting winter snow feeds a major period in spring through to early summer. In the second, a period occurs in late autumn driven by rain storms. Figure 5 shows a typical year of the flow records for a gauging station in the catchment.

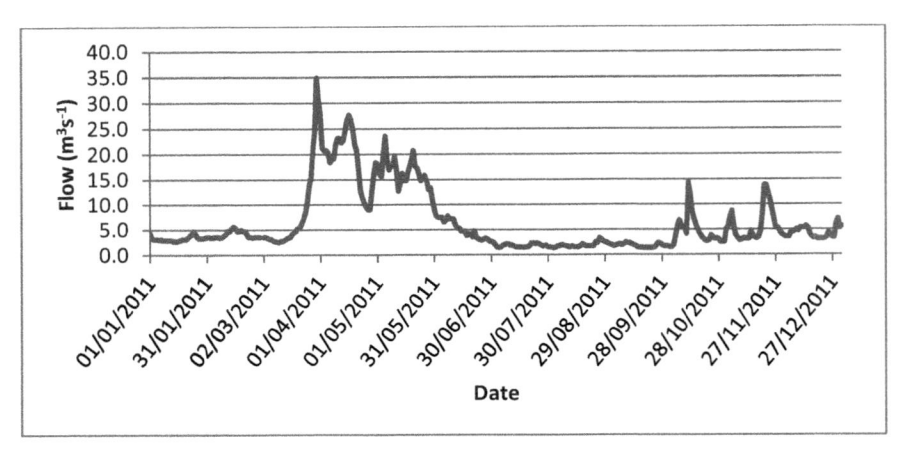

Figure 5. Typical annual hydrograph

Design flows for the site were assessed using various methods including the development of a rainfall-runoff model and gauging station frequency analysis. The final design flows for Didachara Dam site are shown below.

Table 1. Didachara Dam Design Flows

Return Period (1 in x years)	Flow (m³/s)
100	450
1,000	700
10,000	950
PMF	2390

Criteria were then adopted based on general guidance presented in ICOLD Bulletin 82 in order to evaluate the design return periods. These criteria, used to select design flows, are shown below.

Table 2. Design Flood Criteria

Item	Criterion
Diversion design flood	10% probability flood (1 in 10 year)
Design Flood (DF)	0.1 % probability flood (1 in 1,000 year)
Safety Check Flood[1]	Probable Maximum Flood (PMF)

1. Safety check floods may be reduced to design flood if dambreak analysis shows that impact of breach wave on flood wave is nominal (< 10% increase and no additional property affected)

As the PMF event remained high, dam breach analysis was undertaken in order to ascertain the likely increase in risk to downstream populations. The results of this assessment allowed the lowering of the SCF return period from PMF to the 1 in 10,000 year event.

Site Sediment Regime

Adjaristsqali River Conditions
The Adjaristsqali river system is hydro-morphologically active with erosion, transport and deposition of sediment creating a wide range of channel morphology along the length of the river and tributaries. Figure 6 shows an example of the river morphology on the Adjaristsqali.

Figure 6. River morphology at the dam site

The sediment being carried by the river arrives as a result of run-off from precipitation, both snowmelt and rainfall run-off, and by the river washing debris and landslip materials from along the river banks. In its upper reaches, such as the location of Didachara, velocities can be high in storm run-offs and the bed load can include cobbles and boulders

As a result of the sediment load carried within the river its careful management is required in order to minimise the impact on:

- the turbines, as sediment passing through the turbines could potentially cause wear and other damage to the runners and associated parts; and
- available storage, as the sediment trapped in the reservoirs would reduce the ability to regulate available flows to the power station.

PROPOSED SEDIMENT MANAGEMENT SYSTEMS
As noted previously, the two primary design challenges for the Shuakhevi scheme were maximising diurnal storage and managing the heavy sediment loads within the Adjaristsqali River.

During normal operation of the scheme, Didachara reservoir is expected to remove a large proportion of the coarse sediment from the river inflows. Thus the reservoir is able to provide effective sediment management with respect to protection of the turbines, but will also gradually fill over the course of a year reducing the available diurnal storage. In this way it is necessary to remove this sediment through low level flushing to ensure effective operation of the reservoir.

Flushing would be expected to occur annually during the high flow periods in early spring. This would ensure sufficient flow was available to permit effective clearing of reservoir sediment. Based on the local hydrology a flushing flow of 25m³/s was selected as a flow rate that could be achieved for a minimum period of two days during the month of April. As such a typical deposition cycle could be assumed to commence at the beginning of May and ending in April. At the selected flow rate a flushable channel width was assessed to be around 40m with a depth of flow of 0.3m.

To give an indication on the impact of the sedimentation within the reservoir the storage characteristics of the reservoir are now described. The gross available storage at the time of initial reservoir impounding at Didachara is 2.3Mm³, yet operational conditions will limit the variation of water level below Top Water Level (TWL) to 10m. As such the maximum available diurnal storage is limited to 1.3Mm³. However due to unavoidable loss of storage away from the flushable channel the available long term live storage was assessed be limited to just 0.63Mm³. Figure 7 presents the stage storage curves at initial impounding and the long term case where storage has been lost to sediment deposition outside the flushable channel. TWL and the minimum reservoir level are also indicated showing the range which equates to the 'live' storage available.

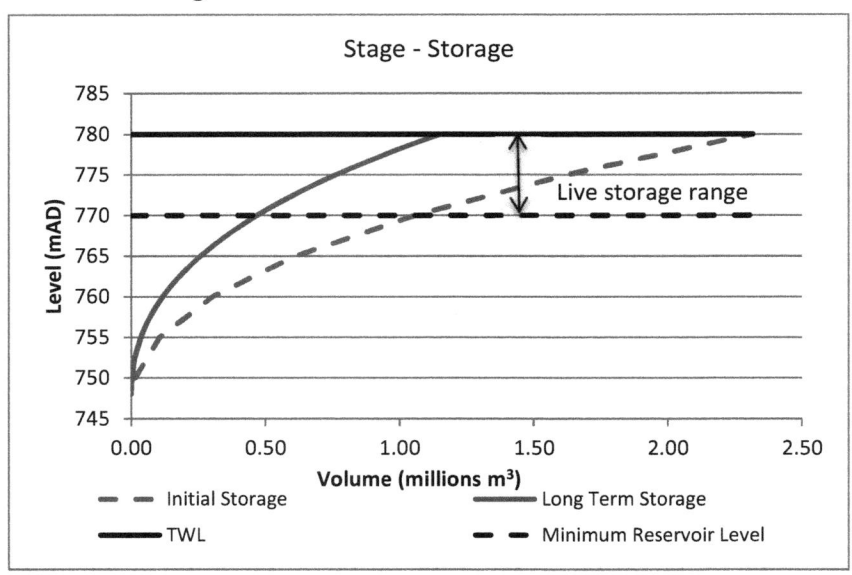

Figure 7. Stage-Storage Curve

In addition to the long term loss of storage, the annual short term storage loss had to then be added. The estimated volume of live storage available at Didachara at the beginning and end of a deposition cycle along with the average quantity of sediment being carried by the river are shown in Table 3. As can be seen it is expected that one-third of the available live storage will be lost in the course of just one year, demonstrating the need for annual flushing.

Table 3. Available storage and sediment inflows at Didachara Dam site

Live Storage Volume at Start of Deposition Cycle (Mm³)	Average Quantity of Sediment Entering (T/year)	Live Storage Volume at End of Deposition Cycle (Mm³)
0.63	327,000	0.42

DAM HYDRAULIC ARRANGEMENT

Based on the constraints outlined above the principal functions of the Didachara spillway are:

- Pass and still the Design Flow;
- Pass the Safety Check Flood with no overtopping of the dam;
- Maintain a flood rise below observed colluvial deposits on the right abutment located at approximately EL. 785;
- Limit flood water level control using low level gates;
- Provide capacity to lower the top water level to allow open channel flow in order to flush out sediment from the reservoir;
- Ensure that the intake to the headrace tunnel is kept clear of sediment with negligible intake of the bedload in transport.

In addition, further operational requirements had to be considered with regard to the long term safe operation of the dam, including:

- Permit full opening of the low level sluice gates;
- Use of the low level sluice gates during extreme flood conditions;
- No high level gates located on the fixed crest spillway;

As such the design infrastructure needed to meet these functions were:

- Low level outlets to allow successful flushing of the reservoir;
- A fixed crest overflow to pass floods and control water levels at TWL reducing the need to operate the low level gates;
- Stilling basins to allow effective control of flow releases over the fixed crest and sluice gates;

- Local sluice gates at the intake to allow flushing of the area in its immediate vicinity.

Flood Control Measures

One main concern in regard to use of the gates during flood events was the required stilling arrangements to control the downstream flows. This was particularly important in the light of the need to limit excavations due to the geological conditions on the right bank and whether the required arrangement was practical.

One of the principal concerns in this regard was the need to control any energy dissipation at the toe of the dam in order to limit the risk of inducing landslips within the area whilst retaining any proposed infrastructure within the confines of the site topography.

This was achieved using roller buckets for the sluice gate stilling arrangements, which allowed the control of the high velocity flow without excessive basin lengths. However, in order for these to be effective the depth of the bucket could not exceed the required level of excavation for the dam. Although this was reasonable due to a 12m deep layer of alluvial deposits, this did require the limiting of unit discharge from the gate to 40m^3/s per metre run.

As a result of this initial design work it was felt that the use of the low level sluices during flood events could be incorporated into the overall arrangement, and as such was adopted within the final design.

Spillway Design

The design of the flood spillway had to consider the balance between the use of the low level gates and the fixed crest spillway. The principal concerns in regard to the design were:

- Minimising the need to control flood flows through the gates due to the high head to which the gates would be subject;
- Minimising the relative increase in depth of flushing flows at the sluice gates relative to the required maximum depth of 0.3m;
- Managing the flood rise and associated unit discharges through the spillway;
- Allowing for effective control of operating water levels

One of the principal difficulties was that associated with minimising the use of the low level gates during flood events due to the high head associated with this operation. Table 4 shows the flow rates over the fixed crest weir for different gate numbers and the approximate associated return period.

Table 4. Weir Flow v Gate Number

Number of Gates	Flow over Weir (Design Flood)(m³/s)	Approx. Return Period
0	750	1 in 1,000
1	750	1 in 1,000
2	640.8	
3	531.7	
4	422.8	1 in 100
5	314.2	
6	206.1	

Based on the evidence above it became clear that a logical argument for the dam arrangement was to use the 1 in 100 year event for the sizing of the fixed crest spillway, above which the gates would be used.

This arrangement allowed for the provision of four gates, thus minimising the increase in depth of flow relative to the required depth for flushing at the dam location. This in turn would maximise the flushable channel size within the reservoir, thus maximising the overall diurnal storage throughout the year.

As such this arrangement also allowed the retention of the flood rise below EL. 785 and resulted in unit discharges which were within a manageable range.

Final Arrangement

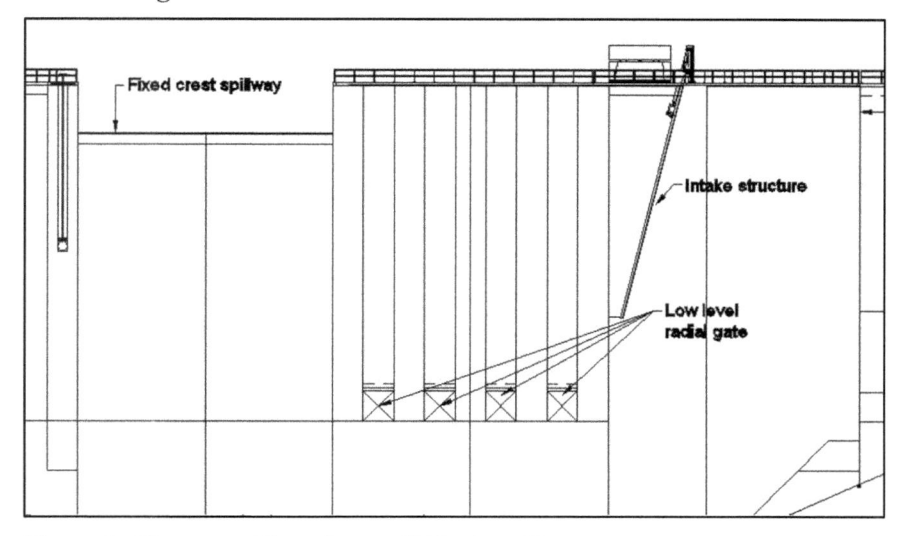

Figure 8. Upstream Elevation on Didachara Dam

In the final arrangement a 25m wide fixed crest spillway was provided along with 4 no. 3m wide x 2.4m high sluice gates located 29m below reservoir top water level. The intake was positioned immediately adjacent to the bottom outlets to allow for effective sluicing in its immediate vicinity (Figure 8).

CONCLUSIONS

The principal design constraints which influenced the optimisation of the hydraulic arrangement at the Didachara Dam site were:

- providing the maximum available diurnal storage available within the confines of the site;
- managing the large inflows of sediment load carried by the Adjaristsqali river.

The need to provide sufficient diurnal storage led to the adoption of the overall location of the dam, however the need to retain this storage in light of the high sediment loads within the Adjaristsqali river meant there was a need to provide for low level sluicing.

In addition to these concerns a variety of other factors had to be accounted for when considering the overall hydraulic arrangement. These led to the incorporation of a fixed crest spillway, the use of the low level outlets during flood conditions and a need to retain energy dissipation at the toe of the dam. The consequences of incorporating all these constraints into the final design would define the final outcome of the hydraulic arrangement and associated civil works.

The result was to provide a high level fixed crest weir with 4 no. low level sluices with associated roller bucket energy dissipaters.

SECTION 3:
MECHANICAL COMPONENTS OF DAMS

Maintaining the Safety of our Dams and Reservoirs
ISBN 978-0-7277-6034-0

ICE Publishing: All rights reserved
doi: 10.1680/mdam.60340.175

Refurbishment of the Portora Sluice Gates, Enniskillen Northern Ireland

K J McCUSKER, URS, Belfast, UK
G A COOPER, URS, Belfast, UK

SYNOPSIS URS was commissioned in 2005 by the Department of Agriculture and Rural Development – Rivers Agency, on behalf of the funders ESB, to design and install new gates at the Portora Sluice Structure.

The Portora Sluice Structure was completed in 1956 by Cementation Ltd and is located in County Fermanagh, Northern Ireland. It is essential to balancing the environment and flood protection requirements in Upper Lough Erne, including the town of Enniskillen. It plays a vital role in sustaining a navigable waterway in Upper Lough Erne. The Portora Sluice Structure comprises four gates each weighing 14 tonnes, which regulate water levels thus reducing flooding and providing huge benefits to the local area. The Portora Sluice Structure also plays an important role in relation to the generation of hydroelectric power at two large installations downstream.

In 2005 an Emergency Report highlighted significant health and safety hazards with the structure, including the possibility of the sluice gates falling under their own weight and the resulting impact on operational safety. Subsequent refurbishment works were commissioned by the Department of Agriculture and Rural Development – Rivers Agency, including:

- Replacing the 4 No. 13.17m long by 4.26m high existing sluice gates
- Mechanical and electrical refurbishment works to gear boxes and control systems and repainting the structure.

This paper describes the practicalities and health and safety issues surrounding the refurbishment works, the design of the replacement sluice gates and how the strict tolerance levels were achieved. In addition, this paper explains how the new gates were successfully installed in the structure, in under four months to enable it to continue to play a vital role in controlling winter water levels in Lough Erne.

INTRODUCTION

In 1942 the high demands for electricity within the Republic of Ireland led the Electricity Supply Board (ESB) to begin planning for the construction of hydroelectric generating stations on the Erne catchment between Belleek in County Fermanagh and Ballyshannon in County Donegal. The scheme was entitled the Erne Drainage and Hydroelectric Development.

This scheme would see the both Governments working together on the most important and complex cross-border initiative undertaken at that time. It would also lead to the construction of three key structures; the Cathaleens Falls Dam, Cliff Dam, and the Portora Sluice Structure. The two dams are located in County Donegal and the Portora Sluice Structure is located in County Fermanagh. Figure 1 shows the locations of the key structures.

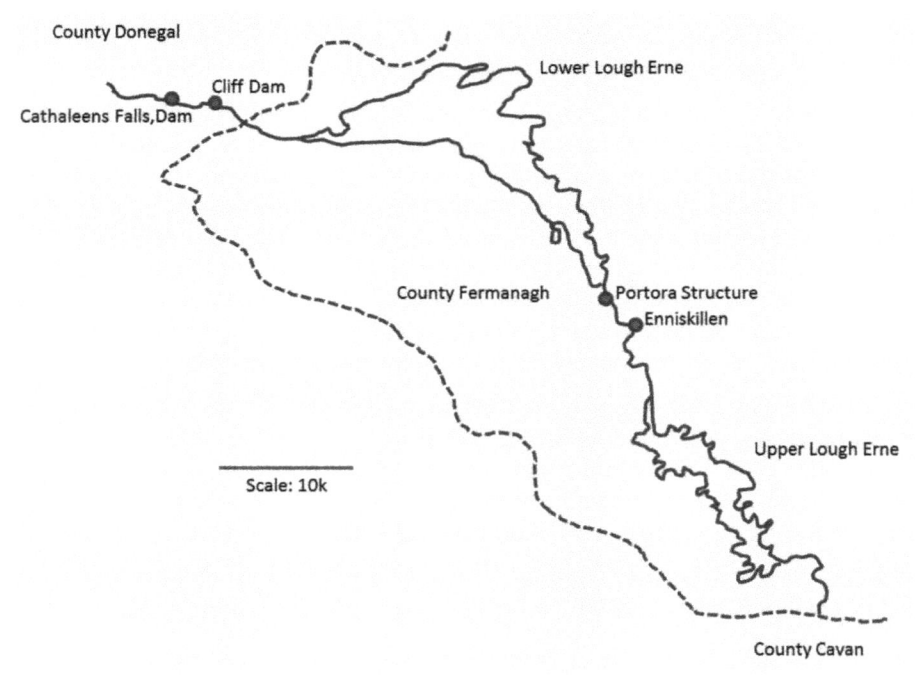

Figure 1 – Location map of key structures on the Erne Drainage and Hydroelectric Development

Today the Portora Sluice Structure is operated by the Department of Agriculture and Rural Development, Rivers Agency (in Northern Ireland) in partnership with ESB (in the Republic of Ireland). In 2005 URS was appointed by Rivers Agency on behalf of the funders ESB to complete a condition inspection of the Portora Sluice Structure.

URS, with specialist support from KGAL, undertook a condition inspection of the Portora Sluice Structure. The inspection identified a number of deficiencies and recommended the replacement of the existing sluice gates and refurbishment of the winch system that operates the gates. The construction of the new gates was completed in 2008 by Deane Public Works under the NEC Engineering and Construction Contract Option A.

THE ERNE DRAINAGE AND HYDROELECTRIC DEVELOPMENT

Erne Drainage and Hydroelectric Development Overview
The Erne Drainage and Hydroelectric scheme was seen as being of benefit to both Governments. The Government of the Republic of Ireland would benefit from the construction of two large hydroelectric dams that would supply power to the existing ESB network. The Government of Northern Ireland would benefit from flood control and river improvement works.

The Portora Sluice Structure allows Upper Lough on the Erne River system to act as regulated storage. The sluice gates control the river between the Upper and Lower Loughs. This allows the drawing down of the Lower Lough for the production of power from the hydroelectric dams while maintaining the water level in Upper Lough Erne. This creates a level difference between the Loughs for flood storage. As flooding from the winter storms begins in the areas surrounding the Upper Lough Erne the gates are opened and the flood storage is utilised.

Erne System
The River Erne is approximately 95km long and has a catchment area of over 4000 km^2. The River Erne flows North from County Cavan through Upper Lough Erne. It then flows in a north-westerly direction, past the town of Enniskillen into Lower Lough Erne. The Lower Lough is significantly the larger of the two Loughs. The Portora Sluice Structure is located just downstream of Enniskillen. On exiting the Lower Lough the River Erne flows through the two hydroelectric dams before discharging into Donegal Bay and finally the Atlantic Ocean.

Cathaleen's Falls and Cliff Dams
The Cathaleen's Falls Dam is located upstream of the town of Ballyshannon, County Donegal. The dam is of concrete gravity construction and is 250m long by 28m high. The dam incorporates two turbo-alternators each with an output of 22.5MW. The two turbines can generate enough power to supply 26,000 homes.

The Cliff Dam is located upstream of Cathaleen's Falls Dam, and is also of concrete gravity construction and is 210m long by 18m high. The Cliff Dam has two turbo-alternators with an output of 10MW. The two turbines can generate enough power to supply 12,000 homes.

Portora Sluice Structure

The Portora Sluice Structure takes the form of a reinforced concrete bridge with four arches (figure 2). The concrete structure has a deck width of 4.5m and is 65m long with four piers. The structure incorporates fish and elver passes within the middle and end piers. The western abutment accommodates a lock structure for navigation. The concrete bridge structure houses four steel up-and-over sluice gates. When open each gate is held in the horizontal position under the concrete deck (figure 3). As the gates close they are lowered into the vertical position under the concrete deck.

Figure 2 – Portora Slice Structure (Facing upstream from western abutment)

The sluice gates are vital to the management of flood storage within the Erne River System, as they allow the hydroelectric stations at Cliff and Cathaleen's Falls to draw down water from the Lower Lough. This permits the use of the storage in the Lower Lough without impacting on the water levels in the Upper Lough.

Upstream of the Portora Sluice Structure is the town of Enniskillen where the majority of buildings are founded on sensitive clay and many are close to the River Erne. The sluice gates maintain the upstream water level to prevent any clay shrinkage or river bank slippage.

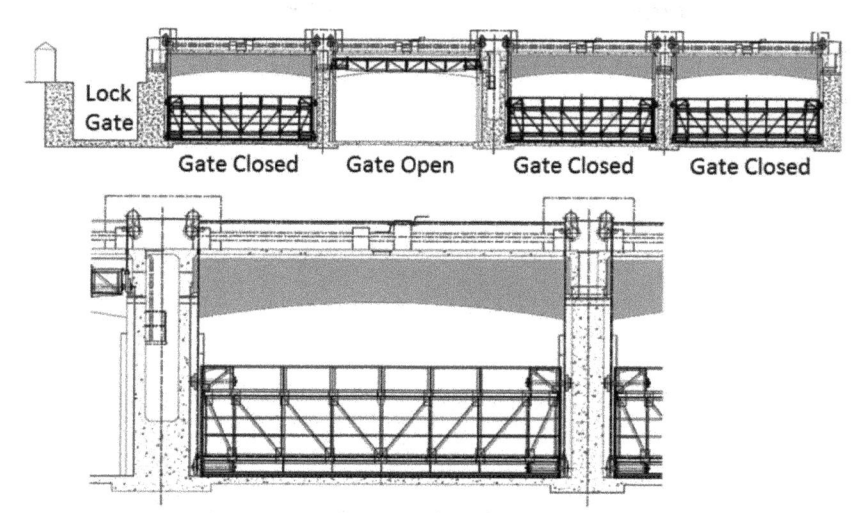

Figure 3 – Upstream view of Portora Structure with Gates in the Open and Closed position

The Upper Lough is an important recreational waterway and also contains numerous islands in close proximity to each other. The sluice gates help to maintain the width and depth of navigation between these islands, and also help in the management of wetlands, shallow rivers and the associated wildlife.

PORTORA SLUICE GATES

Original Gates
The original gates were constructed by Arrol from channel and angle sections riveted together with a skin plate on the downstream side, which is unusual. See figures 4 and 5.

There are four drive wheels supported within a box arrangement on each corner of the gate. The mass of each gate was approximately 14 tonnes and they measure 13.170m long x 4.257m high.

Existing Winch and Operating System
The winch operating system is located on the concrete deck and consists of a central braked electric motor driving two shafts through a fluid coupling, worm gearbox and electro-magnetic brake.

At each end of the shaft is a gearbox driving the wire rope drum. In turn the steel rope is connected from the drum directly to the front / lower wheels on the sluice gate.

Figures 4 & 5 – Portora Sluice Gate with bracing upstream (gate is in the "half closed" position)

As each rope drum rotates, the steel rope lowers the front / lower wheels. Steel tracks are cast into the side of each concrete pier to guide the wheels. As the front / lower wheels track downwards into the water the upper / back wheels follow the same route. The gate then comes into the vertical position and lowers onto a concrete apron on the river bed. There is also a hand operated gearbox in the system to allow manual gate operation in the event of electrical failure.

Condition Survey

A detailed condition survey of the sluice gates, winch and operating system was undertaken in 2005.

The existing steel skin plate was in poor condition and extensive "pitting" had taken place. Many of the rivets joining the skin plate sections together had rusted, particularly on the upstream side, where some have been reduced to approximately 50% of their original size.

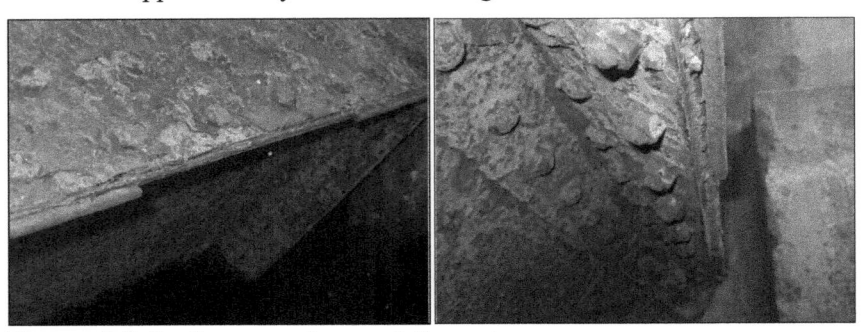

Figures 6 & 7 – Lower seals on existing gates

The gate seals were all found to be absent or in a poor condition and provided little or no hydraulic restriction to water flow. See figures 6 and 7.

The existing wheels also had extensive surface "pitting" and the wheels were no longer freely turning. This caused the wheels to jam, resulting in a shuddering movement as the gates were being lowered into the water. (Figures 8 & 9).

Figure 8 (Left) - Front / Lower wheel with steel rope attached
Figure 9 (Right) - Upper / Back wheel.

The winches and operating system were generally in good condition. However, the protective housing and covers were no longer providing protection to the workings of the winch system.

Consideration was given to the possibility of refurbishing the existing sluice gates. However, due to the poor condition and reduced steel thickness of the gates, it was recommended that all four steel gates should be replaced with new gates. The housing for all mechanical works was also recommended to be replaced.

NEW PORTORA SLUICE GATES DESIGN

New Gates
The original gates were constructed from channel and angle sections riveted together with a skin plate. Due to the constraints imposed by the wheel guide rails being cast into the concrete piers, it was decided to design the new gates to largely follow the original layout and operating regime since that arrangement had operated successfully for 50 years.

The original gates were constructed using imperial riveted sections, but with the advancement of welding and metric steel sections, the new gates would replicate the original gates but would be lighter.

The new gates were to be welded and to have bolted fixings only for the seal clamps. This was to allow for the future replacement of the seals as and when necessary. Due to the extent of the skin plate and amount of bracing, care needed to be taken during the welding process to prevent the skin plate from warping.

Four lifting points were also included in the gate fabrication. The lifting points were vital during installation and will assist in future maintenance of the gates and the lifting ropes.

SITE CONSTRAINTS
The site had a number of constraints that required to be overcome, both before and during the construction process.

Access
The existing site is located adjacent to the 17th century Portora Castle, which stands adjacent to a steep bank only 10m from the western abutment of structure. The site is only accessible along this western abutment via a concrete footpath which is mainly used by personnel operating the lock gates. Access to the eastern abutment is limited to a single water-bound gravel road. The geometry of the road and ground conditions did not permit large craneage plant to access the site. Therefore, all access for any removal or loading of gates had to be completed by floating plant.

Environment
Portora Structure has fish and elver passes accommodated within the middle and end piers. This is to allow fish passage at times when the gates are in the closed position.

The site is located within an Area of Special Scientific Interest (ASSI) and qualifies under Article 4.1 of EC Directive 79/409 on Conservation of Wild Birds. This placed a number of constraints on both the timing of the works and the construction methods employed by the contractor. The contractor also had to develop detailed anti-spill and emergency pollution plans.

Programme
The primary function of the gates is to provide flood control within the Erne system. The maximum discharge from the Erne System is 330m³/s. The maximum recorded flow into the system was 1007m³/s in 1991. This highlights the need for the gates and the importance of upstream flood storage during the work.

Both inflow and outflow of the system needed to be closely monitored throughout the contract and the outflow from the hydroelectric plant adjusted to maintain water levels to facilitate the works. This was required to allow any floating plant to pass under the structure when removing and installing gates.

CONSTRUCTION

The construction programme was set with a completion date of October 2008. This was the forecast date for when the gate would have to be closed to allow the Lower Lough to be drawn down. This would produce flood storage within the Lower Lough.

Installation of Gates

Due to the importance of the sluice gates for flood control, only one gate was permitted to be out of operation at any one time. The new gates were manufactured in Derrylin, County Fermanagh and transported individually to site.

The first stage of installation process was the removal of the old gate. It is likely that the existing Portora Sluice Structure was constructed within a coffer dam in two phases and this option was obviously not available. Removal of the existing gates was achieved using a barge and winch system. Four holes were cored through the concrete deck in a location to match the new gate lifting points. Corresponding holes were then cut in the old gate.

A temporary winch system was erected on the deck over the four holes using four winches, two beams and a set of "A" frames. Steel chains were then passed through the formed holes in the gate and bridge deck and attached to the winches. The winches supported the load of the gate allowing the steel ropes to be detached. The four wheel axis were then cut and removed, allowing the gate to be lowered onto a barge below and removed. Installing the new gate involved attaching the chains to the new lifting points and then winching the gate up into position. The wheels were then fitted and steel ropes attached. The winches then lowered the gate onto the steel ropes. The winch system was then dismantled and erected for the next gate. See figures 10 & 11 for installation of new sluice gates.

Testing of Gates

The new gates were lowered into position and an underwater survey was completed to check the seal of each gate. The gates were also raised and lowered to check for alignment and the running of the wheels.

Winch and Operation System

The housing and electrical works were refurbished by the Department of Agriculture and Rural Development - Rivers Agency prior to the commencement of the gate replacement.

Figure 10 – New sluice gate being moved into position on barge.

Figure 11 – New sluice gate being moved into position under structure prior to been winched.

SUMMARY

The replacement of the Portora Sluice Gates required the cooperation of a number of parties to manage the challenges presented by this project.

The scheme was completed in October 2008 and the gates were lowered two weeks later in order to start providing control.

The Portora Sluice Gates Structure continues to be used in its original role and has provided flood protection and control for hydro power to the public since 1956.

The scheme illustrates how dams can assist in a dual role with both flood management and power generation.

With the possibility of climate change resulting in drier summers and wetter winters this installation will provide flexible control and resilience to adapt to changing conditions.

ACKNOWLEDGMENTS

The authors would like to thank: DARD Rivers Agency, ESB, KGAL and Dean Public Works for information on the Portora Sluice Gates.

REFERENCES

Kennedy, M (2006), *The Realms of Practical Politics: North-South Co-Operation on the Erne Hydro-Electric Scheme, 1942-57*. Institute for British-Irish Studies, University College, Dublin

Porter, D & McCusker, K (2008), *Lough Erne System & The Portora Sluice Gates*, The Chartered Institution of Water and Environmental Management.

Dunalastair Dam Floodgate Replacement

R J DIGBY, KGAL, Poole, UK.
M A NOBLE, Scottish and Southern Energy plc, Perth, UK.

SYNOPSIS   Scottish & Southern Energy plc commissioned KGAL Consulting Engineers to carry out the detailed design for the replacement floodgates at Dunalastair Dam, which diverts water into the supply aqueduct for Tummel Power station, part of the Tummel Valley hydroelectric scheme. The major challenges to the replacement of the vertical lift bow string girder roller gates were the combination of remote location, the lack of any existing stoplog system to isolate the gates and the requirement to utilise the existing civil structure and counterweights to operate the replacement gates. The floodgates were exactly as originally installed at the time the scheme was first constructed.

The mechanical contractor, Enterprise Engineering Services Ltd of Aberdeen, participated in the early site discussions with both KGAL and SSE. Key issues such as stoplog and stoplog guide handling and installation and the removal and replacement of the gates were therefore agreed prior to the commencement of the design works. Removable stoplog guides, stoplogs and a multi-part replacement gate were subsequently designed. In addition, permanent dedicated lifting equipment for the deployment of the stoplogs, guides and gate was also designed. All of the equipment was finally installed on site and successfully commissioned.

INTRODUCTION
The decision-making process behind the need to replace the existing gates mainly resulted from concerns over the age of the current asset (installed circa 1932), its general aged condition and the specific risk profile associated with the gates at this site.

In line with the accepted terminology for dams the safety profile of the gate was reviewed. It was determined that any failure of one or other of the two flood gates could result in significant rapid flooding immediately downstream of the dam and its environs which could impact on the local village. Any potential for loss of life was considered unacceptable.

Figure 1. Photograph of Dunalastair Dam taken from upstream

Based on this assessment, the client decided that the design of replacement gates to modern standards was required to reduce the risk of gate failure to a tolerable level.

ASSESSMENT OF SITE CONSTRAINTS

The major challenges to the replacement of the original two vertical lift bow string girder type roller gates were the combination of remote location, the lack of any existing dam board or stoplog system to isolate the gates and the client requirement to utilise the existing civil structure and counterweights to operate the replacement gates. The gates themselves were exactly as originally installed at the time the scheme was first constructed and have not been changed since that time.

Early Contractor Involvement (ECI) in the preliminary site discussions with both designer and client was crucial to the process of ensuring that key issues such as site access, limitations on size of components transportable by road, stoplog and stoplog guide handling and installation and the removal and replacement of the vertical lift gates were fully agreed prior to the commencement of the design works.

The only access route to the site was from the road adjacent to the supply aqueduct, which passes over a restricted single carriageway bridge. The road is narrow and has several tight bends and corners making access with low loaders or cranes challenging.

Notwithstanding this limitation, it was necessary to design individual components to be able to be handled using the largest crane that was capable of reaching site.

An area was prepared adjacent to the dam, where it was possible for the transport delivering components to site to stand and to manoeuvre. The area was also prepared for the deployment of the crane.

The low-level bridge running across the dam was identified as the proposed route for the movement of the components arriving on site. The load capacity of the bridge was confirmed as being adequate for this purpose prior to the commencement of design in order to ensure that this was a viable option.

The client was keen to ensure that any lifting equipment needed to install components on site was rated to lift the heaviest individual item, whether that was part of the enabling, maintenance or permanent works.

A further requirement was that such lifting equipment should also be dedicated to the site i.e. permanently installed and available without any need for separate craneage. This was a key constraint based on the identified site access limitations.

Based on previous experience of the same client, contractor and designer team at Rannoch Weir, the client agreed that the methodology for deconstruction of the existing gates and subsequent installation of the replacement gates would be on a section-by-section basis. This would serve to limit the individual mass to be lifted on site.

For the deconstruction phase, this would involve cutting the gate horizontally into pre-determined sections ready for removal.

DESIGN OF STOPLOGS, TEMPORARY WORKS AND RELATED MECHANICAL LIFTING EQUIPMENT

A permanent dedicated lifting frame for the deployment of the stoplogs, their guides, the sill beams and the gates was designed first to accommodate the site lifting of all other components. The frame was built as an assembly of parts and was attached permanently to the overhead civil works gantry.

Figure 2. Photograph of dedicated lifting frame with traversing beam

The lifting frame was designed to lift components brought to the gate bay on wheeled trolleys over the access bridge. Once lifted from the trolley, the lifting frame included a traversing beam assembly, which allowed the component to be moved from the access bridge side to the 'wet' side of the gate bay. Once in place above the final intended position, the component could then be lowered and fixed securely. This method of installation covered the installation of the gate bay stoplog steel sill beam, two side stoplog guides, stoplogs, new gate sections and the removal of the existing gate sections.

The gate bay stoplog sill beam was designed as a hollow steel fabrication, shaped to fit on top of the existing civil works sill in order to provide an efficient sealing surface for the new stoplogs and also to provide an optimal shape for the discharge of water when the gate was raised. Once it was bolted into position the hollow sill beam fabrication was filled with non-shrink cementitious grout to prevent long term internal corrosion and provide structural stability when subjected to the mass of the stoplog stack.

Site-specific equipment was subsequently designed including removable stoplog guides, stoplogs and a multi-part replacement gate.

The principle of the removable stoplog guides was included by the client as a specific requirement, given that leaving the guides in place would impede the discharge flow capacity of the flood gate apertures. For this reason, female threaded flush inserts were drilled and fitted into the civil works ready for the attachment of the guides. These inserts could then be left in place without detriment to the flow capacity. The fabricated side guides were then galvanised, brought to site, lifted into position and bolted into the female inserts.

The stoplogs were a set of horizontal steel fabricated assemblies, each of which carried its own integral seals and locating dowels. In this way, the stoplogs could be lowered into the water within the stoplog guides and would seal on top the sill beam and against the side guides and mating stoplogs.

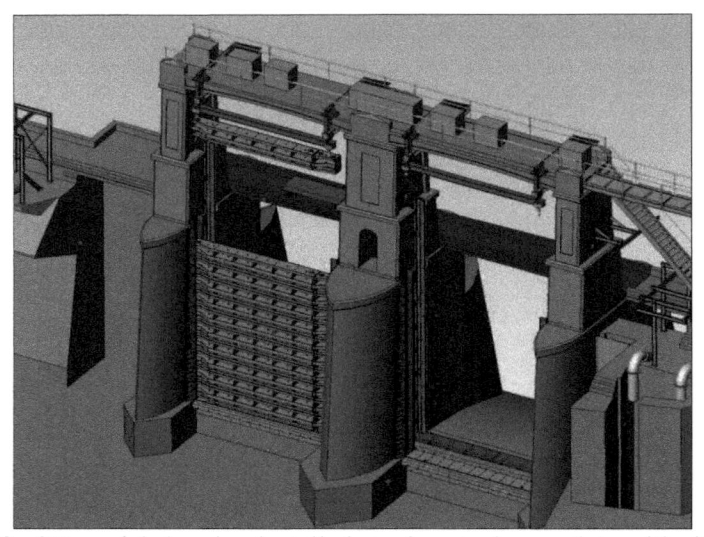

Figure 3. 3D model showing installed stoplog stack complete with sill beam and side guides.

Once the stack of stoplogs was installed in one gate bay, it was possible to simply lift the associated gate and release the water from the void between stoplogs and gate. Thus the dry working area adjacent to the gate was created.

DESIGN OF REPLACEMENT GATES
The bow string girder design principle for the existing gates was considered to be equally appropriate for the design of the replacement gates, given the advantageous strength-to-mass ratio available. On this basis, the design of the replacement gates started with the premise that, wherever possible, the new design would mimic the original design.

In this way, the multi-sectional approach was adopted and utilised the levels of the bow string girders themselves to act as the splitting planes for the new gate design. The new gate sections would then be location dowelled and bolted together using high strength bolts.

The vertical end beams were also designed as separate fabricated parts to which the central sections were also bolted to form the overall gate assembly. The roller chains had been replaced a few years previously, were fully functional, and were simply incorporated within the new design.

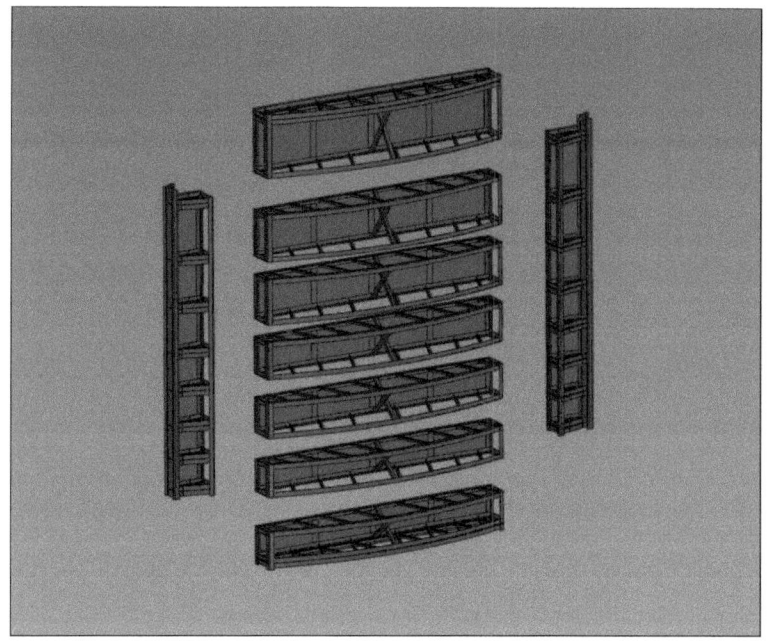

Figure 4. Exploded 3D model of gate showing central and end beam sections

The gate design was carried out using a combination of design codes and finite element (FE) based analysis. Extensive use of both traditional steel design codes such as BS 5950 and hydraulic steel structure design codes such as DIN 19704 were made during the design process, particularly when sizing the sections utilised for the bow string girders and the skinplates.

The design process involved the construction of a 3D solid model of the proposed multi-part gate design within the Solidworks Computer Aided Design (CAD) software program and the subsequent transfer of this solid-based geometry into ANSYS, a general purpose FE Program. The model was then 'populated' with a FE mesh and self-weight, hydrostatic, frictional, seismic and impact loads were applied in various load combinations in order to determine the worst or 'design' case.

Due to the specific seismic requirements placed upon the gate design it was necessary to undertake computer intensive and time consuming modal and response spectra analyses to determine the suitability of the new design. A 'Pseudo Static' approach using the formula for increased hydrostatic pressure due to seismic influence proposed by C.N.Zangar[1] was used to generate a conservative analysis, against which the results of the response spectrum analysis were later compared.

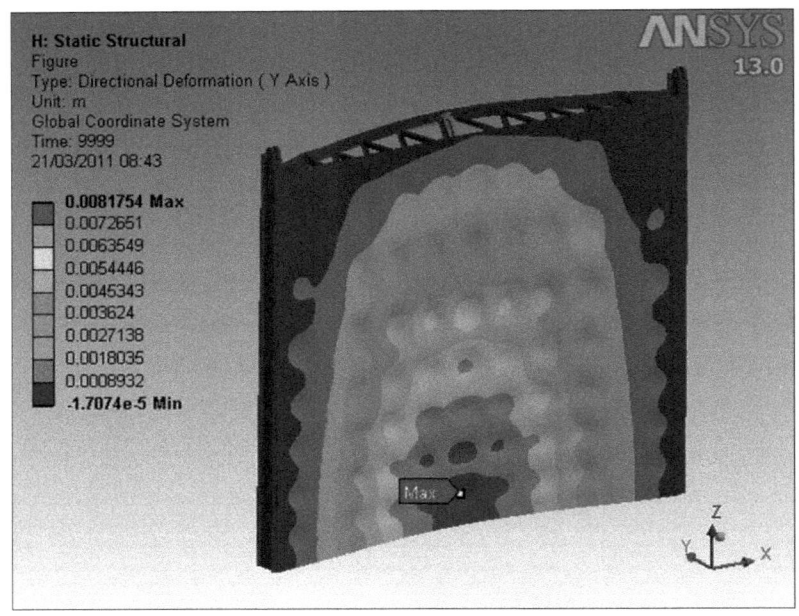

Figure 5. Results of finite element analysis showing deflected shape of gate under load

This process highlighted certain parts of the girder assemblies as requiring increased sectional properties to resist the loadings applied.

An additional design constraint was the client's desire for the existing gate civil structures and counterweights to be retained in their entirety, i.e. for the replacement gates to be completely interchangeable with the existing gates without the need for operating equipment modifications.

This constraint involved the performance of the existing gates, headgear and counterweights to be modelled using retrospective predictive calculations in order to ensure that an operating benchmark was established prior to the finalisation of the replacement gate mass.

In the final analysis, the design of the replacement gates in accordance with the relevant codes and the finite element analyses resulted in a gate that was approximately 10% heavier than the existing gate. This resulted in the need to increase the counterweight mass accordingly and proved to be a small restriction to the overall gate opening capability.

INSTALLATION & COMMISSIONING OF REPLACEMENT GATES

With the temporary works installed and the selected gate bay dry, the process of first disconnecting the gate from the overhead operating equipment and roller chains, and the subsequent cutting of the gate into manageable horizontal sections could begin.

The mass and centre of gravity of each section was pre-calculated in order to ensure that each cut section of gate could be lifted safely using the overhead lifting frame and transported to the crane on a wheeled trolley for lifting on to the low loader.

Once the existing gate had been completely dismantled and removed, the process of erecting the replacement gate in its place could begin.

The two gate end posts were lifted into their correct positions and then the process of lifting the horizontal centre sections began. Starting with the lowest, each section was bolted first to the two end beams and then to the next horizontal section lowered on top of it. In this way, the construction of the gate was completed.

Figure 6. Photograph showing "dry" commissioning gate tests

The retained roller chains and the overhead operating equipment were then attached to the new gate. Once this had been successfully achieved the gate was raised and lowered several times to ensure free travel and good alignment.

The gate was then closed and water pumped back into the void between the stoplog stack and the gate in order to test the effectiveness of the sealing system. Once this had been achieved the gate was then deemed to be serviceable.

The entire sequence was then repeated for the remaining gate bay with the end result that both replacement gates were completely installed and commissioned to the satisfaction of the client.

REFERENCES

Zangar, C. N. (1952). *Hydrodynamic Pressures on Dams due to Horizontal Earthquake Effects*. Engineering Monograph No.11, United States Department of the Interior, Bureau of Reclamation, Denver, Colorado, USA.

Maintaining the Safety of our Dams and Reservoirs
ISBN 978-0-7277-6034-0

ICE Publishing: All rights reserved
doi: 10.1680/mdam.60340.195

Rudyard Reservoir (Staffordshire) Safety Related Works

D M WINDSOR, Canal & River Trust, Fazeley, UK
M COOMBS, Hyder Ltd, Plymouth, UK

SYNOPSIS   This paper details works carried out by the Canal and River Trust at Rudyard Reservoir. It highlights the work recently completed at the reservoir, which has been undertaken to improve the inlet control, operation of the scour valves and impacts of the PMF.

Specifically this has involved the installation of a modified inlet structure with a new hydraulically operated penstock, new eccentric plug scour valves (operated from 10m above on a new access platform within the valve tower) and a new boom across the spillway to prevent boats and debris restricting the flow.

INTRODUCTION
The Canal and River Trust (the Trust) is the undertaker for 72 'large raised reservoirs' (as in the Reservoirs Act 1975 definition) in England. They have an average age of 195 years, ranging from the 238 year old Pebley Reservoir to the 123 year old Winterburn Reservoir.

Hyder Consulting (UK) Ltd is the framework design consultant for the Trust. Hyder is an international multidisciplinary civil engineering consultancy and has undertaken design work on various Trust reservoirs, with Rudyard being the most recently completed on site.

Rudyard Reservoir was constructed between 1797 and 1800 by John Rennie and with a capacity of 2.6Mm^3 is the Trust's second largest reservoir. The 11m high dam was constructed primarily to feed James Brindley's Trent and Mersey Canal which had experienced water supply problems from its opening in 1777.

The reservoir was always used for leisure and gave its name to Rudyard Kipling whose parents first met at a party by its shores in 1863. It was actively developed by the North Staffordshire Railway Company as a leisure destination for the Potteries, Midlands and North West. Enormous numbers of visitors were carried to and from the small station at Rudyard with 88 trains and 20,253 passengers alone in one day in June 1913. The reservoir continues to provide an ideal location for boating, bird watching, fishing, sailing and walking and recently the lake has become a national centre of excellence for sailing for those with disabilities.

RUDYARD RESERVOIR

The 155m long and 11m high dam was completed in 1800 with some remedial sealing works required in the following year. The embankment dam is constructed of earthfill, with a puddle clay core, 3.6m wide at the bottom and 1.8m wide at the top. The crest of the dam is wide, varying from 14m to 18m at the right abutment. The reservoir water level was raised in 1905, with wooden planks installed at the spillway to increase top water level by 600mm. The general layout is shown in Figure 1.

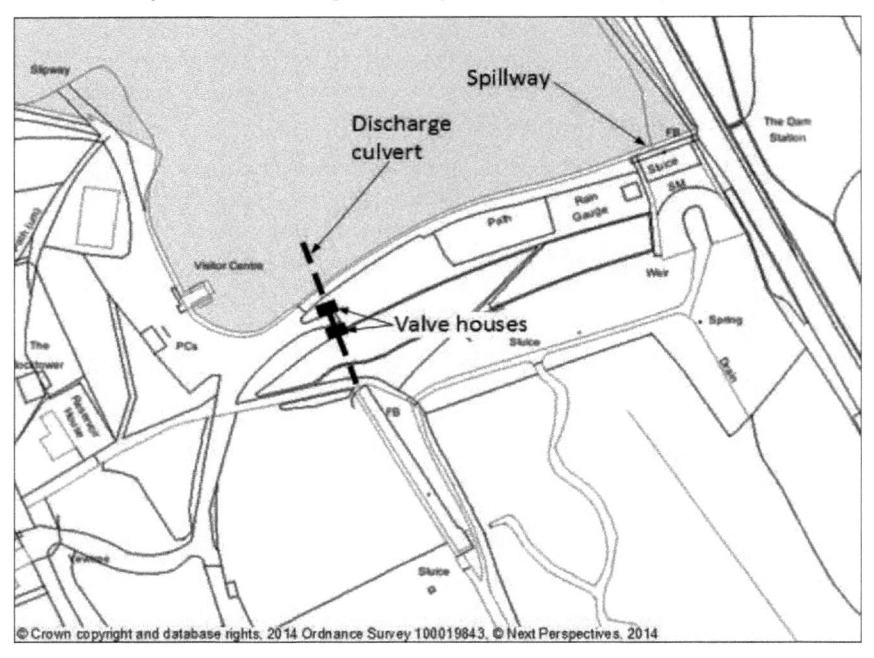

Figure 1. Rudyard Reservoir Dam Layout

A single discharge/draw-off brick culvert, 900mm in diameter, passes under the dam. This was lined in 1992 with 825mm GRP pipe, and was taken through the core in twin 300mm diameter ductile iron pipes which were grouted in place. The upstream timber paddle valve of the draw-off was also replaced at this time, with a sliding cast iron penstock installed at the slope angle and operated via a solid stainless steel rod from within the original valve tower on the wet side of the crest. The downstream control is provided by twin 300mm valves, operated from the base of the other valve tower at the dry well as shown in Figure 2.

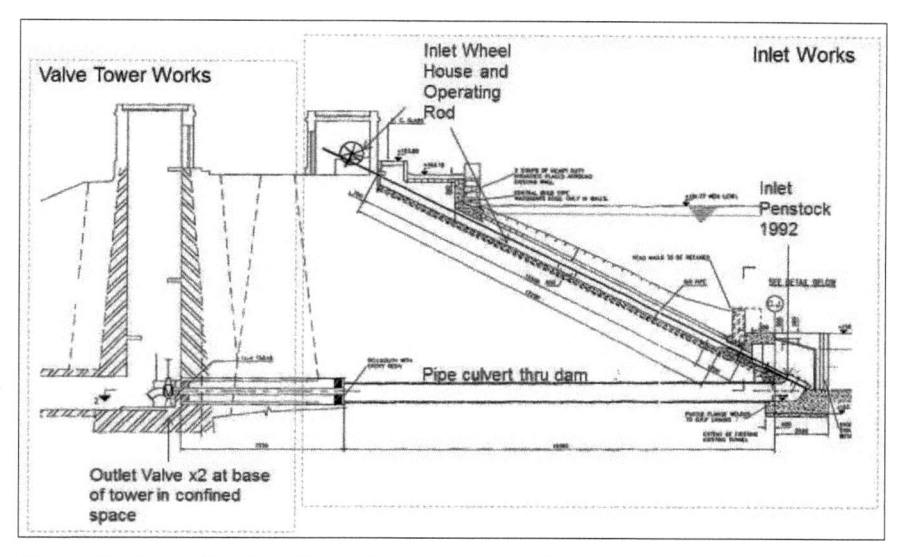

Figure 2. Cross Section through Dam pre Works

Inspections

The most recent S10 inspection was carried out on 10 August 2004 by Andrew Rowland of Black and Veatch.

The main points were:

- the reservoir was re-categorised A having been B;
- the reservoir could be lowered at 0.1m/day by the use of the scour (canal feeder discharge) and when the reservoir is full by removing the stop boards on the spillway weir (discussed but there needed to be a means of doing this safely).

There were no recommendations in the interests of safety, however the following were recommended:

- preventing drifting boats blocking the weir;
- replacing the spillway footbridge beams;
- provide a safe means of removing the stopboards on the spillway.

In addition monitoring of leakage into the upstream GRP lined draw-off culvert was recommended to take place every three months. This required the inlet penstock to be closed and the outlet valves opened such that any leakage could be measured.

In 2007 the Trust's Supervising Engineer, Paul Howlett, raised initial concerns that the inlet penstock was becoming difficult to operate. The penstock could not be closed at all by his S12 visit in November 2011.

Having lost the facility for inlet closure and unable to carry out monitoring of leakage of the upstream draw-off culvert, major works were scheduled and funded to take place in 2013.

The Supervising Engineer also noted that the downstream valves were becoming difficult to operate and it was evident that there were other works required at the reservoir, of which some might have become measures to be taken in the interests of safety at the next S10 inspection, due in 2014, or were required to improve operation and operational safety at the reservoir.

It was then decided for efficiency of scale to combine a number of work items into a single contract for delivery by the Trust's framework consultant and contractor.

The three main items of work, the brief, constraints and design options are discussed separately under the following headings:

- Inlet Control Penstock
- Outlet Valves and Safe Operation
- Spillway Boom

The works were delivered by the Canal and River Trust's framework contractor, Kier May Gurney Ltd and commenced in November 2013.

INLET CONTROL PENSTOCK
The penstock installed in 1992 had ceased to operate by the end of 2011, seemingly because the force required to overcome the friction at the sliding surfaces could not be generated using the solid metal rod, which was displacing intermediate support brackets as it buckled during attempts to close the penstock. It could be fully opened but only partially closed, which meant there was supply water for the canals and emergency drawdown capability, but no means of protecting the dam if there had been a failure of the outlet culvert.

Existing Details
Drawings and some photographs of the 1992 works were available, and the likely reason for the deterioration of the penstock's sliding surfaces was apparent. The penstock was mounted at an angle parallel to the face of the dam, as shown in Figures 3, 4 and 5 below, and not in the more traditional vertical plane. This sloping orientation of the penstock would be more likely to allow accumulation of debris and silt on the sliding surfaces. To compound matters, the large fixed screen did not allow for maintenance or inspection access.

The inlet control rod was installed in a box culvert (approx. 850mm by 600mm by 17m long) with a concrete and rip rap layer over the top of it.

There was then no means to inspect or maintain the control rod system, which was only accessible at the top from the valve tower.

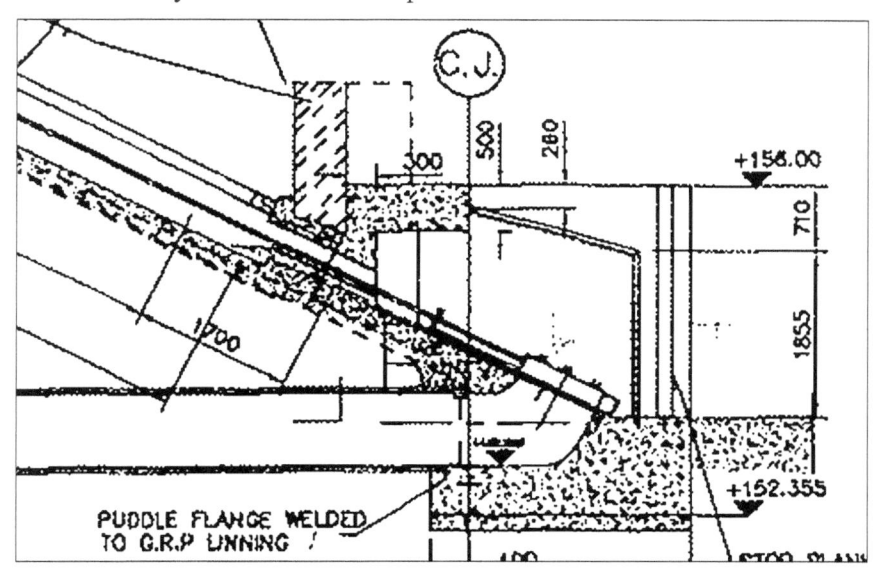

Figure 3. 1992 Inlet Works

Figure 4. 1992 Headwall Works   Figure 5. 1992 Penstock

Brief and Constraints for Replacement
The brief was for a reliable inlet control that could be maintained and replaced without needing to draw the water down.

The main constraint for the works was the level, or volume, that the reservoir could be drawn down to. This was restricted by:

- The desire to maximise the chances of a full refill prior to the 2014 boating season;

- Minimising disruption to leisure activities, particularly since the Sailability Centre was officially opened earlier in 2013 by Her Royal Highness, the Princess Royal;

- Avoiding a fish rescue on such a large and publicly accessible reservoir.

The 1992 works included a concrete headwall with stop plank grooves that on first inspection seemed to provide a facility to carry out works. Those works were carried out behind a cofferdam, unfortunately removed on completion, with water levels down to around 10% of holding. This is the generally accepted minimum holding to avoid the need for a fish rescue, provided that other factors such as fish stock levels, time of year and temperatures are favourable. There were, however, reports of fish deaths for up to four years following the 1992 works. The concrete headwall was obviously intended to be used for future works but would have needed the water volume to be less than 5% of total volume, effectively draining the reservoir.

Options and Design Solution
The first consideration was whether or not to install a cofferdam, typically to be left in place for future maintenance access. This would have been costly and disruptive and with no control of inflows from the 17km² catchment would have needed considerable overpumping to be on site to reduce, but not remove, the risk of the works being inundated.

This meant that one key factor, in terms of design and installation, was that water levels would not be fully reduced and therefore submerged installation using divers was stipulated.

In order to replace the penstock with a more reliable opening system a number of potential solutions were considered:

- Installing a new, or refurbished penstock on the same arrangement with the existing inclined rod operation;

- Installing a replacement penstock or valve, on a new arrangement with an improved system of manual operation (i.e. a vertical penstock with bridge access);

- Installing a replacement penstock or valve, but with remote operation (i.e. hydraulic operation).

Considering the site factors and the requirements the latter two options were considered the most viable.

Installing a replacement, or refurbished, penstock using the same arrangement as the existing would require not only removal of the existing inoperable penstock, but also removal of the rod within the concrete box culvert and replacement of the associated brackets and fixings.

Ideally any new penstock/valve would be installed vertically to promote the best operating conditions. In order to achieve this, in a way which would allow installation by divers, the best method was considered to be to utilise the existing headwall structure. A front and top panel arrangement would allow the structure to be sealed with a new penstock mounted on the vertical front panel.

The system of operation would ideally be a vertical spindle arrangement accessed by bridge but, due to the distance away from the main embankment of the inlet structure, this would have been a costly option. Cable, actuated or hydraulic operated alternatives were then considered. With no permanent power supply available in the adjacent valve tower, where any form of operation would need to be housed, the system would need to be manually operated.

The new 450mm x 450mm penstock would be mounted on a stainless steel plate bolted to the front face of the existing concrete inlet chamber. The chamber had to be sealed and this was achieved with a stainless steel top plate and fluoroelastomer gaskets. Both plates were approximately 2m x 2.5m, 15mm thick and had additional welded beams providing structural support.

To prevent any blocking of or damage to the penstock by debris a new inlet grille was required, approximately 2.1m x 1.1m x 0.7m deep, with large front opening doors to allow diver access for maintenance of the penstock and top doors to allow future replacement of the penstock.

Fabrication and assembly of the two plates, penstock and grille were completed prior to installation, minimising the time spent working underwater. Dive surveys were undertaken to verify the condition of the existing concrete and to confirm all dimensions at pre-design and pre-fabrication stages.

The new penstock was provided with a submersible hydraulic ram system operated from the valve tower manually or alternatively by using a hydraulic power unit and a portable generator. Hydraulic hoses were ducted from the submersible ram on the penstock, via the existing concrete culvert, into the valve house.

The submerged environment and inaccessible nature of the structure meant that longevity was essential and the materials used reflected this, with stainless steel and fluoroelastomer gaskets throughout.

OUTLET VALVES AND SAFE OPERATION
The discharge outlet valves are located at the bottom of a 10m deep shaft which, being designated as a confined space, requires at least two people to be in attendance and the use of a gas detector whenever the valves are adjusted, typically twice a day or more during the summer. Access and egress is via a narrow helical staircase. The valve operation had become increasingly difficult, adding to the operator's exertions and potential problems. A view down the tower prior to the works is shown in Figure 6.

Figure 6. Outlet Valve Tower

Brief and Constraints
The brief was to replace the valves and bring the valve operation out of the confined space.
Good conservation practice for such a historic structure is that there are no fixings into the face of the ashlar stone walls. The towers, or indeed any other part of Rudyard Reservoir are not listed, but are clearly of heritage value and one of the Trust's principles is that "The Trust will give all its heritage assets, whether designated (i.e. legally protected) or non-designated, the same level of beneficial treatment".

Options and Design Solution
In order to operate the valves outside of the confined space, a system would be required that could be accessed at crest level allowing day-to-day operation to be undertaken safely by a single operative. The following options were therefore considered;

- A cable system,
- Actuated valves (though no power present within the tower),
- Hydraulic system,
- Spindles, extended to operating level.

The existing valves are operated frequently to feed varying levels of flow to the canal system, so the valves were changed to eccentric plug valves to provide finer flow control. The valves were specified with IP67 rated gearboxes to cater for the damp nature of the tower and to aid operation and life expectancy.

The new valves were relocated more centrally within the base of the tower by extending the existing outlet pipework. This allowed a straight spindle to extend from each valve to a new access platform at crest level. The access platform was hung from the roof structure, thus avoiding fixings into the wall faces. The spindles were supported over the 10m length from a central column fixed only at the base and top (the platform), again avoiding fixing into the masonry walls.

SPILLWAY BOOM
The overflow spillway is a semi-circular masonry drop structure, with a footbridge crossing the approach channel. The overflow and footbridge is shown in Figure 7.

The operation of the reservoir was also being changed. Working practice had, for many years, been to hold the reservoir down such that it did not spill, to mitigate the perceived risks of flooding occurring downstream. The timber boards, installed in 1905 to increase capacity, had not held water back for many years as a result of this practice. The top water level had effectively been some 600mm lower than the theoretical level, which had led to some properties encroaching onto land that would have been submerged at top water level previously. The decision was taken to remove the timber boards, lowering top water level by 600mm, but allowing the spillway to spill as it had been designed to do when constructed, removing the need for regular human intervention. There was some ditch clearance required downstream to reduce flooding risk, though some problems persist. Allowing the reservoir to spill also highlighted the need to provide a safety boom to prevent out of control boats from passing over the spillway. The need for a boom was then twofold, a need in terms of reservoir safety but also a need for user safety

Figure 7. Spillway

Brief and Constraints for Replacement
The requirement was to prevent boats reaching the spillway and to provide something that would not be too visually intrusive. There was no requirement for planning permission, since the works could be done under permitted development rights.

Options and Design Solution
Any form of boom would have to withstand the impact of a boat breaking free during the PMF. The options for the system were:

- Wooden or concrete posts driven into the bed at narrow spacings to form a barrier (like a large trash screen).
- As above with wider spacings and a floating boom/barrier.
- A boom (fixed into the reservoir walls at each end).
- A floating boom (fixed at each end but on sliding rails).

Whist the post option provides a more rigid line of defence its installation is more disruptive to the reservoir with the posts needing to be driven into the upstream embankment and the water level lowered to provide access.

A decision was made to provide a cable based floating boom set back from the spillway, fixed into a reinforced concrete wall and the masonry spillway wall. The boom was installed over a length of 25m, fixed to a stainless steel rails at either end to allow the boom to raise and fall with the water and flood level. Pull out tests were conducted in both walls prior to fixings being made to ensure the required loadings were achieved.

CONSTRUCTION
The works were delivered by the Canal and River Trust's framework contractor, Kier May Gurney Ltd.

Water levels had been reduced during the boating season and into late summer to aid access for installation of the hydraulic hose duct through the riprap and into the concrete trough. This work was programmed for the first few weeks and once complete the valves were closed to allow diving to take place without any differential head. The reservoir was then allowed to refill, with all other works within the reservoir carried out by divers.

Inlet Control
All of the inlet works were carried out by divers, with as much prefabrication carried out on the surface as practical. The dive team operated from a pontoon which had a hydraulic arm mounted on it, to manoeuvre the large steel plates and heavier components as shown in Figures 8 and 9.

Figure 8. Diving Pontoon   Figure 9. Front Plate and Penstock

Spillway Boom
The spillway boom was installed whilst the water level was still low, enabling it to be floated across the reservoir but the plates to be fitted to the walls in the dry. Figure 10 shows the boom at installation.

Figure 10. Spillway Boom

Other Works
Other works were carried out within the same contract as the items that have been detailed within this paper. These other works included masonry repairs to the compensation weir (the "coffin weir"); the installation of an adjustable sluice to further regulate the compensation flow; installation of steps and a viewing platform for the spillway; stop plank groove installation to the feeder channel; safety chains and signs to the spillway overflow. A condition survey and load assessment of the footbridge structure was carried out and the capacity of the bridge beams was shown to be adequate, hence only minor refurbishment of the footbridge was carried out.

The works commenced in November 2013 with an overall cost including design and management costs of the order of £700k.

CONCLUSION
The works have brought the operation, safety, inspection and maintenance requirements of the reservoir to a standard which should ensure its continued success as a critical water resource for the canal system, a sports and general public amenity and a key visitor attraction for the Canal and River Trust.

Sliplining Bottom Draw-Offs

J P WALKER, Mott MacDonald Ltd, Leeds, UK
M T TIETAVAINEN, Mott MacDonald Ltd, Leeds, UK

SYNOPSIS Unlined bottom draw offs in embankment dams are undesirable due to the high risk of internal erosion that may occur should the conduits fail structurally. There are a variety of lining techniques available, the experience of these usually coming from the sewer and water main rehabilitation industries. Although sliplining is a well-known technique, it needs to be carefully planned and executed when being used on embankment dams.

The paper will describe two case studies at reservoirs in Yorkshire. Both had unlined conduits passing through earth embankment dams with puddle clay cores that were constructed over 150 years ago. Inspections, under Section 10 of the Reservoirs Act 1975, recommended that the conduits were lined to reduce the risk of internal erosion. The practical experiences of these projects is explained, focussing in particular on the annulus grouting procedure, why it needs to be planned in detail, and what the areas of good practice are.

INTRODUCTION
Unprotected outlet pipes or conduits are quite common in UK embankment dams. It was found that as many as half the total number of dams could have these (Millmore and Charles, 1988) with a large majority of private dams having no upstream control.

The risk of internal erosion caused by leakage from the pipe or conduit is one of particular concern to dam owners. The worst event, and possibly best known example, of this is Warmwithens reservoir, Lancashire, UK, where erosion behind a concrete lined tunnel eventually led to the catastrophic failure of the embankment (Wickham, 1992). There have also been a number of incidents which have required an emergency draw down of the reservoir due to the concern that the embankments could fail. At Upper Rivington reservoir in 2002 increased leakage behind a 1.8m diameter brick culvert instigated an emergency response and lowering of the reservoir using pumps. Upon inspection it was apparent that a large void had been eroded behind the culvert (Gardiner et al, 2004).

Lining pipes or conduits with modern materials is a method of extending their asset life and reducing the risk of leakage of water into the embankment. A large variety of lining techniques, that have mainly been development from the water distribution and sewerage rehabilitation industries, is available. However, a number of these techniques need to be analysed and in some cases modified before use on embankment dams. This can be due to the limited access for repairs should something not go to plan during installation, and also because in most cases the conduits could be exposed to the external pressures from full reservoir head. Appraisals of the various techniques can be found in the Sewer Rehabilitation Manual (WRc, 2001) and CIRIA Guide 147: Trenchless and minimum excavation techniques (CIRIA, 1998).

SLIPLINING

To be able to assess the appropriate lining method and decide the liner pipe diameter, the host pipe or tunnel diameter needs to be verified. For non-man-entry situations this can be achieved with a laser profile survey using a CCTV tractor unit. The survey also provides information on the existing condition of the host conduit. Tunnel deformations or partial collapses may require the selection of a smaller liner pipe. The liner is usually within 10% of the host pipe diameter (FEMA, 2005).

The pipe selection for sliplining is governed by the various pressure cases the liner could be subjected to. These include:

- the internal design pressure;
- the external pressure of water;
- earth pressure if the risk of future conduit collapse is expected;
- temporary pressure generated from annulus grouting.

The assessment should make allowances for potential deformation of the liner during installation and due to flotation of the lining to the crown of the host tunnel. According to Sewer Rehabilitation Manual (WRc, 2001) the maximum allowable deformation of the pipe of 6% can nearly halve the external pressure capacity of the pipe. This increases the potential for buckling of the pipe both during temporary grouting operation and in the long-term load condition. Table 1 shows the maximum permissible external pressures for Polyethylene (PE) pipe of different thicknesses (Standard Dimension Ratio (SDR) value) and how this vary considerable with the initial deformation. The manual recommends a minimum design pressure of 35kN/m^2. This means if the annulus between the host pipe and the liner is to be grouted, pipes with SDR greater than 17 (i.e. thinner walled) are unsuitable.

Table 1. Short term permissible external pressures on PE pipes (reproduced from Sewer Rehabilitation Manual, 4th Ed, 2001)

Initial Deformation	0%	1%	3%	6%
SDR	Permissible External Pressure (kN/m²)			
33	13	12	10	7
26	27	25	20	15
17	96	88	73	59
11	417	383	317	242

The maximum external head that the liner pipe can withstand is also of critical importance in the use of sliplining in embankment dams. Whereas in water main rehabilitation the external water pressure is usually low (as groundwater pressures are unlikely to be high unless the mains are buried particularly deep) in embankment dams, the impounded head of water needs to be considered. Table 2 shows the maximum permissible head for different SDR values and percentage deformations and shows that thin wall pipes (high SDR value) will only be acceptable for embankment dams with fairly low impounded depths.

Table 2. Maximum safe head of water for PE pipes (reproduced from Sewer Rehabilitation Manual, 4th Ed, 2001)

Initial Deformation	0%	1%	3%	6%
SDR	Safe head of water (m)			
33	1.9	1.8	1.5	1.1
26	4.0	3.7	3.1	2.3
17	15.4	14.2	11.7	8.9
11	63.1	58.0	48.0	36.6

The values are derived from the Timoshenko buckling equation and include an enhancement factor of 4 that accounts for situations where the annulus may not be fully grouted. The values also include a Factor of Safety (FOS) of 2 and are based on PE 80 pipe. If using PE 100 pipe, the values for the flexural modulus should be adjust for both the short-term and long-term condition.

The annulus grouting has potential to cause liner flotation. The flotation can be partly controlled with filling the pipe with water and providing end restraints to the pipe, especially for shorter pipe runs. If the pipe needs to be kept central to the host pipe, spacers should be used.

CASE STUDY 1: REDMIRES UPPER RESERVOIR

Redmires Upper Reservoir is the top of a cascade of three reservoirs constructed over Wyming Brook, a tributary of the River Rivelin, west of Sheffield. The reservoir is retained by an earthfill dam 15m high and 686m long, forming a reservoir of approximately 1.6Mm^3 of water. The embankment has a puddle clay core with puddle filled cutoff trench and was an early example of a "Pennine" dam, being constructed by 1854.

The original draw off system consisted of a 12 inch (300mm) diameter penstock (4.3m below overflow level) fed into a cast iron valve tower from which 2 no. 14 inch (350mm) diameter pipes passed under the embankment. A 2ft 6 inch (750mm) diameter tunnel in the upstream shoulder, some 12.8m below overflow level fed through an 18inch (450mm) cast iron lower draw-off sluice valve into the same valve tower. The pipes through the downstream shoulder were lined with a cured-in-place liner 9mm thick in 1977.

Following statutory inspections of the Upper and Middle Reservoirs in March 2006, the Inspecting Engineer included the following recommendations in the interests of Reservoir Safety:

(i) The valve shaft and tunnel should be inspected and a Panel Engineer should be asked to review the results of the inspection.

(ii) The toe drain should be re-laid along the length of the embankment.

(iii) The two pipes should be lined with a structural liner.

Investigation of the valve shaft was carried out by Mott MacDonald Bentley Ltd (MMB) and the results presented to the appointed Qualified Civil Engineer (QCE). The investigation highlighted that the structure could not be safely accessed for maintenance and inspection purposes. This, together with the poor condition of the draw off valves, resulted in a "follow on" recommendation to "provide a modern draw off facility at the reservoir".

During the optioneering phase it was also identified that the existing scour arrangement did not have sufficient capacity to lower the reservoir by 1m per day, which was identified as a desirable target. However, it was decided that the immediate threat to the embankment was the condition of the scour pipework and subsequent risk of internal erosion should the pipes leak.

The options for structurally lining the draw off pipes were mainly dictated by the external pressure and the need to maintain as much flow capacity as possible. Because it was not known how the existing pipes were constructed through the clay core, it was assumed that any lining could be subject to full reservoir head of about 13m. This ruled out some rehabilitation methods such as "rolldown" or Cured in Place Pipe (CIPP), as the liner thickness required to withstand the external head would have been too great a compromise for the flow capacity. Access to each end of the scour pipes was also limited.

Construction works

The remedial works combined an improved access to the upstream end of the scour pipes, together with an element of future proofing so that the issue of draw down capacity could be addressed at a later date. The existing cast iron valve shaft was removed and replaced with a segmental concrete shaft, constructed using a reinforced concrete support ring on the upstream shoulder. The new shaft was then excavated and shaft segments installed and grouted behind as the excavation progressed, as shown in Figure 1.

Figure 1. New draw off shaft under construction at Redmires Upper.

The new shaft included a concrete plug that would be sufficient to resist uplift, should the shaft be made water retaining in future. However, the proposal was to leave the shaft "wet" by allowing the upstream culvert to discharge into it through a new ductile iron pipe connection.

The agreed option for structurally lining the scour pipes was sliplining using High Performance Polyethylene (HPPE) pipes. The selected pipes were 355mm OD SDR 11, which had sufficient wall thickness to withstand the external head whilst maximising the internal diameter.

The existing pipes were cleaned using a jet-vac, and then surveyed using CCTV with a laser profiler. This confirmed the internal diameter and level of the pipes. It showed that both pipes had settled vertically by about 300mm in the vicinity of the clay core. There was a concern that this would prohibit the new pipes to be pulled through as the bend would be more than the allowable bending radius of the pipes. It was agreed to manage the risk by monitoring the winching forces at the upstream end. If they exceeded the set limits, the installation would be suspended.

The HPPE pipes were strung out on the downstream side of the embankment in the old spillway. The pipes were butt fused together and internal and external beads removed. External bespoke "lugs" were welded onto the outside of the pipes to allow them to be spaced centrally within the host pipes. The lugs were tapered at both ends to aid installation or removal if problems were encountered.

Both pipes we successfully pulled into the host pipes without exceeding the maximum pulling force on the winch, which was located in the new upstream draw off shaft. The annulus between the host and new HPPE pipes was then sealed with a Linkseal. Pre-drilled grouting points into the annulus were connected with standpipes where the grout would be pumped in. Water was first pumped into the annulus which would act to lubricate the internal surfaces in preparation for grout pumping. The volume of water pumped in was approximately equal to the total annulus of the pipe. The grout was then mixed in a grout pan using a proprietary grout product. Both pipes were grouted in one day.

The following day, a CCTV survey was carried out of both pipes internally. A large intrusion was noticed on one of the pipes, approximately 15m from the downstream end.

On further investigation, it was discovered that the pressure of the grouting had been sufficient to collapse the HPPE pipe. Various options for remedial works were then identified which involved various methods of excavation in the downstream shoulder to access the damage pipe. It was decided that a sheet piled excavation would be lowest risk to the embankment. Other options included headings of pipe jacking along the lengths of the pipes. The excavation in the downstream shoulder was approximately 6.5m deep at the upstream end. As the embankment was excavated it was possible to confirm the original construction details of the cast iron pipes. They had been surrounded in clay, and although it was not recognisable as puddle, it did appear to have formed a good seal around the outside of the pipes.

The section of damaged pipe was located and cut out using a keel cutter. It could be seen that the HPPE pipe had buckled due to excessive grouting pressures, as the annulus both in the damaged section and upstream was well grouted (Figure 3). The undamaged upstream section of pipe was carefully cut so that the host pipe could be removed and a stub of the HPPE pipe left to weld on a new section of pipe. This was done and then sleeved inside a new section of ductile iron pipe. The annulus of the pipe was then grouted and the excavation backfilled.

Figure 3. Section of damaged pipe showing buckled HPPE pipe

CASE STUDY 2: REDMIRES MIDDLE RESERVOIR
Redmires Middle Reservoir was constructed before Redmires Upper Reservoir, and was completed by around 1836. It is therefore considered to be one of the oldest embankment dams in the UK. The reservoir is formed by 12m high 894m long earth embankment with a storage capacity of 852,000m³ at the top water level (TWL). The original construction includes a masonry tunnel which contains a 12 inch (300mm) diameter cast iron draw off and scour pipe. The scour pipe passes into the tunnel at the base of a valve tower at the toe of the upstream shoulder. The outlet discharges directly into Redmires Lower Reservoir. Access to the tunnel is very difficult since the outlet is nearly 5.0m below the top water level of the Lower Reservoir. Additionally, the tunnel itself is only 900mm in diameter which, given the presence of the scour pipe, has prevented previous inspections. A statutory inspection of the reservoir was carried out in 2006 in accordance with the Reservoirs Act 1975. A recommendation was made to inspect the tunnel. Various attempts to CCTV survey the tunnel were made with conventional tractor units, a remotely operated underwater vehicle (used when the tunnel was submerged by the lower reservoir) and a bespoke remotely controlled tractor unit mounted on the scour pipe. The bespoke unit was partially successful but only managed to survey about two

thirds of the tunnel. Because of the inherent difficulty in inspecting the tunnel, its age and knowledge of the embankment, the QCE therefore made a recommendation for follow on works to structurally line the tunnel to prevent future collapse.

Pipe Removal and Insertion
Rather than grouting the annulus of the tunnel with the old scour pipe in place it was decided that the pipe had to be removed as its condition was considered to be poor, and it would be difficult to prevent the pipe from floating during grouting, as it was not strapped to the tunnel invert. A sheet piled cofferdam was constructed at the upstream end of the tunnel to allow for pipe removal and lining of the tunnel.

The existing cast iron pipe was removed by extracting it from the outlet end, breaking it in sections as it was removed. A 19T directional drilling rig was used to insert 75mm diameter drilling rods into the cast iron pipe. A drilling head was attached at the upstream end, and then the rods were pulled back.

Figure 4. Spacers and grout pipes attached to pipe prior to insertion

With the old pipe had been removed a laser profile survey of the tunnel was undertaken. This indicated that the tunnel had at a minimum 850mm diameter throughout and allowed a decision to be made to use 710mm outer diameter HPPE pipe, the maximum possible diameter that could be pulled in the tunnel (see Figure 4).

Spacers were attached to the outside of the pipe which increased the overall liner diameter by little over 100mm, which meant that there was less than the recommended space available around the pipe left for the pipe installation. To calculate the pulling forces required, an assessment needed to be made of the frictional resistance of the pipe in the conduit, including the spacer arrangement. Paggioli & Turkopp (2011) state that common friction factors are within a range of 0.3 – 0.5. The actual factor applied at Redmires Middle based on the pipe weight and pull force used for the pipe was within this range.

Grout Tests
Grout trials were carried out prior to the main grouting operation to establish stiffening and setting times for the specified grout mix. The test times then formed a benchmark of grout consistency for the annulus grouting operation. The grout mix was tested using a grout flow cone compliant with BS EN 445. BS EN 447 indicates the time to empty the cone needs to be less than 25 seconds to ensure workability of the grout mix.

It was estimated that the full grouting operation could take up to 10 hours (including contingency for equipment failure) whilst the manufacturer estimated setting time for the grout to be eight hours. Furthermore, the grout trial on stationary grout indicated that the grout would start to stiffen considerably at around four hours. Due to the length of the grouting operation it was seen imperative to maintain fluid mix throughout the grouting to ensure complete void filling of the tunnel annulus. It was therefore decided to use two large (2,000 litre) grout pans for the main operation to maintain continuous pumping of grout. Grout tubes were inserted in to the annulus to discharge at third points along the length of the pipe, to correspond with the grout setting times.

Annulus Grouting
The dry grout was also specified to be high flow with relatively high water content (solids ratio of 0.4). The mixing was staggered in such a manner that while one of the pans was being pumped into the tunnel, the grout mix was prepared in the other pan. It is worth noting that to produce the required amount of grout (approximately 19.5m^3) 936 no. 25kg bags of grout and 9360 litres of water had to be provided for the mix. The storage and movement of such a quantity of materials on site before and during grouting can be a logistical challenge depending on the volume and method.

Each pan of the grout mix was tested using the Marsh cone before the grout was pumped into the tunnel. The grout injection pressure was monitored continuously with a pressure gauge. The maximum allowed pressure at the injection point within the tunnel was 50kN/m^2 (0.5 bar) following the guidance in the Sewer Rehabilitation Manual (WRc, 2001). The grout pressure at the start of the test was monitored to establish a baseline pressure due to friction losses in the grouting hoses. The maximum allowable pressure was then taken to be the baseline pressure plus the 0.5bar.

Figure 5. Valve tower, coffer dam and grout pans during grouting

There were two grout pipes which were strapped to the HPPE liner when this was pulled in to the tunnel (Figure 4). The different length of pipes minimised the risk of blockage, as the grout in different parts of the tunnel was expected to stiffen at different rates. It was found that as long as the initial grout line was active and the grout was kept in the agitated state, it was still operating within the pressure limits.

The full length of the grouting operation took six hours. Most of the grout batches were pumped into the tunnel in less than 15 minutes, which equates to a rate of just under 2 l/s.

CONCLUSIONS

The two case studies have illustrated the need for careful planning and control of the sliplining and, in particular, grouting operations when working on lower draw-offs on impounding reservoirs. The lessons learned from grouting the annulus of the sliplined pipes at Redmires Upper Reservoir were tested during similar construction works at the Middle Reservoir, and proved to be invaluable. The remedial works at the Upper Reservoir were time consuming and costly. Whilst excavating into the downstream shoulder was difficult, there may have been more risk in attempting alternative solutions, such as auguring out the damaged pipe. This is essentially the method adopted at Poka Beck Reservoir in Cumbria (Wickham, 1999) and was unsuccessful. In that case the scour pipework had to be fully grouted up and abandoned.

Although some excellent literature exists from the sewer and water main rehabilitation industries, there is less information on good practice for procedures such as annulus grouting. The following 10 "Golden Rules" are therefore suggested that a designer and contractor can adopt when carrying out similar works:

1. Ensure the maximum **external pressure rating** of the sliplined pipe is sufficient for both the maximum head of the reservoir and the grouting pressures.

2. **Fill (but do not pressurise) pipe with water** to help reduce flotation and reduce the risk of buckling. If the pipe is pressurised beforehand, it should be allowed to rest for at least 24 hours.

3. Ensure the liner pipe is inserted using **spacers** to control floatation and ensure good coverage of grout around the annulus.

4. A **grout mix** with slow set time, good flowability and low solids content should be used. It may be necessary to pre-wet the annulus to prevent the water in the grout mix being absorbed and the mix becoming too stiff at an early stage.

5. Grout **pumping length** should be reduced as much as possible, with multiple injection points if possible. This reduces the risk of having to abandon the operation should one of the lines block or if the grout starts to set in part of the annulus during pumping.

6. Carry out **grout testing** using a grout flow cone in real time during the grouting operation. Grout trials beforehand are also recommended to set a benchmark and for site personnel to familiarise themselves with the equipment and the expected grout consistency.

7. Ensure the grout is **pumped in a continuous operation** to prevent blockages if grout starts to stiffen in any part of the annulus or delivery pipework. This may mean having multiple grout pans to ensure uninterrupted mixing and pumping.

8. Ensure a **pressure gauge** is used to monitor the grout delivery pressure and ensure it does not exceed the stated maximum at the injection point. Adjustments should be made for the friction losses in the grout delivery hoses or pipework.

9. Ensure the **site logistics are well planned in advance**. This includes how the grout is mixed, how the water is supplied, whether there is

there a suitable washout area, methods of communication to either end of the conduit, and numbers of personnel required on site.

10. Ensure **detailed records** are taken during the operation, especially volumes of water and dry grout added to the pan. These should be checked against the pumping rate and calculated volume of the annulus.

ACKNOWLEDGEMENTS

Special thanks to Yorkshire Water for giving their permission for this paper to be written, JN Bentley for their share of the design and construction works, and Dr Andy Hughes who was the appointed QCE for the project.

REFERENCES

BSI (2007), *BS EN 445: 2007, Grout for Prestressing Tendons – Test Methods*. British Standards Institution, London, UK

BSI (2007), *BD EN 447:2007, Grout for Prestressing Tendons – Basic Requirements*. British Standards Institution, London, UK

CIRIA (1998), *Trenchless and minimum excavation techniques: planning and selection*. CIRIA, London, UK

FEMA (2005), *Technical Manual: Conduits through Embankment Dams, Best Practices for Design, Construction, Problem Identification and Evaluation, Inspection, Maintenance, Renovation and Repair*. US Dept of Homeland Security, Bureau of Reclamation, Denver, Colorado, USA

Gardiner K D, Hughes A K and Brown A (2004). Lessons from an incident at Upper Rivington reservoir - January 2002. *Dams & Reservoirs, vol 14, no 2, September, pp 17-20*. British Dam Society, London UK

Millmore J P and Charles J A (1988). A survey of UK embankment dams. Reservoir Renovation. *Proceedings of BNCOLD Conference, Manchester, Technical Note 1*. British Dam Society, London UK

Paggioli, K and Turkopp, R (2011), *Sliplining and Microtunnelling of Large Diameter Pipelines*, Arema, Lanham, Maryland, USA

Wickham D B (1992). Collapse of an earth embankment dam. *Dams & Reservoirs, vol 2, no 3, October, pp 18-19*. British Dam Society, London UK

Wickham D B (1999). *Poaka Beck Reservoir - failure of a cast iron outlet pipe. Dams & Reservoirs, vol 9, no 2, August, pp 13-14*. British Dam Society, London UK

WRc (2001), *Sewerage Rehabilitation Manual, 4th Ed., Volume II Sewer Renovation*. Water Research Centre, Swindon, UK

Section 4:
Risk analysis and reduction measures

Maintaining the Safety of our Dams and Reservoirs
ISBN 978-0-7277-6034-0

ICE Publishing: All rights reserved
doi: 10.1680/mdam.60340.221

Improving serviceability through Portfolio Risk Assessment

O J CHESTERTON, Mott MacDonald Ltd
I M HOPE, Severn Trent Water Ltd
T J HILL, Mott MacDonald Ltd
R L GAULDIE, Mott MacDonald Ltd

SYNOPSIS   Severn Trent Water (STW) strives to achieve a high degree of confidence in the serviceability of its reservoirs. Ahead of any statutory drivers, a key component to achieving this strategic objective is Portfolio Risk Assessment (PRA) of both statutory reservoirs and those that are likely to become statutory following legislation in 2010, together with elevated sludge lagoons.

The PRA included recommendations for a programme of capital works schemes that further improved reservoir safety. It identified where further portfolio wide studies could improve the understanding of reservoir risk and help to bring other deficiencies to the fore. The PRA also recommended that the assessment process be a live one and periodically revisited. Following process review STW has now undertaken a second portfolio risk assessment. To further improve resilience this has included more detailed quantitative assessments on selected reservoirs which has provided greater confidence in serviceability.

Capital works were reviewed, ranked and initiated between the assessment periods. While the reservoir risk ranking was informative, the prioritisation of the works was more heavily led by works programming to effect construction cost efficiencies.

To further improve resilience, a portfolio-wide study to assess drawdown was undertaken and applied a set of engineering and risk criteria across all of STW's reservoirs. This study has helped bring certain deficiencies to light and provided vital information for Emergency Action Plans.

The second PRA benefitted from the presence, at an early stage, of the contractor responsible for implementation and has confirmed many of the expected movements in the risk ranking resulting from works and studies undertaken in the intervening period.

Two reservoirs have been subject to detailed fault tree analysis and the understanding gained from this process is compared to that of the PRA exercise undertaken as a screening assessment. The fault tree analysis has also shown promise in identifying and prioritising mitigations, further emphasising the importance of planned interventions such as monitoring and surveillance.

INTRODUCTION

STW's portfolio of reservoirs includes structures built from the 1850s. At 86 years, the average age of the portfolio is lower than most other water company owners. However, arguably some of these structures are working beyond their financial asset life. In seeking to establish and maintain infrastructure resilience STW has implemented a pro-active strategy to ensure the continued reliability and serviceability of these aging structures working to modern standards.

The goal of resilient infrastructure is achieved by the ability of the asset, networks and systems (crucially this involves people) to anticipate, absorb, adapt to and/or rapidly recover from a disruptive event. The following diagram places key supporting components to this goal in context:-

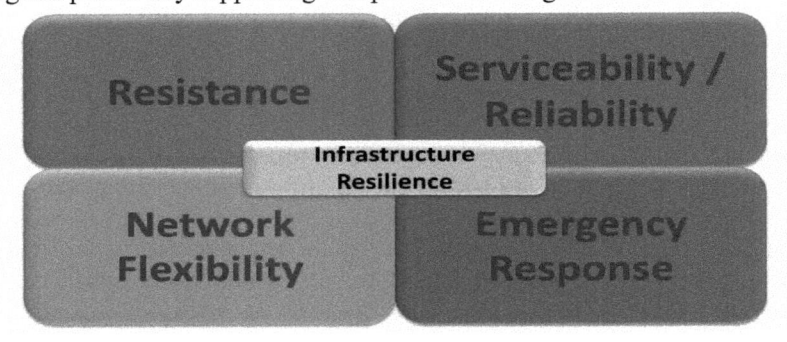

Figure 1: Components of infrastructure resilience (Cabinet Office 2011)

Serviceability/reliability is achieved through applying portfolio risk assessment. This focuses on achieving reliability when operating under the full range of conditions confronting the reservoir. This paper outlines the measures taken to implement the recommendations of PRA to achieve this objective.

In 2011 STW undertook a Portfolio Risk Assessment of its reservoir portfolio including both Statutory Reservoirs and Non-Statutory reservoirs and elevated sludge lagoons. The review was a qualitative summary review only, and in appraising potential risks served to provide STW with a relative risk ranking of reservoirs. From this schedule, a list of capital works was identified.

In adopting a proactive approach, ahead of any regulatory drivers, the assessment was driven by STW's strategy to understand its exposure to reservoir risk and ensure the portfolio was properly managed in order to guarantee serviceability. This strategic approach determined the allocation of funding for AMP5 (2010-2015) and identification of AMP6 (2015-2020) and AMP7 (2020-2025) funding needs and, crucially, start to gather evidence for these and future funding cycles. To meet its objectives STW selected a pragmatic approach to assessment, going to the level of detail appropriate to the need.

Following PRA 1 (the first assessment), STW implemented a £2.8M programme of capital works at 36 reservoir sites.

In late 2013, ahead of a 2nd PRA review (PRA 2), a lessons-learnt review was carried out. This was conducted as part of STW's ethos of continuous improvement and is a key strand to the "Safer Better Faster" (Hope 2012) way of working. In implementing PRA 1 the contractor had been provided with a schedule of works. Efficiencies were achieved through allowing the contractor to determine his programme of work. A key change to the PRA 2 process was the full involvement of the contractor in the detailed PRA review of each reservoir site. The contractor was able to feed into the workshops the learning from his knowledge of the sites and provide a more informed view on the estimated costs and proposed methodology of implementation of the planned works.

IMPLEMENTATION OF RECOMMENDATIONS
From the consultant's perspective, the initial PRA revealed a strong awareness of risk within STW and endorsed its proactive approach to reservoir safety. The perspective gained through the PRA was used to further build this awareness.

Recommendations from PRA 1 included:

- Adopt a regular PRA review period and methodology
- Address gaps in studies
- Build on previous PRAs and increase level of detail
- Improve data gathering and management systems

Not simply working to fix a list of deficiencies, STW has worked to apply the recommendations and findings of the PRA on a number of fronts which include:

- Prioritised and focused risk-based investment planning including up-to-date PRAs and site specific risk assessments.
- Efficient capital works procurement through block spending on minor items to guarantee serviceability and head off potential emerging problems.

- Developing information and reducing uncertainty through proactively commissioning studies where deficiencies may occur
- Improving the inspection regime through the continued use of pre-statutory inspections (these occur two years ahead of the 10 yearly statutory inspection)
- Strengthening monitoring and surveillance by the provision of a suite of professional training materials supported by an assessed training course for operators.

Capital works schemes
In 2011, PRA 1 exercise was applied to the list of capital investment needs identified. Individual work items were assigned a severity rank by the Supervising Engineer (the engineer appointed by statute to the respective reservoir) and scored based on this and the PRA rank.

Works were then ordered in two ways; by reservoir and across the portfolio.

Works ordered by reservoir were listed by reservoir PRA rank and investment totalled per reservoir. In preparing to mobilise to site, the view was taken that it would be more efficient to complete all necessary works identified at each site in one go. While this may have resulted in investment targeted toward reservoirs that were perceived to be more risky, it did not take into account the level of risk reduction achieved per investment or per reservoir. Hence, with this method, minor works may be prioritised on a reservoir with more inherent risk over measures on less risky structures even when the latter could result in greater portfolio risk reduction. It was important to recognize that the driver was a higher level of serviceability and not purely safety measures that were re-ranked.

Works were also ordered across the entire portfolio based on their individual score which corrected for this problem and against which a total cumulative investment was assigned. The aim was to prioritise severe problems on the riskiest reservoirs first and continuing in this order to the level of funding available for this period.

These final ranked works on the 36 reservoir sites were then put forward for implementation.

Programming and prioritisation
As part of the privatisation of the water industry in the UK in 1989, the Water Services Regulation Authority (Ofwat) was established. Ofwat regulates the industry through five-year Asset Management Plan (AMP) cycles which set limits on water prices and thereby determines the level of capital investment for the period.

In common with most water companies, for each of these five-year cycles STW appointed consultants and contractors onto a framework to partner in delivering capital works schemes. Previously this had led to a cyclical and inefficient approach to resource deployment, often referred to in the industry as the "bow wave effect" and the overall implementation of works. To remove this cyclical approach and drive further economies STW reviewed and extended the tenure of its implementation partners to a 10 year period covering both AMP 5 and 6.

Following PRA 1 in 2011 the capital works schemes identified were estimated to range in cost from £100,000 to £1,000 with an average investment of approximately £25,000. Works varied from remediation of leakages to cleaning and painting valve and pipe work spread over 57 sites across STW's area stretching from Wales in the west to Leicestershire in the east.

Given the varied nature and locations of the capital works schemes identified it was considered more efficient to implement the works as a "block" rather than to develop a scheme for each reservoir or work type. A framework contractor, North Midland Construction, was brought on board to deliver the work, proposing a target cost with pain/gain share which was developed and approved.

The framework partner had visited the sites to assess the work. Costing revealed early on that the total investment needed across the portfolio was well within the level of funding required, barring a number of larger schemes that justified a special project. It was also obvious through the process of reviewing, ranking, and costing that to address individual items a simple order of risk reduction was not the most efficient. However, in order to drive efficiencies during implementation the contractor was given the freedom to develop an efficient programme and prioritisation of works.

Implementation
At the end of 2013 approximately 55% of the works identified were underway or had been completed. During the two year implementation programme several items were removed from the scope as no longer needed or because they subsequently became parts of larger capital schemes at the same site.

Access to service reservoirs has also proved to be a challenge because they require emptying and rendering off-line to facilitate the construction of works. Identifying closedown periods for inspection, survey and implementation of works currently remains a problem for certain clean water service reservoirs. A further scheme to provide improved versatility in the supply network to resolve this issue is currently under review.

Portfolio Wide Studies

While risk assessments may help a reservoir owner to apportion funding, traditional engineering assessments provide the basis for evaluating a reservoir against various failure modes. The reservoirs in STW's portfolio were built over a span of more than 164 years, during which engineering guidance and standards have developed significantly. Ensuring reservoirs meet and are maintained to an appropriate standard is vital to their continued safety and serviceability.

Under current GB legislation, a reservoir is independently reviewed every 10 years against current guidance. This guidance is continually reviewed against international and ICOLD standards. However, for certain critical areas, no specific standard currently exists and when engineering judgment is applied this may vary depending on the Inspecting Engineer. Additionally, full engineering assessments may not have been revisited for quite some time. For some older structures, given engineering practice and understanding at the time, certain assessments considered necessary today were not carried out during design. For example, earth embankments more than 70-80 years old did not benefit from modern advances in soil mechanics and in many cases, the internal geometry and soil properties were not recorded. Inspecting engineers apply judgement to their decision making and may not require assessments / intrusive examination if the structure has performed adequately for over 70 years.

Through the application of the PRA process STW is now proactively reviewing the status of engineering assessments across its portfolio. In many cases, studies have been initiated that would be anticipated in the next statutory Inspection.

For STW and most similar undertakers a key metric to demonstrate reservoir safety is to have zero mandatory works following a statutory inspection. The key means to achieve this, portfolio wide evaluations, were made of the following engineering assessments that may be required in coming years.

- Dam break
- Risk Assessments
- Drawdown Capability
- On-site Plans (Flood Plans)
- Hydro-Mechanical Assessments
- Flood Studies
- Seismic Hazard Assessments
- Stability Analysis
- Ventilation (Service Reservoirs)

In some cases, industry wide standards/guidance exist enabling the portfolio to be evaluated against them. In other areas, such as Emergency Reservoir Drawdown, guidance has yet to be developed.

Portfolio Drawdown Evaluation

Reservoir drawdown, the lowering of the water level in the reservoir in case of emergency, was an area identified for further study.

While discussions about appropriate levels of drawdown have been ongoing for a number of years, UK industry guidelines have not yet been developed. Hence decisions regarding the need for improvement of drawdown facilities have been based on the judgement of the Inspecting Engineer. Works can be both intrusive and potentially expensive to achieve the objectives required. In certain instances when resolution could only be addressed by multimillion pound schemes a legal challenge by the owner has resulted.

GB legislation gives the Inspecting Engineer the authority to make 'recommendations in the interest of safety' that the owner must, by law, comply with in the specified time. This can prove costly, for example, should a significant increase in drawdown capacity be required. Furthermore funding may not be included in the owner's asset management plan.

Therefore, to understand the company's potential exposure to such recommendations, STW has undertaken a portfolio wide engineering assessment of drawdown.

The following methodology was adopted as follows;

- Firstly, information was gathered on the drawoff and scour facilities of all STW's statutory reservoirs.

- Secondly, a recommended drawdown capacity was developed. This comprised a twofold approach. The Hinks formula (Hinks, 2008), giving a drawdown capacity that is dependent on the reservoir surface area, height and inflow, was compared with a simple time to drawdown to 50% loading which varied from 14 days for higher consequence reservoirs (Category A and B) and 30 days for the rest. The most conservative of these two approaches was adopted as the recommended drawdown across the portfolio.

- Thirdly, drawdown capacity calculations were undertaken for all reservoirs using the information gathered. These included network analysis of draw off pipework to develop an expected installed capacity for each reservoir. Finally, the reservoir portfolio was tested against the recommended capacities. The review highlighted certain reservoirs that required further study and some reservoirs that may require upgrading.

As part of the Emergency Action Plan, the assessment has usefully informed the level of additional temporary pumping that should be planned for potential on site emergencies.

The UK government co-sponsors a programme of research and development in order to provide reservoir safety guidance. This is overseen by an Institution of Civil Engineers (ICE) sponsored Reservoir Safety Advisory Group (RSAG). As part of this programme of work, which STW and other leading water companies are co-sponsoring, drawdown guidance will be available for use in 2015. With a thorough understanding of its own portfolio, STW will quickly understand the impact of any new guidance planned and be a strong contributor to the discussion.

Serviceability Scour Releases
To ensure the serviceability of scour release facilities, every bottom outlet valves should be operated periodically to ensure it is fully operable in an emergency and is not clogged with debris or compacted sediment.

Recognising this, STW has begun a pilot study to investigate the feasibility of opening these valves at sites where no current consent exists. This has included investigating the best way to manage the downstream impacts and obtain consent for periodic operation from the environmental regulator, the Environment Agency.

RISK ASSESSMENT REVIEW (PRA 2)
Eighteen months ahead of the commencement of AMP6, towards the end of 2013, as part of its strategy to evenly spread annual capital investment, but earlier than recommended, STW has also revisited the Portfolio Assessment.

A similar process to PRA 1 was undertaken incorporating lessons learnt following the first exercise, and a coarse screening methodology adopted for a second time. The approach was similar to a Tier 1 assessment now recommended in the recent "Guide to risk assessment for reservoir safety management" (2013). This high level of assessment was considered appropriate as it was shown during the previous assessment that this delivered the required outcomes without the significant level of detail - and crucially cost - needed to undertake a Tier 2 or Tier 3 assessment.

The methodology employed included a data gathering phase followed by interviews undertaken for each reservoir. The interview panel comprised an independent Inspecting Engineer, STW's Dams and Reservoirs Manager, the Programme Manager, NMC's Contracts Manager and a technical secretary together with the respective Supervising Engineer and Reservoir Technician, both of whom shared an in-depth knowledge of the sites in question. Interviews included reviews of data gathered and an update on the current reservoir condition together with works required by the Supervising Engineer as part of his annual inspections. Failure modes were reviewed and a screening evaluation agreed. After completing the interviews, data was collated and a revised risk ranking produced.

Gap Analysis

The database of reservoir documentation held by STW was interrogated and a gap analysis undertaken to identify areas where information or studies may be lacking.

To determine where gaps existed or studies had expired, known changes in assessment methodology were used as markers, and where no recent studies could be found, this was flagged. Inspection reports were also interrogated for assessments carried out by the inspecting engineers that may have addressed any study shortfalls. In some cases it proved difficult to determine if older studies should be considered insufficient. A general limit of 30 years was applied to engineering assessments, beyond which they would be subject to detailed review.

Results

The completion of PRA 2 has shown the movements in reservoir rank that would have been expected.

The rank methodology itself was updated to represent a more balanced weighting between consequence score and likelihood score. For example, age was included in, and affected, the reservoir rank. Age was considered to be related predominantly to the likelihood of failure given the uncertainty surrounding the construction of older structures. This needed to be balanced by a parameter that was related to consequence, such as reservoir volume, which was also included in the rank score.

Non-Statutory reservoirs (Volume < 25,000m^3) were also considered and ranked but with less detail due to the limited information available. These reservoirs, which will be regulated in the future, are likely to require significant investment to bring them to a standard on a par with the currently managed portfolio. A large number of capital works schemes were identified at these sites and STW intends to improve their safety well in advance of regulatory requirements.

FAULT TREE ASSESSMENT

As a component of the Birmingham Resilience programme, but separate to the PRA, STW required an assessment of the likelihood of failure at the Bartley and Frankley Reservoirs in order to substantiate evidence in their AMP 6 business plan to the regulator and to better inform emergency planning.

These assessments coincided with PRA 2 and provided a timely comparison to the screening assessment.

Methodology
A detailed failure mode assessment was undertaken for the two reservoirs evaluated using fault tree analysis.

Fault tree analysis is a visual top-down deductive failure analysis, in which an undesired state of a system (in this case, a reservoir failure causing catastrophic uncontrolled release of water) is analysed using Boolean logic to combine a series of lower-level events.

In order to perform the fault tree analysis and generate a risk assessment of the reservoirs, the following procedure was adopted:

A site visit was undertaken to understand the operating procedures and specific reservoir details. Two All Reservoirs Panel Engineers and both operational and engineering representatives from STW attended each visit.

Following the site visit a risk workshop was held to draw upon the operational and technical experience of those present. An open discussion was undertaken to evaluate the likely failure modes at each reservoir.

Failure mode determination: Based on a combination of experience and research, potential failure modes were defined. This process referred to the current "Guide to risk assessment for reservoir safety management" published by the Environment Agency / Defra in March 2013.

Construction of the fault tree: A fault tree was constructed on specialist Fault Tree Analysis software.

Evaluation of results: The results from the fault tree analysis were assessed to determine how the failure probability compares to previous assessments at these sites and at other reservoirs.

Analysis of the failure paths was undertaken to assess the riskiest route to the top event to provide an indication of where risk management measures might be best focused.

Qualitative assessment of likelihood
Given the age of the reservoirs (Frankley 1904 and Bartley 1930) and the difficulties associated with determining the likelihood of various initiating events, a series of subjective probability bands were developed and applied to each event during the workshops.

The following bands were adopted.

Table 1. Qualifying Subjective Probability Bands

Category	Annual Likelihood	Guiding Qualifier
Realised	1	Process/ incident is already on-going
Extreme	0.5	Process/Incident likely to occur annually
Very High	10^{-1}	Direct experience of the incident at the site under consideration by past or present
High	10^{-2}	Occurrence is common within a local portfolio of reservoirs
Medium – High	10^{-3}	Occurrence is common within the UK portfolio of reservoirs
Medium – Low	10^{-4}	Occurrence is unusual within the UK portfolio of reservoirs
Low	10^{-5}	Occurrence is very rare within the UK portfolio of reservoirs
Very Low	10^{-6}	Process is plausible but very unusual based on national and international experience
Negligible	10^{-7}	A theoretically plausible process but with almost no precedents in international experience

The use of qualitative probability bands proved successful and showed that fault tree analysis does not need to be as data intensive as some may suggest but can be carried out and prove informative on the basis of expert judgement and the knowledge available in standard reporting, so long as the limitations are understood. Once more data is acquired the fault tree probabilities or structure could then be updated accordingly.

Results

The contribution to the overall likelihood of failure made by each of the initiating events was ranked by their relative importance to the overall likelihood of failure. The relative importance of each initiating event is calculated by assigning a probability of zero to the event and assessing the change in overall failure likelihood as a result. This ranking helped to identify the key risks to the reservoir and where future and existing mitigation measures should be focused.

The assessment further demonstrated that the regular, routine monitoring and surveillance carried out by STW's operational staff is a crucial measure in reducing the risk that initiating events propagate to catastrophic failure. From an independent viewpoint STW's reservoir monitoring is currently considered amongst the best in the UK and continued diligence on this front should be considered as the vanguard of its effective reservoir safety management regime.

The process identified several gaps in knowledge emanating from certain aspects of the 110 year old construction features. Beyond conjecture at the workshops, further investigation of archives is planned.

Contrasting benefits of Screening and Qualitative Assessments
While the pros and cons of Qualitative and Quantitative assessments have been discussed elsewhere, the process highlighted the difference in practical outcomes seen.

Identification of deficiencies through screening
The screening assessment was geared toward identifying deficiencies and succeeded in identifying reservoir safety works that may be required. This was due to the calibre of those interviewed and stresses the importance of interviewing the correct individuals to access their bank of knowledge. The quantitative assessment was more thorough but did not identify any works that had not been picked up during screening.

Identification of priorities through fault tree analysis
The fault tree analysis gave a likelihood of failure in line with what had been expected. Hence it would seem that the real value of the analysis was in its ability to rank initiating events and enable the prioritisation of mitigations per reservoir. While capital works may be easy to justify to the business, analysis of a fault tree shows the real impact of monitoring and surveillance. This highlights the impact small investments such as the recent production of the reservoir training manuals could make.

The increased awareness gained by the team in debating and considering credible vulnerabilities on each structure contributed to an improved knowledge bank for the client.

Serviceability as the main driver
The client author holds a strong view that focus and resource is best concentrated on the reservoir site itself. The level of consequence off-site expressed qualitatively as "high" or "low" is deemed sufficient information to achieve this focus and prioritise resource. Ranking of works themselves is better informed by considering the impact on serviceability rather than developing F N curves.

The credibility of any company after an embankment failure would be called into serious question. As such, an owner works to maintain a reservoir portfolio to a level of serviceability, not simply to address safety concerns as they arise.

CONCLUSIONS

The current PRA Screening process has assisted in assuring the serviceability of STW's reservoir portfolio by the early identification of capital works and studies required. It has also identified those areas where capital would be best employed and highlighted the importance of surveillance and monitoring.

A comprehensive programme of risk assessment across the Severn Trent Water portfolio of reservoirs has informed the current five-year capital investment programme (AMP 5 (2010-15)) and established the needs for the AMP 6 cycle, allowing resources to be focused and prioritised to further reduce risk to tolerable levels and extend asset life.

REFERENCES

Bowles D S (2006) Portfolio Risk Assessment to Portfolio Risk Management, *Proceedings ANCOLD 2006 conference, Dams: the challenges of the 21st century*, Manley, Sydney

Cabinet Office (2011), *Keeping the Country Running: Natural Hazards & Infrastructure A Guide to improving the resilience of critical infrastructure and essential services*, Civil Contingencies Secretariat, Cabinet Office

Chesterton O J, Hill T J, Hope I M, Airey, M, (2012) A pragmatic approach to Portfolio Risk Assessment at Severn Trent Water *Proceedings of the British Dam Society (BDS) 17th Biannual Conference, Leeds*, ICE Publishing, London, UK

Environment Agency (2013) *Risk Assessment in Reservoir Safety Management*, Volumes 1 & 2, Horison House, Deanery Road, Bristol, United Kingdom

Hinks J, *et al* (2008) Low Level Outlets for Dams, *Dams & Reservoirs Vol 19, No 1, March 2009,* Thomas Telford, London.

Hope I M (2012) The Implementation of Severn Trent Water's People Plan to be recognised as the Best in Great Britain at Managing Reservoir Safety *Proceedings ICOLD Conference, Kyoto 2012*, Kyoto, Japan

Hughes A K *et al* (2000) CIRIA C542 - *Risk Management for UK Reservoirs*, R & D Technical Summary W5B-028/TS, Rio House, Waterside Drive, Aztec West, Almondsbury, Bristol, United Kingdom

A Practical Application of UK Guidelines for the Public Acceptability of the Risk of Dam Failure

K D GARDINER, Atkins, Warrington, UK
C BROWN, Manchester Airports Group, Manchester, UK

SYNOPSIS   The recently published "Guide to Risk Assessment for Reservoir Safety Management" (Environment Agency 2013) sets out methodologies for the assessment of the probability that a dam will fail. However the acceptability of those risks to the general public, and to the HSE, in the wake of a catastrophic dam failure is by no means obvious. In the aftermath of a dam disaster the first question that will be asked is "Was a risk assessment carried out?" and the second question might be "Was the risk acceptable by UK standards?" If the answer to either of these questions is "no" the dam owner may be at risk of incurring a considerable fine or even a custodial sentence if fatalities have occurred; so the implications of the correct application of the published guidance are profound. But what is an acceptable risk?

The authors have reviewed the published UK guidance and formulated a hierarchy of risk evaluation. This starts with the probability that dam failure is "intolerable" if even a single fatality occurs, then assesses societal risk, where the number of people likely to be killed is compared with the probability of the fatal event occurring, and finally calculates the individual risk that can be weighed against the cost of reducing the failure probability by carrying out remedial works.

INTRODUCTION
The authors have attempted to present the current guidance that has been published in the UK on what constitutes an acceptable risk to the population living or working downstream of a dam. The principal source of information on the acceptability of risk used in this country is the Health and Safety Executive's publication "Reducing Risks Protecting People" commonly known as R2P2 (HSE 2001). This gives somewhat confusing and even contradictory advice, although other published guidance by the HSE and others does give an indication of the factors to be considered when justifying expenditure on risk reduction measures on dams.

THE BOUNDARIES OF ACCEPTABILITY.

Tolerable risk

"HSE believes that an individual risk of death of one in a million per annum for both workers and the public corresponds to a very low level of risk and should be used as a guideline for the boundary between the broadly acceptable and tolerable regions." (R2P2 para 130) (HSE 2001)

This suggests that a risk of 1×10^{-6} per annum or lower is acceptable.

Intolerable risk

The HSE limit of intolerability is 1 in 10,000 for members of the public who have a risk imposed on them 'in the wider interest of society' (R2P2 para 132), so it has been argued that this level should be adopted as long as at least one life is in jeopardy.

"... in our document on the tolerability of risks in nuclear power stations, we suggested that an individual risk of death of one in a thousand per annum should on its own represent the dividing line between what could be just tolerable for any substantial category of workers for any large part of a working life, and what is unacceptable for any but fairly exceptional groups. For members of the public who have a risk imposed on them 'in the wider interest of society' this limit is judged to be an order of magnitude lower – at 1 in 10 000 per annum." (R2P2 para 132)

This suggests that a risk of 1×10^{-4} per annum or greater is judged intolerable.

As Low As Practicable

Between an intolerable risk and an acceptable risk lies a "tolerable region" within which the risk must be reduced to As Low As Reasonably Practicable, known as the ALARP region.

"Health and safety legislation made under the Act may be absolute or qualified by expressions such as 'practicable' or 'reasonable practicability'. The latter expressions provide duty holders with a defence against a duty. They are therefore used for instances where HSC/E would like duty holders to have such a defence,..." (R2P2 Appendix 3 para 2)

It is therefore important to know what might constitute such a defence.

"Of particular importance in the interpretation of SFAIRP [So Far As Is Reasonably Practicable which can be equated to ALARP] *is Edwards v. The National Coal Board (1949). This case established that a computation must be made in which the quantum of risk is placed on one scale and the sacrifice, whether in money, time or trouble, involved in the measures necessary to avert the risk is placed in the other; and that, if it be shown that there is a gross disproportion between them, the risk being insignificant*

in relation to the sacrifice, the person upon whom the duty is laid discharges the burden of proving that compliance was not reasonably practicable." (R2P2 Appendix 3 para 4)

Edwards vs NCB established the principle of disproportion that introduces an element of cost vs benefit into risk management: but how disproportionate must the cost be before it is grossly disproportionate?

The HSE guidance document "Guidance on ALARP decisions in COMAH" (HSE 2010) gives an indication:

"All RRMs [Risk Reduction Measures] *will involve a cost to the Operator. Equally, an RRM is intended to reduce risk from an operation and this reduction will bring about a benefit (in lives saved, etc.) which can be expressed in monetary terms. The ratio of the costs to the benefits can be described as a "proportion factor" (PF). This factor is also referred to as the "Gross Disproportion Factor" or GDF – the terms are interchangeable."* (Para 40)

"... In the individual risk intolerable region, RRMs must be implemented almost regardless of cost, implying a very high or infinite PF (though it is recognised that CBAs and gross disproportion are not applicable in this region).

The difficulty lies in defining the upper limit of PF and the way PF increases with risk. An upper value for PF of 10 has been suggested, but the way PF changes with risk is still unclear. However, the basic principle is shown in Figure 5." (Paras 42 and 43)

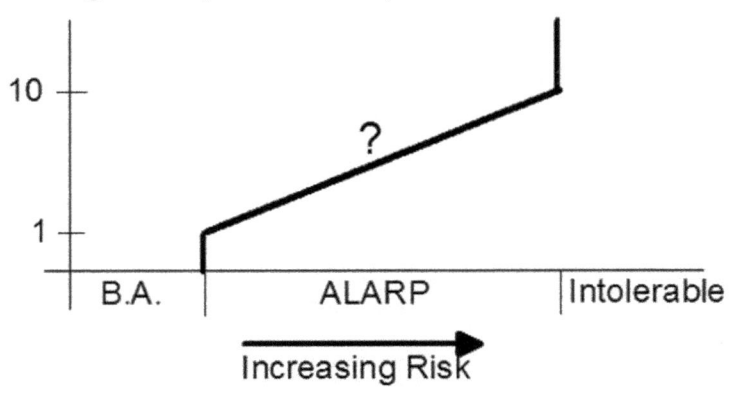

Figure 1. Figure 5: Change of Proportion Factor with Risk (in HSE 2010)

Thus we can infer that at the ALARP/Intolerable boundary (1 x 10^{-4} pa) the proportionality factor need not exceed 10.

Figure 1 above (Figure 5 in HSE 2010) also indicates that at the boundary of the Broadly Acceptable Risk/ALARP the proportionality factor need not exceed 1. In a private discussion with the HSE they indicated that this need not exceed 3 so this has been used in Figure 2, but it actually makes little practical difference.

Thus within the tolerable risk region a cost benefit measure can be used to justify, or not justify, expenditure on risk reduction measures. The cost/benefit measure chosen by the authors is that proposed by Dr D S Bowles which is Disproportionality. (D S Bowles 2003)

DISPROPORTIONALITY

The disproportionality ratio (R) is obtained by comparing the cost that would be incurred if an embankment failed catastrophically with the cost of applying a risk reduction measure.

It has been defined as:

[Annualised Cost of Risk Reduction Measure - Annualised Economic Benefit of Risk Reduction Measure] / Annualised Life Safety Benefit of Risk Reduction Measure.

in which:

Annualised Economic Benefit of Risk Reduction Measure = Economic Loss x Reduction in Probability of Dam Failure estimated for Risk Reduction Measure

Annualised Life Safety Benefit of Risk Reduction Measure = Value of Preventing a Fatality x Estimated Number of Fatalities x Reduction in Probability of Dam Failure estimated for Risk Reduction Measure

"*Principles and guidelines to assist HSE in its judgements that duty-holders have reduced risk as low as reasonably practicable*" (HSE Dec 2001, para 15), suggests that the x axis in Fig 2 should be based on the risk to a statistical individual and not the probability of asset failure. In the event of a reservoir failure, a number of people may be killed, but the relevant factor is the risk to any one individual rather than the probability of at least one person dying.

Figure 2 shows the boundaries that have been used by the authors based on the foregoing.

The x values of the points in Figure 2 represent the sum of the probabilities of failure at a dam multiplied by the probability that the dam failure will cause loss of life to the most exposed individual, before any risk reduction measure is carried out. After any particular risk reduction measure is carried out, the residual risk can be calculated and the dam plotted on the chart in a new position.

To plot a dam on Figure 2 the total probability of failure at a dam before risk reduction measures are carried out needs to be known. The estimation of this probability is beyond the scope of this paper and reference should be made to the RARS Guidance document (EA 2013). Also the risk to the most vulnerable individual that could lose their life in a catastrophic dam failure must be estimated.

Interpretation of HSE Guidance

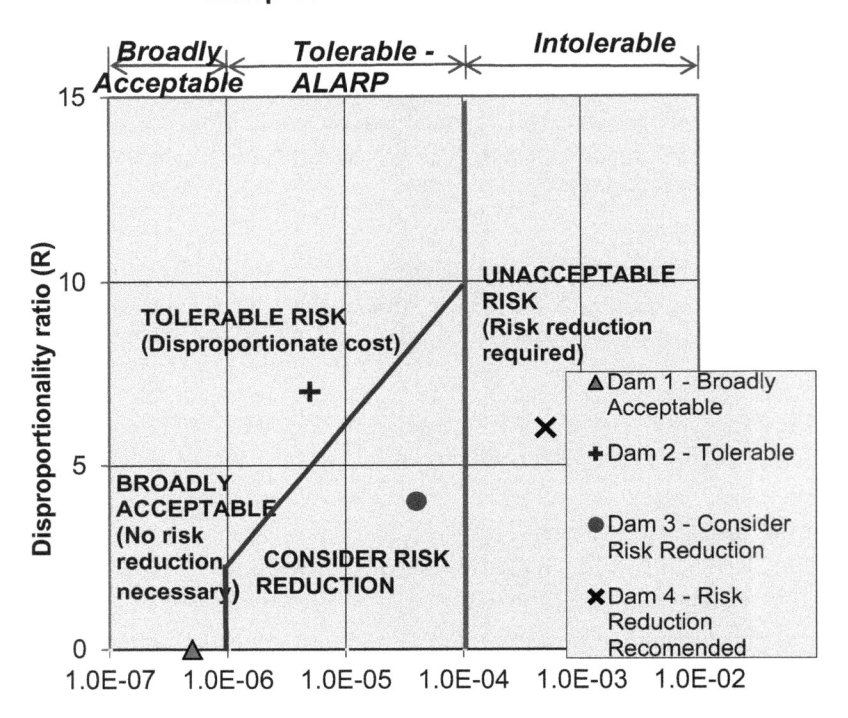

[Chart: Disproportionality ratio (R) vs. Probability of Loss of Life, showing zones: Broadly Acceptable, Tolerable - ALARP, Intolerable. Regions labelled: TOLERABLE RISK (Disproportionate cost), UNACCEPTABLE RISK (Risk reduction required), BROADLY ACCEPTABLE (No risk reduction necessary), CONSIDER RISK REDUCTION. Data points: Dam 1 - Broadly Acceptable (▲), Dam 2 - Tolerable (+), Dam 3 - Consider Risk Reduction (●), Dam 4 - Risk Reduction Recomended (✗). X-axis: 1.0E-07 to 1.0E-02. Y-axis: 0 to 15.]

Figure 2: Probability of Loss of Life for Most Exposed Individual per annum.

Risk to the most exposed individual
R2P2 Appendix 1 sets out the approach for assessing risk in terms of a 'hypothetical person', normally taken to be the theoretical individual most exposed to the hazard (HSE Dec 2001) paragraph 15;

"Risk should be assessed in relation to a hypothetical person, e.g. the person most exposed to the hazard, or a person living at some fixed point or with some assumed pattern of life."

In the context of reservoirs, the authors have considered the likely risk groups within society as a whole and suggest the following:

- Residents of the property that is closest to the embankment's downstream face, which would be hit by the flood with least warning. This assessment would take account of the likelihood of residents being present at the time of an incident, taking into consideration time away from the property for work, holidays, shopping or other social engagements. These would be typical for society rather than specific to current residents since residence is transient but reservoir safety is being considered for the long term.
- A dedicated rambler. Someone who will walk the area for recreational purpose, although even a frequent user for this reason will have quite a low total exposure time compared to a resident. Such an individual is only likely to be the most exposed where there is no permanent residence in the flood zone.
- The nearest catchment group who are likely to have high residence time, such as those in care homes. Such establishments are likely to remain in the same location for a long period so that it is reasonable to include them in this assessment.

All of these, plus specific cases for each location, should be calculated to identify the 'hypothetical individual most at risk'.

An Example.

Here is an example of how the Disproportionality value changes as the risk and consequence parameters are varied.

[The calculation uses a discount rate of 5.1% to annualize the costs and a Value of Preventing a Fatality of £1.58M (This figure is published annually by the Department of Transport.)]

Table 1 Effect of varying the risk/consequence parameters on the Disproportionality Ratio

Cost of risk reduction measure	£200k	£200k	£200k	£400k	£200k	£200k	£200k
Probability of LOL	9×10^{-5}	9×10^{-5}	9×10^{-5}	9×10^{-5}	7×10^{-5}	9×10^{-5}	9×10^{-5}
LOL*	50	20	10	20	10	10	5
Financial Consequences	£100k	£100k	£100k	£100k	£100k	£1k	£100k
P of LOL after Works	5×10^{-6}	5×10^{-6}	5×10^{-6}	5×10^{-6}	5×10^{-6}	5×10^{-6}	5×10^{-6}
Disproportionality Ratio =	1.51	3.78	7.56	7.57	9.89	7.57	15.13

As can be seen from the table, the Disproportionality Ratio is changed little by a large change in the Financial Consequences of the dam failure, however there is a marked change when the Number of Fatalities, the Cost of the risk reduction measure, and the Probability of Loss of Life are varied.

SOCIETAL RISK
In addition to the Intolerability and Disproportionality tests, HSE have also suggested that a different consideration should be made to address "societal risk" where many lives would be lost in a single incident, although whether this is just a suggestion or a requirement is unclear. They also say that this type of assessment may not be valid for a dam burst.

"HSE has adopted the criteria below for addressing societal concerns arising when there is a risk of multiple fatalities occurring in one single event. These were developed through the use of so-called FN-curves. The technique provides a useful means of comparing the impact profiles of man-made accidents with the equivalent profiles for natural disasters with which society has to live. The method is not without its drawbacks but in the absence of much else it has proved a helpful tool if used sensibly. These criteria are, however, directly applicable only to risks from major industrial installations and may not be valid for very different types of risk such as flooding from a burst dam or crushing from crowds in sports stadia." (R2P2 para 135)

Other guidelines are much more clear-cut, saying that societal risk should be assessed in all cases.

"We believe it is right that, in all cases, the judgment as to whether measures are grossly disproportionate should reflect societal risk, that is to say, large numbers of people (employees or the public) being killed at one go. This is because society has a greater aversion to an accident killing 10 people than to 10 accidents killing one person each"(para 34) (HSE Dec 2001)

Until this can be clarified with the HSE it is proposed that the principles of societal risk are applied since the weight of advice pushes in this direction.

If we are to address this issue, it is necessary to establish a limit at which societal risk is considered intolerable. The societal risk boundary is defined by the relationship between asset failure probability and the number of deaths caused by a failure.

The HSE (HSE 2001) provides one anchor point in this relationship, being an event causing 50 deaths and having a probability of 1 in 5000. They give no direct guidance on extrapolating this for other numbers of deaths but do supply a reference to a document which discusses this issue. (Ball and Floyd, 1998)

Quote from page ix: *"The second parameter fixing an FN line is the slope. Most FN criteria are drawn with slopes of between -1 and -2 on log-log diagrams. A slope of -1 is commonly regarded as 'risk neutral' in that the weighting in preference of large accidents is proportional to N, and not some higher power of N as in what is commonly referred to as 'risk averse'* or *'multiple fatality averse' formulations. Most of the criteria which have been published in the UK and Hong Kong have gradients of -1."*

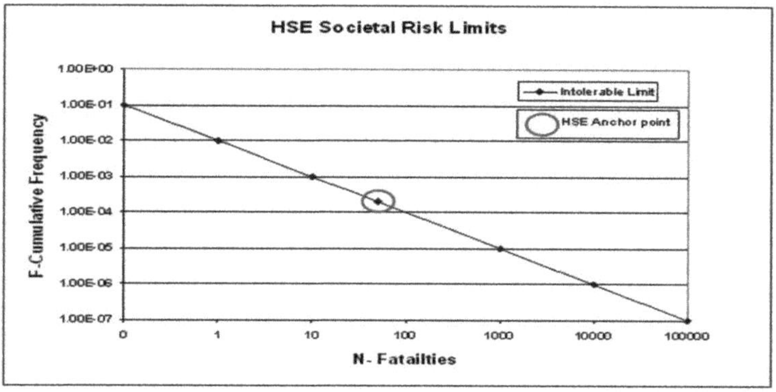

Figure 3. Societal Risk Diagram

Any dams plotting above the line have an Intolerable Societal Risk but there is also the over-riding criterion that the probability of loss of life for the Most Exposed Individual should not exceed 1 x 10^{-4}.

It is instructive to note that for higher numbers of fatalities, for some dams it may not be possible to reduce the probability of failure low enough though risk reduction works to meet the societal risk limit, in which case other measures such as increased monitoring and surveillance, reducing the reservoir water level or even demolishing the dam will need to be considered.

PRIORITIZATION OF RISK REDUCTION MEASURES.
Based on the guidance summarized above, risk reduction measures at dams can be prioritized on the following basis.

1. If the probability of life loss of the Most Exposed Individual per annum caused by dam failure is greater than 1 x 10^{-4}, it is intolerable and works therefore need no further justification.

2. If the probability of failure of the dam per annum plots above the line on the F-N diagram in Fig 3, it is also intolerable on the basis of Societal Risk and therefore works are justified.

3. If the probability of life loss of the Most Exposed Individual and the Disproportionality of the risk reduction measure plots below the line in Figure 2 the expenditure is justified and if it plots above the line not carrying out the works is justified.

CONCLUSION

A logical and defendable method of prioritizing and justifying expenditure, or justifying no expenditure, on risk reduction works at reservoirs, as far as can be ascertained from guidance published in the UK, is to first consider whether the risk is intolerable either to individuals or to society and, if it can be shown to be tolerable, to use a cost benefit approach. Unfortunately, if there were to be a dam disaster, the courts will decide if sufficient care had been taken by the dam owner.

REFERENCES

Ball D J and Floyd P J (1998). *Societal risks.* Risk Assessment Policy Unit, HSE, London, UK

Bowles DS (2003) Alarp Evaluation: Using Cost Effectiveness And Disproportionality To Justify Risk Reduction, *Institute for Dam Safety Risk Management*, ANCOLD 2003 Conference on Dams

Environment Agency (2013) "*Guide to Risk Assessment for Reservoir Safety Management*" Environment Agency, Bristol, UK

HSE (2001). *Reducing Risks Protecting People.* Health and Safety Executive, HMSO, London, UK

HSE (2001) *Principles and guidelines to assist HSE in its judgements that duty-holders have reduced risk as low as reasonably practicable* (Added to HSE website December 2001), HSE, London, UK

HSE (2010). *Guidance on ALARP decisions in COMAH.* HSE, London, UK

… **Emergency planning for mining waste facilities in England**

M CAMBRIDGE, Cantab Consulting Ltd, Ashford, Kent
T J HILL, Mott MacDonald, Cambridge
P HARVEY, Mott MacDonald, Cambridge

SYNOPSIS   Mineral extraction and processing operations generate significant volumes of coarse and fine wastes, which need to be stored in safety in purpose-built mine waste facilities. The finer wastes comprise particulate materials which, together with industrial water, are generally deposited into stage-constructed storage reservoirs. In the UK such facilities are regulated by Health and Safety Legislation and there are additional regulatory requirements, which include emergency planning, for those facilities characterised as Category A under the Extractive Waste Directive.

The hazard potential which would arise should the confining embankment fail in such a manner that a breach were to develop and lead to an uncontrolled outflow of any water and solids which have been impounded therefore needs to be assessed during the design and construction of any UK mine waste facility. A standard method of determining impacts arising from such a breach was developed for all high-risk reservoirs in England and Wales in 2010 but was never extended to cover Category A mine waste facilities.

This paper describes the methodology used to determine the most credible failure mode for two mine waste facilities in England, and how conventional flood mapping techniques, originally developed for large raised reservoirs on behalf of the Environment Agency, have been adapted in order to define the downstream impacts.

BACKGROUND
Mineral extraction and processing operations result in a significant volume of mine wastes being stored in purpose-built mine waste facilities (MWF). A MWF needs to accommodate this waste in safety, and be designed, constructed, inspected and certified in accordance with statutory requirements (Cambridge, 2008). The initial design and all subsequent modifications need to be approved by the Health and Safety Executive (HSE) and the operation and performance undertaken in accordance with

both UK legislation and the EU Extractive Waste Directive (EWD) (EC, 2012).

During the design of such facilities it is good practice to assess the hazard potential which would arise should the confining embankment fail and a breach develop, leading to an uncontrolled outflow of any water which may be impounded. Such an assessment was required as part of the permitting procedures for two MWFs in England, namely Clemows Valley Tailings Dam at Wheal Jane near Truro in Cornwall and the Hemerdon Mine Waste Facility near Plymouth in Devon.

LEGISLATION
An emergency response plan is required for all MWFs in accordance with good practice and generic emergency preparedness, as well as with the following:

HSE standards of equivalence
The HSE is unequivocal concerning the definition of lagoons which hold liquid/fluid waste or "waste with the potential to flow" from the extractive industries and fall under the Mines and Quarries (Tips) 1971 or Quarries Regulations 1999. However, for flood provision the HSE applies the "nearest equivalent standards" principle, and requires the adoption of similar standards to those of the Reservoirs Act 1975. For any mine waste facility which has the ability to store more than 25,000m^3 of water, a suitably qualified civil engineer will need to be engaged to undertake any necessary hydrological assessment, be responsible for defining the necessary flood standards to be applied and for defining the critical breach scenarios and of assessing downstream impacts (Cambridge, 2010).

The Extractive Waste Directive
The EWD requires that an emergency response plan be prepared for all Category A MWFs as follows:

"Article 1
1. A waste facility shall be classified under Category A in accordance with the first indent of Annex III of Directive 2006/21/EC if the predicted consequences in the short or the long term of a failure due to loss of structural integrity, or due to incorrect operation of a waste facility could lead to:
    a) non-negligible potential for loss of life;
    b) serious danger to human health;
    c) serious danger to the environment."

"Article 4
1. Member States shall assess the consequences of a failure due to loss of structural integrity or incorrect operation of a waste facility in accordance with …"

The EWD was transposed into legislation in England and Wales through the Environmental Permitting (England and Wales) Regulations 2010. The subsequent guidelines (EPR6.14, 2011) specified additional permitting requirements for a Category A waste facility and, in particular, the requirement for both internal and external emergency plans.

The Civil Contingencies Act
The Civil Contingencies Act 2004 describes a civil contingency as:
> **"an event or situation which threatens serious damage to human welfare in a place in the United Kingdom** – where serious damage is defined as 'loss of human life or injury; homelessness; damage to property; ….".

The guidance indicates the need to identify the risks and the instances of possible major accidents which are reasonably likely to happen and could cause significant harm and disruption in the UK. The types of emergency which meet the definition given in the Act are presented in the National Risk Register and include, under the heading *Overview of the main types of civil emergency*, major industrial accidents, contamination and technical failures, and cite equivalent installations to that of a mine waste facility, i.e. a water dam and a waste water treatment plant. The aim of the emergency planning is that "in the event that an industrial accident involving hazardous materials does take place, there is a well-developed capability among the emergency responders to deal with it."

BASIS FOR EMERGENCY PLANNING FOR A MWF
In 2010/2011, the Environment Agency initiated the preparation of the mapping for use by local resilience forums in the development of emergency plans for all large raised reservoirs in England and Wales. A standard approach had been developed in 2009 in concert with Mott Macdonald (Reservoir Inundation Mapping – Specification (Defra, 2009)) to develop the necessary mapping tools and inundation data. However, despite the evident anomaly, MWFs were not included in this study and therefore, when regulation necessitated the preparation of emergency plans for two of the UK's largest facilities, there was no precedent. Inundation maps had therefore to be prepared from first principles in order to obtain the necessary environmental permits for the Clemows Valley and Hemerdon facilities. In developing the plans consideration was first given to the nearest equivalent example, i.e. the methodology adopted for large raised

reservoirs, which assumed overtopping to be the critical condition due to the age and construction of the associated waterways (Figure 1).

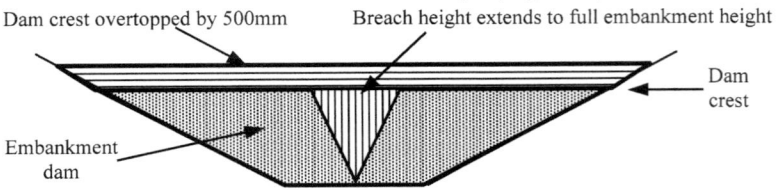

Figure 1. Failure model for high risk reservoirs

The MWFs referenced in this paper are of modern design, are geotechnical structures and are stage-constructed using either the mine waste itself or a combination of mine waste and local fills. The factors of safety against failure of the confining embankments under both normal operating and extreme conditions exceed the accepted standards as defined in ICOLD and national guidelines. In addition, the flood standards adopted for these facilities are in full accordance with the Reservoirs Act 1975 and cater for extreme flood events (ICE, 1996), with emergency spillway provision made (Cambridge, 2010). There is also a strict requirement for inspection, instrumentation monitoring and performance data collection, in parallel with independent expert inspection, in order to ensure that the facility remains safe, stable and meets all design objectives (EC, 2012). The inundation model adopted for large reservoirs was therefore considered to be inappropriate for these structures since overtopping would be the least likely mode of failure due to the stringent flood standards applied.

The development of an alternative failure model which accommodated the principal design and construction features of a MWF was needed. Initially an assessment was made of the differences between reservoir and MWF failures from an analysis of historic failures. Events such as those at Bafokeng, Merriespruit, Aznacollar and Kolontar (Cambridge, 2013) were compared with reservoir failures such as that at the Teton Dam (Snorteland, 2013).

Figure 2. Teton Dam post failure    Figure 3. Kolontar post failure

Water supply reservoirs are prone to a dam breach developing to near foundation level and for the basin to empty rapidly (Figure 2). A failure in a MWF such as Kolontar (Figure 3), containing both water and settled fine particulate materials may, dependent on the characteristics of the depository, result in a partial breach through the dam wall and the rapid evacuation of the fluid portion, but a release of only the more mobile fraction of the mine waste. Thus a failure of any water supply reservoir is dictated by hydrodynamics (Figure 4) and that of a MWF by geotechnics (Figure 5).

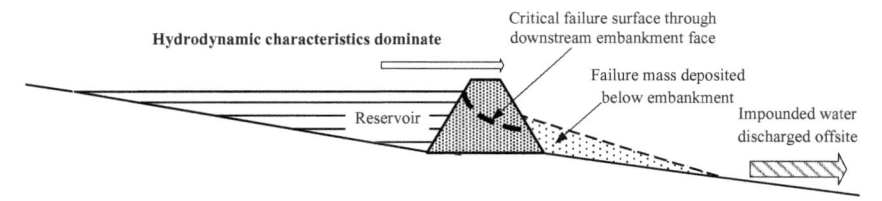

Figure 4. Single-phase hydrodynamic model for a water supply reservoir (e.g. Teton 1976)

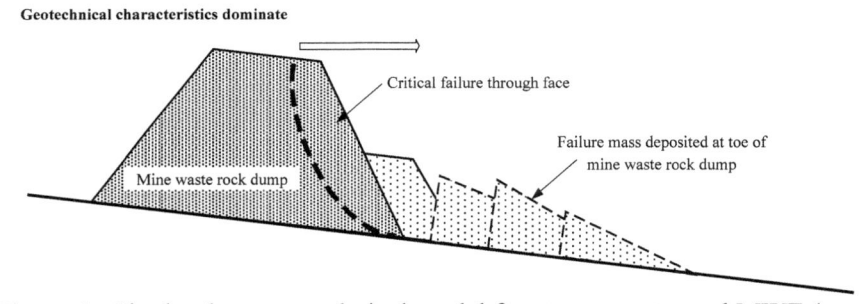

Figure 5. Single-phase geotechnical model for stage constructed MWF (e.g. Aberfan 1966)

This assessment led to the postulation of a two-phase breach model for a MWF using the basic characteristics as a means of both determining the failure mode and prescribing the event outcome. This model is predicated on the volume of free water stored on the facility at failure and separates the event into upstream and downstream phases. The failure characteristics, as outlined in Figure 6, are determined as being geotechnical upstream of the downstream toe of the initiating failure surface and hydrodynamic downstream of, and beyond, the embankment toe. This two-phase model, which enables a realistic assessment of the volumes of both solid and liquid waste involved in the failure, was used as the basis for developing the breach mode for the Clemows Valley Tailings Dam and for the Hemerdon MWF. The breach mechanisms were developed so that the conventional hydraulic model adopted for reservoirs could be used for preparing downstream catchment inundation maps.

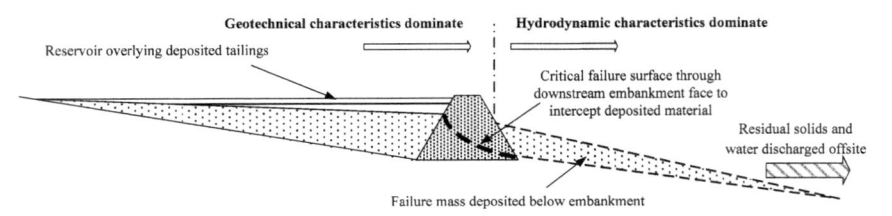

Figure 6. Two-phase model for a stage constructed MWF (e.g. Kolontar 2010)

DEVELOPMENT OF THE CRITICAL FAILURE MODE

An engineering study was undertaken to establish the worst credible event(s) which could lead to the development of a dam breach in these MWFs and a subsequent downstream flood. This required the use of basic risk assessment methods in order to identify and model the most likely failure mode and outcome. The Environment Agency's generic approach, using tiered assessments, was used and included the development of the following:

- credible failure modes – Tier 1 assessment;
- the source – volume of solids and water released;
- the pathway - the critical failure and release mechanism;
- probability rankings for failure modes - Tier 2 assessment;
- the receptor - comparative study of inundation extent for credible failure modes;

Failures of reservoirs confined by embankment dams are predominately caused as a result of embankment instability, untoward or uncontrolled seepage, or by overtopping, and the initial Tier 1 risk assessment was used to determine the worst-case critical breach scenario. Though the failure of a correctly designed and constructed MWF is considered unlikely, it is necessary to assess which modes should be addressed in developing a critical state for use in a breach analysis. Paradoxically, failure of a stable confining dam has to be considered in order to allow an emergency off-site plan to be prepared and the potential downstream impacts to be identified.

Design basis for a MWF

The design of a MWF involves the identification of all potential hazards, not only during operation but post closure (Adam, 2004). This enables the designer to mitigate all risks during the design and construction process. The key issues to be addressed, in addition to those normally associated with dam design, are the geotechnical and geochemical characteristics of the mine waste, the site water balance, the site-specific local hydrology and the robustness of the facility under seismic loading. The threat to life and the environment downstream must be identified in order to assess the risk category of the facility and thus the appropriate factors of safety to be used in the design. Again, these assessments must include an evaluation of the potential for long-term geotechnical/geochemical deterioration of the materials stored in the depository or used to confine the waste product.

The final design stage for both the Clemows Valley and Hemerdon MWFs included full design risk mitigation for embankment stability under all anticipated conditions, as well as flood provision. Further, there is a statutory requirement for ongoing inspection and monitoring of these facilities, as well as for annual independent inspections. Thus, in the case of these properly designed, constructed and operated MWFs the initiation of failure leading to a breach is considered to be extremely unlikely.

Initiating event (Tier 1) assessment
As previously explained, despite the safety and stability provisions it remains necessary to assess the hazard potential arising from embankment failure and an uncontrolled outflow of water and solids. The Environment Agency approach (HMSO, 2011), as defined below, was adopted for the purposes of this assessment:
- Source: material disturbed/released by the failure mode
- Pathway: release of material from the designed position towards a potential receptor
- Receptor: the downstream catchment and, in particular, the river/estuary system

The MWFs under consideration were at different stages at the commencement of the permitting process. The Clemows Valley Tailings Dam was fully developed and the Hemerdon Mine Waste Facility was at final design stage. These MWFs were the first facilities to apply for environmental permits under the 2011 Regulations.

Clemows Valley Tailings Dam
Since underground operations commenced at Wheal Jane in 1970 the tailings derived from mineral processing had been deposited into this purpose-built stage-constructed reservoir. The facility was designed and constructed in accordance with statutory (HSE) requirements and the construction and performance monitored throughout and certified by the Competent Person during annual statutory inspections (Cambridge, 2008).

The Tier 1 assessment for this 50m-high facility involved a review of potential failure modes and of both geotechnical and hydrological safety provisions. It was concluded that the most recent stability assessment, supported by some forty years of instrumentation data, could be adopted as the principal failure mode, the flood standards and the associated spillway design making overtopping extremely unlikely. The critical failure surface, with a factor of safety against failure of 1.72, intersected the tailings deposit.

The instrumentation records indicated the beach deposits to be effectively drained and, in the extreme event that the embankment were to be breached, the volume of material likely to liquefy and flow would be minimal. The Tier 1 assessment undertaken in order to define the downstream impacts concluded that this failure surface should be used to derive the breach mode. Further, the estimate of the quantity of sediment which might be released should be based on limited liquefaction.

Hemerdon Mine Waste Facility
Mineral extraction and processing operations are to be undertaken by Wolf Minerals (UK) Limited at the Hemerdon Tungsten Project on Crownhill Down in south Devon. These operations will result in a significant volume of mine waste being stored in a purpose-built mine waste facility in excess of 100m-high. The fine waste (tailings) will be deposited into a storage reservoir confined by a stage-constructed rockfill embankment comprising coarse mine waste from the open pit. The facility will be constructed in accordance with statutory (HSE) requirements and is to be monitored throughout and certified by the Competent Person. The MWF is to be operated in accordance with both UK legislation and the EWD in order to meet good standards of inspection and monitoring, and will be closed to an acceptable end point on cessation of mineral operations.

There being no construction or monitoring data for Hemerdon, a range of failure modes needed to be considered in order to ensure that all mechanisms had been assessed. The draft guide (HRA, 2013) was used as the basis for defining all failure modes, assessing the mitigation measures included in the design, and categorising each for further consideration, as appropriate.

Though the failure of this MWF is considered unlikely, it is necessary to assess which failure mode should be addressed in developing the critical state. Again, overtopping is considered to be the least likely due to the flood standards applied and the provision of both storage capacity for the design storm and the construction of an emergency spillway. Further, as no correctly constructed rockfill embankment has failed catastrophically under extreme seismic loading in modern times, an earthquake is not considered to be a credible initiating event other than in causing large crest settlements. The design incorporates measures which are intended to mitigate against failure of the facility at all stages. The critical failure mode assessment had therefore to be based on those parameters which cannot be guaranteed. The critical states which under realistic worst-case conditions might precipitate a failure at Hemerdon can therefore be considered to relate to localised geological unknowns, inadequate construction and material control, and poor management of the facility (Table 1).

Table 1. Credible failure modes – Example of Tier 1 assessment

Initiating event	Mitigation in design	Tier 1 Categorisation	Model No.
MWF spillway			
Overtopping	Spillway designed to pass PMF	Non-Credible	N/A
Spillway blockage	Flood storage volume (1000-year event) No catchment debris	Non-Credible	N/A
Erosion of spillway	Construction quality control	Credible	FM
MWF embankment			
Seismic induced slope failure	Properly constructed rockfill dam	Non-Credible	N/A
Seismic induced settlement	Construction quality control	Credible	FM
Uncontrolled seepage	Factors of safety on filters >10 Construction quality control	Credible	FM
Erosion of underdrains	Erosion of underdrains Construction quality control	Credible	FM
Untoward settlement	Construction quality control	Credible	FM
Foundation competence	Site investigation Geological knowledge	Non-Credible	N/A
Abutment competence	Unknown geological conditions Construction control	Credible	FM

The Tier 1 assessment identified the likely credible mechanisms, and a critical failure surface was developed for each mode which involved the full face of the embankment and intersected the tailings beach below reservoir level. This approach was taken since the development of a breach and the subsequent untoward release of the stored water is predicated on the slide surface intersecting the deposited material such that a pipe develops through the stored tailings mass. The critical failure surface adopted as a result of the Tier 1 assessment was then used to derive the breach mode and to estimate the quantity of sediment which might be released (Figure 7).

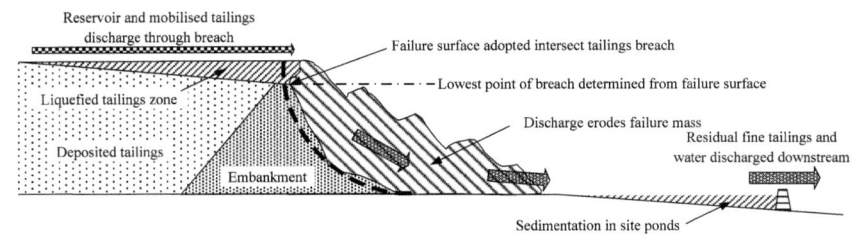

Figure 7. Critical failure mode for a MWF

The critical failure mechanisms

The worst credible tailings impoundment and reservoir conditions were assumed for the critical failure mode in spite of their transient nature. In the case of the Hemerdon MWF, these conditions might be applicable for no more than three months. The assumed failure mechanism for the critical

failure mode leading to an uncontrolled release can be summarised as follows:
- initial slope failure at the point of maximum embankment height;
- retention of coarse failure debris from the slide mass at the toe of the embankment;
- interception of the coarse tailings by the failure surface at a point below reservoir level;
- progressive development of a pipe through the deposit, leading to a breach through the tailings adjacent to the crest;
- discharge of the retained process water through the breach, and erosion of the finer surficial sediments within the reservoir area;
- development of a low-angle conical failure surface immediately upstream of the breach;
- discharge of the release down the embankment face around the failure mass;
- erosion of finer particles from the failure debris;
- attenuation of the release in downstream retention areas;
- discharge of water and some tailings into the downstream environment.

Materials released during the critical event
The critical failure mechanism (Figure 7) was used to determine the volumes of the various materials likely to be involved in a breach and inundation event at the Clemows Valley and Hemerdon MWFs. The material volumes involved in each failure were assessed as follows:
- Rockfill/earthfill: slide mass based on an elliptical projection of the critical circular failure surface;
- Tailings: eroded volume following release, based on the development of a low-angle (equivalent residual shear strength) conical surface in the tailings beach, together with an allowance for erosion of the reservoir surface area during drawdown;
- Rockfill/earthfill fines: fine particles eroded from the debris mass by the released water, the volume being based on the percentage of fines;
- Reservoir: volume of excess process, recycle and storm water impounded on the facility, based on the process water balance and the local hydrology.

INUNDATION MODELLING

Modelling Methodology
The assessment of the downstream impacts was undertaken using the methodology developed by the Environment Agency for category A and B large raised reservoirs. This involved the derivation of an input hydrograph based on that developed for embankment dams as proposed by Froehlich

(1995) and routing this using TUFLOW, a commercially available two-dimensional hydrodynamic model.

The input parameters developed from the critical failure modes used in the reservoir breach analyses were based on the Reservoir Inundation Mapping - Specification (Defra 2009). These inputs included the aforementioned breach hydrograph, the digital terrain model (DTM), based on LiDAR data (Geomatics, 2009), and an assumed mean high-water spring tide level as a downstream boundary condition. The DTM was adjusted, where appropriate, to ensure that the channel depths used in developing the inundation mapping for the downstream catchment were representative. The modelling also assumed that all culverts and bridges less than 1.5m wide were blocked due to debris from the breach event.

The outputs were then post-processed to allow effective presentation of the model results. This allowed an assessment of the number of downstream properties at risk and an indication of the extent of any potential environmental impact.

Model Outputs
Inundation maps were produced and the data used to assess the impact on the downstream catchment. The following were prepared for each upstream condition considered:

- Maximum Flood Event Map – full catchment;
- Maximum Flood Event Map – enlarged to show the upper catchment;
- Both detailed velocity and flood depth maps for the full catchment, with a separate enlargement to show the upper area;
- Inflow and outflow hydrographs for the downstream retention areas;
- Comparison of inundation extent, peak velocity, maximum inundation depth and potential properties at risk for each catchment sub-area.

Furthermore, the hydrographs developed for each model assessed the impact of any downstream facilities in attenuating the flood discharge. The results from the modelling were used to determine the extent of inundation in all scenarios developed for these MWFs. The maximum total area of inundation, together with velocity and depth profiles, in relation to at-risk properties was thus generated for use by the local resilience forum in developing the emergency plan for each MWF.

Results from the modelling showed, in general, relatively small amounts of inundation. This was predominantly a function of both the conveyance of the modelled watercourse as well as the quantity of associated infrastructure such as road bridges and culverts. As a result of such structures fairly significant attenuation of the peak flows was found. This is particularly applicable given that such structures were generally assumed to be blocked.

This resulted in reduced flooding in the downstream reaches, at the expense of upstream areas as water is stored.

Sedimentation

The process water, mixed tailings and fine silt discharged downstream will, at both facilities, initially be captured in site containment ponds which will attenuate the discharge and reduce outflow velocities, resulting in a proportion of the coarse tailings and the eroded debris material settling-out on site. The inflow and outflow hydrographs and the peak flow were used to assess settling velocities in such areas, and to derive the minimum particle size which will be deposited. On this basis, it was feasible to assess the maximum particle size retained on site and thus to determine the tonnage of solids eventually released into the downstream catchment. These release data were used by the emergency planning team with respect to post-incident rehabilitation.

Table 2 presents an example of breakdown of the total volume of material likely to be involved in the release event. It was recognised during the analyses of release data for both MWFs that, due to the shape of the outflow hydrograph produced by this methodology and the transient nature of the peak velocity, such estimates of the solid fraction released are extremely conservative.

Table 2. Summary of typical release data

Materials involved in the release	Quantity
Rockfill in the embankment slide mass (t)	
Eroded mixed tailings (t)	
Fine silts eroded from the failure mass (t)	
Mixed tailings/silt retained in containment ponds(t)	
Tonnage of fine tailings released downstream (t)	
Volume of water and solids released downstream (m^3)	
Average silt concentration in the discharge (g/l) **Note 1**	

Note 1: Assumes that silt is dispersed evenly within the released water volume.

SUMMARY AND CONCLUSIONS

During the design of any mine waste facility it is good practice to assess the hazard potential which would arise should embankment failure lead to the development of a breach and the uncontrolled outflow of any impounded water and solids. The development of emergency plans was required during the permitting of two UK MWFs, namely the Clemows Valley Tailings Dam and the Hemerdon Mine Waste Facility, in order to satisfy the requirements of the local resilience forum, and involved the following:

- review of historic failures which evidenced the difference in mode between large raised reservoirs and MWFs, particularly in terms of material release, indicating that the initiation of failure in the latter

was determined by geotechnical, as opposed to hydrodynamic, characteristics;
- review of the Environment Agency single-phase (hydrodynamic) model for large raised reservoirs, which indicated it to be inadequate, requiring the development of an alternative two-phase (geotechnical/hydrodynamic) model;
- adoption of a tiered risk assessment approach to develop the failure mode for these facilities, recognising that the probability of failure due to the design, construction and inspection standards adopted was extremely low;
- development from the Tier 1 assessment of a critical breach mode with a failure surface which intercepted the dam crest at its highest point and the deposited tailings below reservoir level;
- development of this failure mode using the source-pathway-receptor approach to derive material disturbed during the breach and released into the downstream catchment;
- development of the parameters for the inundation mapping subsequently undertaken using the model adopted for large raised reservoirs;
- provision of the inundation mapping and sedimentation analyses required by the local resilience forums for the preparation of the emergency plans.

In accordance with the requirements of the EWD the emergency plans will be regularly reviewed and updated throughout the life of these facilities.

ACKNOWLEDGEMENTS
The authors would like to thank Wheal Jane Ltd and Wolf Minerals (UK) Limited for their permission to publish this paper.

REFERENCES
Cambridge M (2003). The use of historic tailings dam incidents in the development of emergency plans. *Proceedings ICOLD Conference*, Stockholm, Sweden.

Snorteland N (2013). Fontenelle Dam, Ririe Dam and Teton Dam – an example of the influence of organizational culture on decision making, *Proceedings ICOLD Conference*. ICOLD, Stockholm, Sweden

HR Wallingford (2013). *Guide to risk assessment for reservoir safety management*. Environment Agency, Bristol, UK

European Commission (2012). *DG Environment, No. 070307/2010/ 576108/ETU/C2, Annexe 2, Guidelines for the inspection of mining waste facilities*. European Commission, Brussels, Belgium.

HMSO (2011). *Environmental Permitting Regulations EPR6.14.* HMSO, London, UK

ICOLD (2011). *Sustainable Design and Post-Closure Performance of Tailings Dams.* ICOLD

Cambridge, M (2010). Flood assessment at UK tailings management facilities. *Proceedings of the 16th BDS Biennial Conference, Strathclyde.* Thomas Telford, London UK.

HMSO (2010), *Floods and Water Act 2010.* HMSO, London, UK

Mott MacDonald, (2009). *Reservoir Inundation Mapping – Specification.* Defra, London, UK

Cambridge, M (2008). The application of the Mines and Quarries (Tips) and the Reservoirs Act. *Proceedings of the 15th BDS Biennial Conference Warwick.* Thomas Telford, London UK.

European Commission (2006). *Directive on the management of waste from the extractive industries, 2006/21/EC/.* European Commission, Brussels.

Adam K and Cambridge M (2004). *Evaluation of Potential Risks and Mitigation Measures:* The Design of a Mining Project; Professor Kontopoulos Memorial Volume.

Cambridge M (2004). *Tailings Disposal in Cornwall – Past and Present.* Professor Kontopoulos Memorial Volume.

ICOLD (2001). *Tailings Dams, Risks of Dangerous Occurrences; Bulletin 121.* ICOLD, Paris, France

ICE (2000). *A Guide to the Reservoirs Act 1975.* Thomas Telford, London

CIRIA (2000) *Risk management for UK reservoirs, C542.* CIRIA, London.

Health & Safety Commission (1999). *Health and Safety at Quarries, Quarries Regulations 1999.* HSE, London, UK

ICE (1996). *Floods and reservoir safety.* Thomas Telford, London, UK

ICOLD (1995). *Dam Failures, Statistical Analysis; Bulletin 99.* ICOLD, Paris, France.

Froelich, D.C. (1995) Peak Outflow from Breached Embankment Dam *ASCE J. Water Resources Planning and Management 121, 1, 90-97.* ASCE, Reston, Virginia, USA

BRE (1991), *An engineering guide to seismic risk to dams in the United Kingdom.* BRE, Watford, UK

HMSO (1971). *Mines and Quarries (Tips) Regulations, 1971.* HMSO, London, UK.

Quantitative risk assessment applied to sludge lagoon embankments

M EDDLESTON, MWH Global, UK
C ROSE, MWH Global, UK
E GALLAGHER, MWH Global, UK
I HOPE, Severn Trent Water, UK
P SUGDEN, Severn Trent Water, UK

SYNOPSIS   Following privatisation in 1989 Severn Trent Water (STW) inherited a number of waste water treatment works from Severn Trent Water Authority and its predecessors. A number of the works contain raised sludge lagoons, some of which date back over 100 years. Ageing, elevated, sewage sludge lagoons can pose a significant risk as the tip failure at Deighton, Yorkshire in 1992 (Claydon *et al*, 1997) proved. In learning from this incident, STW is adopting a proactive approach to reduce the risk of failure of sludge lagoons. Whilst associated risks to life or property are very low and not subject to reservoir safety legislation, the catastrophic failure of a sludge lagoon would have a major impact on business reputation as well as attracting punitive remedial costs and litigation. This has led STW to implement a proportionate response to the potential risks posed; significant investment in investigating the nature of these assets has been made, leading to the design and implementation of risk reduction measures over a number of years. Regular and effective monitoring of the assets is also an important part of this process, assisting in detecting changes in behaviour so that problems can be spotted early before they become potentially unstable and expensive to fix.

The sludge embankments have been in place for many decades without major slope failures. The non-engineered nature of the embankment materials, un-recorded construction methods and lack of knowledge of the nature of the materials retained leaves them inherently vulnerable to internal erosion due to seepage, rotational slips and internal combustion. The current lack of any secondary engineered defence could lead to rapid failure in the event of a serious internal erosion or slope stability incident. The uncertainties associated with the non-engineered and variable nature of the embankment materials means that they are difficult to characterise even with extensive intrusive geotechnical investigations. This makes it challenging to assess the future stability of the lagoons using conventional engineering analysis. This paper will describe how the level of risk associated with the various sections of a large sludge embankment were assessed utilising quantitative slope risk assessment techniques applied to infrastructure embankments.

INTRODUCTION

Severn Trent Water (STW) owns a number of sludge lagoons which contain a mix of sludge and water as a by-product of past sewage treatment processes. Some of the lagoons are above natural ground level supported by embankments. Some older sludge lagoons were not built in the traditional way of creating an impermeable core, or by incorporating an impermeable liner, but simply by using soil, excavated from within the lagoon site. Some of the older embankments were even constructed from waste materials such as refuse and blast furnace slag. This paper describes how a Quantitative Risk Assessment was developed to assist STW in managing and developing mitigating measures for the risks associated with one of its larger lagoons.

The extensive sewage treatment works site, was originally developed by Birmingham City Council, passing to the Severn Trent Water Authority in 1974 and to the present owner, STW, following privatisation. The site has a long history of use for sewage treatment and probably domestic refuse disposal extending over a period of 100 years. The original construction of the lagoons area predates the ownership of STW.

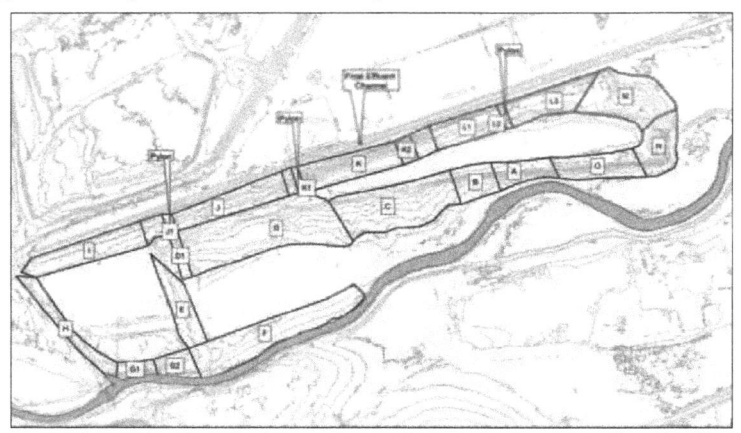

Figure 1. Sludge lagoon site

The study site (Figure 1) has not been operational for recent decades and has been allowed to vegetate naturally. The current lagoon surface is wet, of low bearing capacity, and thus unsafe for access. The site was the subject of a major study in 1996 which included some intrusive site investigations. There have been a series of problems in recent years, with extensive underground fires along the southern embankment and a minor slope failure on the eastern extremity. Each incident lead to further investigation and the slope failure has been monitored on an annual basis.

In November 2010, as part of a portfolio risk assessment review of reservoir safety (Chesterton *et al*, 2012 and Hope, 2012), STW identified that the sewage lagoon sites, though not covered by reservoir legislation and not subject to statutory inspection rigour, potentially posed significant risks to the company and to the environment. STW therefore requested an All Reservoirs Panel Engineer to undertake a preliminary inspection of all sludge storage facilities with elevated embankments in order to quantify the hazard potential and identify necessary remedial works. This inspection recommended that:

- an overview of the stability of all embankments be undertaken, using data from previous investigations where available so that the most up-to-date information can be used to assess the critical embankments and
- a review of the environmental impact of the sites be undertaken, with scrutiny of untoward seepages and migration of potentially highly contaminated material off-site.

In addition, to ensure appropriate management of risks posed by some of the lagoon structures, and in line with Severn Trent's approach to managing risks and reservoir safety, tri-weekly monitoring by trained operators was established.

In seeking a proportionate response to the risks posed a high level site screening risk assessment was undertaken of for all elevated lagoons at the sewage treatment works site in 2012. This enabled STW to assesses and manage site risks to current best practices. The risk profile, covering physical, environmental and social factors, indicated that the potential physical slope instability of the southern lagoon embankment next to the river posed the greatest risk at the site. Following on from this study it was decided that a phased approach to the detailed assessment of the lagoon slopes, and potential detailed design and implementation of risk mitigation measures (e.g. potentially stabilisation measures or improved monitoring), would be the most efficient way to understand and manage long term risks at the lagoons. The three phases are summarised below:

Phase 1: Definition of provinces (sections of lagoon of similar nature, geometry etc.), generic slope stability modelling (based on published data and available site investigation information) and geotechnical risk mapping of the full lagoon perimeter.

Phase 2: Detailed geotechnical slope modelling of the higher risk areas.

Phase 3: To optioneer and undertake outline design of potential risk reduction measures as required.

For the Phase 1 work a Quantitative Risk Assessment of the slopes as part of the geotechnical mapping process was identified as a benefit. This could be undertaken efficiently by the development of a bespoke iPad based site data collection and collation tool (mTool™).

SITE DESCRIPTION

Historical reports on the lagoon embankments indicated that the existing lagoons were first shown on a site plan dated 1922. The main lagoon (the study lagoon) covers the eastern area of the site with two other lagoons in the western portion of the site. The main lagoon was originally a sludge pit but was extended and impounded by embankments on all sides/perimeters. It stores significant volumes of low-density, potentially unstable jelly-like thixotropic sewage sludge (having the ability of becoming fluid when disturbed) as well as potentially significant volumes of water. Water can, following heavy rainfall, pond on the surface of the sludge behind the embankment. Site investigations indicate that the geotechnical consistency of the sludge behind the lagoon embankment increases with depth with a typically very soft layer up to 2m depth followed by "fibrous" type of sludge of increasing consistency, becoming typically firm to a proven maximum depth of 10.7m.

Engineering details are not available for the construction of the embankments. From exposed elements of the embankments they appear to be constructed from unsorted and uncompacted rubble, furnace slag and ash, waste residues and local soils. Anecdotal evidence suggests the embankment placement method may have been of the "upstream" type, consisting of a foundation "dyke" followed by successive layer(s) imposed on the crest of the foundation structure. It is likely that as sludge level reached a top level behind the foundation embankment another section of the embankment would be constructed and filled with sewage sludge until the current level was reached.

The embankments have a history of subsurface fires. Anecdotally these fires appear to erupt every 10 years or so from within the embankment. The complex chemical and physical reactions that take place during a subsurface fire are unknown because they have not been studied in-situ. However, they are believed to be formed by rapid oxidation of carbon based waste material at elevated temperatures causing smouldering. The fires are accompanied by gas and smoke emissions. On the study lagoon the subsurface fires led to the burning of tree root systems, with mature trees becoming unstable. In 2011 and 2012 vegetation on most of the embankments was removed to assist in controlling fires and enable improved monitoring of the embankment condition.

The maximum embankment height is approximately 10m. The slopes of the embankments are steep (typically 1:2 vertical:horizontal), non-uniform, and suffering from local shallow surface movement and tension cracks. The underground fires are one of the likely causes by creating voiding and internal collapse therefore creating near-surface slope instability.

The lagoon embankments were constructed in the floodplain of a main river. The southerly section of lagoon embankment nearest the river is potentially at risk of being undermined by flood waters at the embankment toe. Flood water reaches the toe of the embankment on a regular basis and could cause erosion which would lead to collapse of the embankment. This would allow sludge to escape into the river and cause a significant pollution event. Three electricity pylons cross the site from east to west and are founded over the northern embankments and the sludge lagoon. A brickwork lined open channel conveying treated effluent from the adjacent sewage treatment works runs along the line of the toe of the lagoon's northern embankments.

SITE GEOLOGY AND GROUND CONDITIONS

The geology of the site comprises drift deposits of River Alluvium and then Terrace Gravels overlying solid deposits of the Mercia Mudstone. A number of site investigations have been undertaken in the embankment adjacent to the river where this geology was confirmed and a typical embankment section is shown in Figure 2.

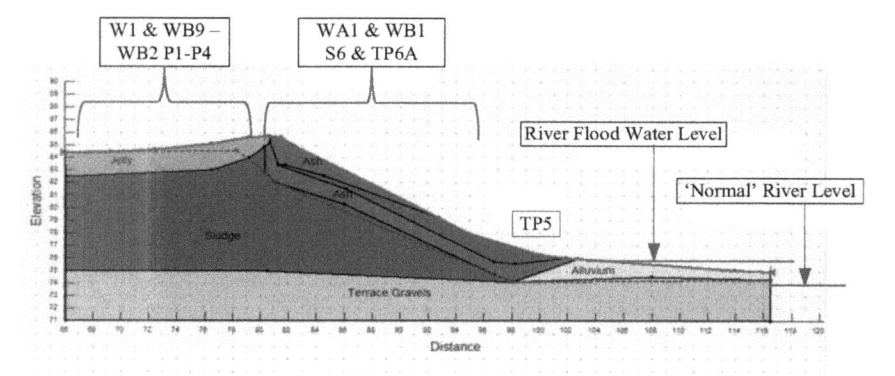

Figure 2. Schematic section through embankment adjacent to the river location of intrusive investigations and river levels

RISK ASSESSMENT METHODOLOGY BACKGROUND

Over the last decade owners of major assets in the UK that can be affected by slope failures, including dams, flood defences, and rail and highway embankments, have developed risk assessment techniques to manage the risk associated with failure. This has enabled the identification, prioritisation and justification of remedial works. Vivid examples such as

Deighton in 1992 (Claydon *et al*, 1997) and Kolontar in Hungary in 2010 (Mecsi J. 2013), demonstrate the destructive impacts of lagoon failure.

Failure at this site means an uncontrolled release of sludge that can have a damaging effect on adjacent assets, including major pollution of the river, blocking of the sewage treatment works final effluent carriers, or damage to the National Grid electricity pylons affecting power supply to Birmingham.

Review of Available Techniques

UK Best Practice
The following slope inspection and risk assessment regimes were reviewed and examined with the aim of finding a suitable inspection method that could be used or modified for the lagoon embankments. These were, in no particular order:

- Seven Trent Water Likelihood of Failure' assessment approach for embankment dams that do not fall under the requirements of the current Reservoir Act (1975)
- Environment Agency - Asset Condition Assessment (Flood Embankments)
- Network Rail - Examination of Earthworks (NR/L3/CIV/065)
- Highway Agency - Maintenance of Highway Geotechnical Assets (HD41/03)
- British Waterways - Asset Inspection Procedures (AIP 2008)
- Defra/EA Performance and Reliability of Flood and Coastal Defences, RD2318/TR1.
- Guide to risk assessment for reservoir safety management (2013), Environment Agency.
- Scottish Road Network: Landslide Study. The Scottish Executive. Winter, M.G., MacGregor, F. & Shackman, L. (2005, 2008).

In general the above approaches produce a semi-quantitative ranking of the likelihood of failure. All the inspection regimes require regular visual inspections of assets, where the frequency of the inspections is dependent on the risk and consequence of failure and condition of the assets. Inspections are generally undertaken by a trained or competent person. When a high risk is identified the Network Rail standard requires a geotechnical engineer to inspect the asset and undertake a much more comprehensive risk assessment which looks at the actual and potential risks of failure of the asset (Manley and Harding, 2003).

Other Best Practice
Discussions with practitioners in Australia and New Zealand have indicated that a similar approach has been developed in the Asia-Pacific region for slope assessments, which produce an assessment of the likelihood of failure in terms of probability of failure (AGS, 2007 a-f, and Transport New Zealand, 2004). The Transport New Zealand method included a methodology to make an assessment of consequences to allow a risk matrix to be devised and give a risk ranking to the potential slope instability threat.

Using best practices a bespoke methodology was derived in order to assess the likelihood of embankment failure and the probable consequences of that failure. The product of the likelihood and consequence equates to the level of risk thus allowing each province of the lagoon embankment perimeter to given a risk rating to assist in prioritising and planning future investigation/remediation/monitoring work at the site.

SELECTED METHODOLOGY
Each step of the derived risk methodology is detailed in the following sections, including key factors to identify and assess, specific tools used, and reference to supporting documentation.

Likelihood (Notional Probability of Failure) Assessment
To make an assessment of potential probability of failure the lagoon embankments were split into provinces. Site data was collected for each province, which was used in the assessment to produce a likelihood score. This score was then converted into a probability of failure. The likelihood assessment developed for the lagoon site uses the general approach used by STW for Likelihood of Failure assessment for embankment dams that do not fall under the requirements of the current Reservoirs Act (1975) amended based on the AGS (2007) and Transport New Zealand Slope Risk Assessment Methodology (2004).

A bespoke software tool (mTool™) was developed for the surveys that was hosted on an iPad (Apple Inc.) for data collection in the field, and a web based data management system (Sharepoint™) where the site data and photographs were uploaded, saved and shared. Once on the Sharepoint site the data was assessed and an estimate of a notional probability of failure was derived (see Table 1). The purpose of developing a project specific software tool was to ensure that the entire process was:

- objective, accurate and consistent;
- repeatable (e.g. when the province reassessed following remediation works or for long-term monitoring records); and,
- auditable, using version control and document management.

Using an iPad with 3G capability also had the advantage that any photographs, province assessments and comments could be located on site.

Each engineer undertook the assessment using the same software tool and gave each of the criteria a score based on knowledge of the site from field observations, measurements and available reports. The criteria considered were as detailed in Figure 3 below. By using the same scoring system each assessment achieved the aims of the process, particularly to be objective and repeatable.

Generic soil parameters used in the slope stability assessment of each province were derived from test results from the various phases of site investigation, back analyses, technical literature (Claydon *et al*, 1997; Klien and Sarsby 1990; Mecsi, 2013) and British Standards.

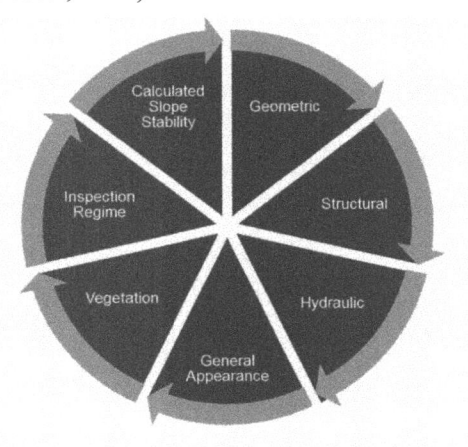

Figure 3. Criteria used in Likelihood of Failure Assessment

Probability of Failure Rating
Likelihood is taken as a rating on a scale based on the detailed likelihood assessment normalised to a scale of 1 to 5. The ratings are converted to a notional probability of failure in accordance with Australian Geomechanics Society and Transport New Zealand Ratings (AGS 2007 a-f, and Transport New Zealand, 2004) as shown in Table 1. The derived probability values can be used to give an indication of the potential need for remedial works to be undertaken by reference to the Health and Safety Executive guidelines given in Reducing Risk Protecting People R2P2 (HSE, 2001) as shown in Figure 4. This recommends a notional probability of failure of greater than 10^{-4} or 1 in 10,000 for an identified hazard potentially leading to loss of life or societal risk. A lower limit of 1 in 1,000 is sometimes reported for lower risk situations where no potential life loss is identified and consequences are minor (Eddleston, 2004, 2012).

However, published data indicates a considerable range of values; therefore balance is needed in assigning both the notional probability of failure and consequence of failure to be used in assessments. There is some consensus that a notional probability of failure of 10^{-4} (1 in 10,000) is considered a generally acceptable criterion for slopes where there is a potential for loss of life. Alonso (1976) equates this to the commonly accepted deterministic factor of safety of 1.5 for new build embankment dams. Also, Christian *et al*, (1994) report probabilities approaching 1 in 1000 for a factor of safety of 1.5.

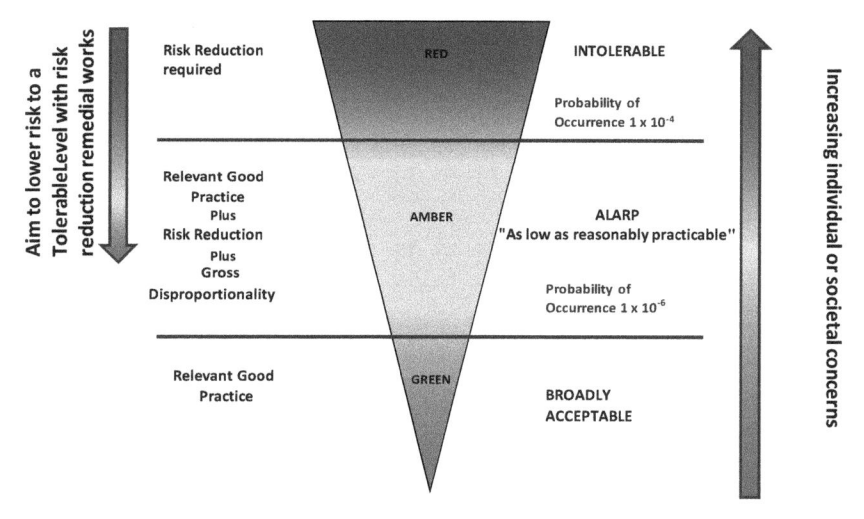

Figure 4 - Health and Safety Executive Likelihood of Failure Guidelines (HSE, 2001).

Consequence Assessment
The assessment of consequences of failure used in STW's Dam Safety Risk Assessment is a quantitative approach based on the general approach for embankment dams that do not fall under the requirements of the current Reservoir Act (1975). This approach was modified to allow for consequences that are important at the site but are not covered in the primarily water supply related method used to assess consequences of reservoirs failure (e.g. damage to pylons). The modified approach also includes a numerical scoring system taken from Transport New Zealand (2004) which considers consequence in terms of damage to image and reputation, structures and installations, the environment and financial consequences in terms of cost and the time required to remediate any damage.

Table 1 - Relationship between Likelihood Rating and Probability of Failure (AGS 2007 a-f, and Transport New Zealand, 2004).

		Description	Indicative Value of Annual Probability of Failure
LIKELIHOOD	Likely (5)	The threat can be expected to occur *or* a very poor state of knowledge has been established on the threat.	10^{-2}
	Quite Common (4)	The threat will quite commonly occur *or* a poor state of knowledge has been established on the threat.	10^{-3}
	Unlikely (3)	The threat may occur occasionally *or* a moderate state of knowledge has been established on the threat.	10^{-4}
	Unusual (2)	The threat could infrequently occur *or* a good state of knowledge has been established on the threat.	10^{-5}
	Rare (1)	The threat may occur in exceptional circumstances *or* a very good state of knowledge has been established on the threat.	10^{-6}

Adopted Risk Classification System

The Risk associated with the identified potential threats is the product of the likelihood of the threat being materialised and the consequences of it occurring, usually expressed in the form of a risk matrix. The matrix proposed to assess the potential threats at the lagoons is shown in Figure 5 below, based on the Transport New Zealand, 2004 approach.

RESULTS OF RISK CLASSIFICATION

The results of the consequence and risk assessments for each province were superimposed on a site aerial photograph (Figure 6).

The risk classifications indicate:

- areas of very high risk associated with slopes adjacent to the river and third party installations,
- high risk areas for slopes adjacent to the treatment works final effluent channel,
- moderate risks on the southern lagoon embankment slopes either side of the bend in the river, and
- an area to the east of the treatment works outfall is shown to be of moderate risk but potentially high likelihood of instability.

	Indicative Value of Annual Probability of Failure	CONSEQUENCES (loss)				
		Negligible (1)	Minor (10)	Medium (40)	Major (70)	Substantial (100)
Likely (5)	10^{-2}	5 Low threat ACCEPT ACTIVELY Enhance systems to minimise potential	50 Moderate threat ACCEPT ACTIVELY Enhance systems to minimise potential	200 Very high threat AVOID Immediate action Enhance systems to minimise potential	350 Extreme threat AVOID Immediate action Cease activity	500 Extreme threat AVOID Immediate action Cease activity
Quite Common (4)	10^{-3}	4 Low Threat ACCEPT ACTIVELY - Enhance systems to minimise potential - Accept	40 Moderate threat ACCEPT ACTIVELY - Enhance systems to minimise potential	160 Very high threat AVOID Immediate action Enhance systems to minimise potential	280 Very high threat AVOID Immediate action Contingency Plans	400 Extreme threat AVOID Immediate action Cease activity
Unlikely (3)	10^{-4}	3 Negligible threat ACCEPT	30 Moderate threat ACCEPT ACTIVELY Enhance systems to minimise potential	120 High threat ACCEPT ACTIVELY OR TRANSFER Immediate action Monitor Contingency Plans	210 Very high threat AVOID Immediate action Avoid Contingency Plans	300 Very high threat AVOID Immediate action Avoid Contingency Plans
Unusual (2)	10^{-5}	2 Negligible threat ACCEPT	20 Low threat ACCEPT ACTIVELY Monitor	80 High threat ACCEPT ACTIVELY Monitor Contingency Plans	140 High threat AVOID OR TRANSFER Monitor Contingency Plans	200 Very high threat AVOID OR Monitor Contingency Plans
Rare (1)	10^{-6}	1 Negligible threat ACCEPT	10 Low threat ACCEPT ACTIVELY Monitor	40 Moderate threat ACCEPT ACTIVELY Monitor Contingency Plans	70 or below Moderate to high AVOID OR ACCEPT ACTIVELY Monitor Contingency Plans	100 High threat AVOID Monitor Contingency Plans

LIKELIHOOD (left axis)

Mitigate whenever possible →
Minimise whenever possible ↓

Figure 5. Risk Classification Matrix

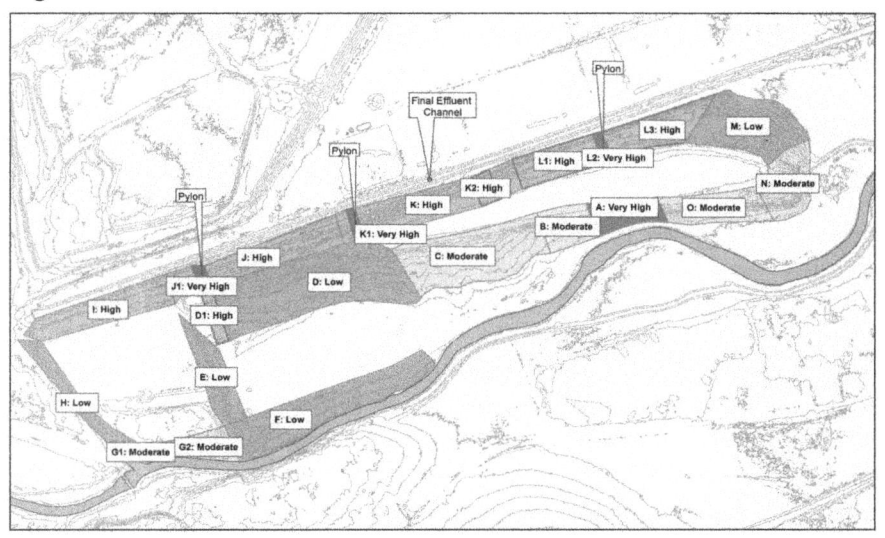

Figure 6 – Lagoon embankments Risk Classifications

RISK MITIGATION MEASURES
The risk profile allowed STW to place into context the risks posed to its assets. It presented the risk in such a way that the concerns could be easily explained to operational and executive management. It also helped prioritise potential risk reduction measures to manage the risks in a proportionate manner. The process indicated that the physical slope instability adjacent to the river posed the greatest risk at the site. This risk is being mitigated by slope remediation and toe protection as part of a major capital project. Other measures are being considered to protect the sewage treatment works final effluent channel and evaluate and monitor third party assets on the sites.

It is also recognised that whilst risk analysis is very useful for understanding asset performance and investment planning, it should be used cautiously, and should not replace judgement-based observations. In fact, as with large embankment dams, careful and frequent monitoring of lagoon embankments is arguably the most important aspect of keeping them safe from a potential failure. As part of its continuous improvement programme for reservoir safety (Hope, 2012) Severn Trent Water has established an assessed, comprehensive training programme with an illustrative manual specifically for maintaining the safety of its sludge and tertiary lagoons (STW, 2013). This reinforces and complements the comprehensive inspection regime already in place at this and other sludge lagoon sites. Other risk mitigation measures include the development of an on-site plan to the Defra template and a planned exercise to test emergency response for later in 2014.

ACKNOWLEDGMENTS
The authors acknowledge the permission of Severn Trent Water to publish this paper.

REFERENCES
Alonso, E E (1976). *Risk analysis of slopes and its application to slopes in Canadian sensitive clays*. Geotechnique Vol. 26, 453-472.

Australian Geomechanics Society. (2007a). *Guideline for Landslide Susceptibility, Hazard and Risk Zoning for Land Use Management*. AGS Landslide Taskforce, Landslide Zoning Working Group, Australian Geomechanics, 42(1), March 2007.

Australian Geomechanics Society. (2007b). *Commentary on Guideline for Landslide Susceptibility, Hazard and Risk Zoning for Land Use Management*. AGS Landslide Taskforce, Landslide Zoning Working Group, Australian Geomechanics, 42(1), March 2007.

Australian Geomechanics Society. (2007c). *Practice Note Guidelines for Landslide Risk Management 2007*. AGS Landslide Taskforce, Practice Note Working Group, Australian Geomechanics, 42(1), March 2007.

Australian Geomechanics Society. (2007d). *Commentary on Practice Note Guidelines for Landslide Risk Management 2007.* AGS Landslide Taskforce, Practice Note Working Group, Australian Geomechanics, 42(1) March 2007.

Australian Geomechanics Society. (2007e). *Australian GeoGuides for Slope Management and Maintenance.* AGS Landslide Taskforce, Slope Management and Maintenance Working Group, Australian Geomechanics, 42(1), March 2007.

Chesterton, O J, Hill, T, Hope, I M and Airey, M, (2012). A pragmatic approach to Portfolio Risk Assessment at Severn Trent Water. *Proceedings of the 17th Biennial BDS Conference, Leeds, pp 219-230.* ICE Publishing, London

Christian, J T, Ladd C C and Baecher G B (1994). Reliability applied to slope stability analysis. *Journal of Geotechnical Engineering, ASCE. 120(12) 2180-2207.* ASCE, USA

Claydon J R, Eadie, H S and Harding C (1997). Deighton Tip – Failure Investigation and Remedial Works. *Proc. of the Nineteenth ICOLD Congress, Florence. Paper Q, 75-R. 19, 233-245.* ICOLD, Paris, France

Eddleston, M, Beeuwsaert, L, Taylor, J and Gardiner, K D (2004). Development of a probabilistic methodology for slope stability and seismic assessments of UK embankment dams. *Proceedings of the 12th Biennial BDS Conference, Canterbury, pp 232-24.* Thomas Telford, London, UK.

Eddleston, M (2012). Exploring the potential use of a risk based approach to assessing the geotechnical well-being of the slopes of old embankment dams. *Proceedings of the 17th Biennial BDS Conference, Leeds, pp 206-218.* ICE Publishing, London

Klien, A and Sarsby, R W (1999). Problems in defining the geotechnical behaviour of wastewater sludges. In *Geomechanics of High Water Content Materials.* ASTM STP 1374.

Health and Safety Executive (2001). *Reducing Risks: Protecting People.* HSE, London, UK

Severn Trent Water (2013) *Maintaining the safety of our sludge and tertiary lagoons - Training Manual.* Severn Trent Water (unpublished)

Hope, I M (2012) Implementing Severn Trent Water's People Plan to become the best in Great Britain at managing Reservoir Safety. *Proceedings of the 17th Biennial BDS Conference, Leeds, pp 43-55.* ICE Publishing, London

Manley, G and Harding, C (2003). Soil slope hazard index as a tool for earthworks management. *Proc. 6th Int. Conf. Railway Engineering, London.*

Mecsi J (2013), Some Technical Aspects of the Tailing Dam Failure at the Ajka Red Mud Reservoirs. *Proceedings of the 18th International Conference on Soil Mechanics and Geotechnical Engineering, Paris 2013, 3309-3312.*

Transport New Zealand (2002). *Risk Management Process Manual AC/Man/1.*

Setting standards for draw-down capability at Scottish Water's Reservoirs

R MANN, Technical Director, Aecom,
J MALIA, Senior Reservoirs Engineer, Scottish Water,
S LOCKETT, Senior Engineer, Aecom

SYNOPSIS   With 270 large raised reservoirs, Scottish Water is the largest owner of reservoirs covered by the Reservoirs Act 1975 in the UK. Although draw-down capability is considered by Inspecting Engineers at statutory inspections, Scottish Water had not established a policy on this matter.

In order to establish best practice, Aecom was commissioned by Scottish Water to study and review current draw-down capacity for their impounding reservoirs, and to propose a standard to adopt for draw-down rates to meet maintenance, precautionary and possible emergency requirements.

This paper reviews current standards for reservoir draw-down proposed and adopted in the UK and overseas. These standards vary over a considerable range. Within this range, draft criteria have been developed for adoption by Scottish Water (SW). These have been determined by considering both risk-based and practical considerations relevant to Scottish Water's reservoirs.

Following an initial review based on limited data, further assessment of draw-down capability is in hand to understand the extent of upgrading required to meet the proposed criteria.

INTRODUCTION
Scottish Water's 270 large raised impounding reservoirs cover a wide variety of type of dam and wide range of dates (late 18th Century to current); impounded volume (to 61Mm³ for Megget Dam and 130Mm³ for Loch Lomond Barrage); height (from very low raisings of lochs to 56m for Megget); and location (from mountainous to lowland areas)

The objective of draw-down provision is to provide a means of reducing the reservoir water level for occasional maintenance purposes, or critically for the rare event of an incident requiring precautionary or emergency draw-down to mitigate the risk of dam failure. Generally this is by means of scour arrangements installed at the dam for this purpose. Alternative or supplementary measures may also need to be employed to achieve a suitable draw-down rate depending on circumstances at the reservoir.

Draw-down provision is currently a matter considered at periodic statutory inspection of reservoirs. Appropriate details incorporated into on-site plans will become a statutory requirement in the near future when the relevant Flood Risk Management (Scotland) Act 2009 and the Reservoirs (Scotland) Act 2011 are brought into effect.

At several dams Scottish Water has carried out measures to improve draw-down capacity following recommendations made by Inspecting Engineers. Measures have included provision of fixed siphons, refurbishment of scours off supply pipes, provision of new scour branches off supply mains, altering capping arrangements to use redundant supply mains for scour purposes, and providing sluices in overflow weirs. All have been designed to achieve draw-down standards determined by the engineer appointed under the Act.

There are various criteria quoted in the literature from UK and overseas practice for the required rates of lowering of a reservoir, with significant variation between them. At each reservoir the adequacy of draw-down provisions has been subject to the considered view of the appointed Inspecting Engineer or the appointed Qualified Civil Engineer. Scottish Water was therefore faced with inconsistent and varied criteria for draw-down requirements considered to be appropriate.

Establishment of a policy and standard for draw-down provisions would allow Scottish Water to:

- Fully understand the issues and adopt a clear, consistent and pro-active approach to draw-down provision and its ability to avert dam failure, in line with best practice;
- decide on the appropriate level of risk that it was willing to take;
- understand the draw-down enhancement required and to make provision for the relevant investment;
- establish a programme for such enhancement;
- communicate SW's drawdown policy to Inspecting Engineers making statutory inspections.

EXISTING CRITERIA FOR DRAW-DOWN OF RESERVOIRS
Various criteria for emergency draw-down within the industry in the UK and overseas are set out in the *Engineering Guide to Emergency Planning for UK Reservoirs* (Jacobs Babtie, Defra, June 2006) and subsequent UK publications. The criteria cover a wide range of target rates of draw-down.

The Defra Guide identifies two Key Points for draw-down to be achieved:

- Initial draw-down to a stated depth below the spillway overflow level using only fixed provisions at the reservoir.

- Further draw-down using both fixed and supplementary provisions to a level where hazard is substantially reduced.

EXISTING CRITERIA FOR INITIAL DRAWDOWN

The Defra Guide refers to a depth of 1m below spillway level (recognising that some internal erosion incidents are located in the upper part of the core) but does not set a specific time for achieving this depth. A previous letter issued to water companies in England (Defra, 2002) commended a lowering of 1m per day (achieved with supplementary pumping) as a factor in averting failure at a serious UK dam incident at that time. A subsequent paper by Hinks (BDS, Dams & Reservoirs, March 2009) states that "... *for many purposes, it is more important to know how far a reservoir can be lowered ... in 24 hours than how long it will take to reduce the volume to, say, 50%*". Hinks gives a target reduction in reservoir level, under moderately wet inflow conditions, of 300mm plus 0.5% of the dam height in the first 24 hours without recourse to imported pumps or other provisions.

Hinks' paper also recognises that a greater initial draw-down up to 1m in the first 24 hours might be regarded as appropriate in some circumstances at high risk reservoirs. Conversely, the paper refers to large flood storage reservoirs where a lower immediate draw-down rate may be applicable.

Target draw-down rates for emergency situations are distinct from maintenance or precautionary draw-down where the rate at earthfill embankments needs to be controlled to avoid risk of destabilising the upstream face. Various rules of thumb exist, one being a limit of 300mm per day for typical fill materials. If the permeability is known, more accurate assessments can be made. Downstream flooding is also a consideration.

EXISTING CRITERIA FOR FURTHER DRAWDOWN

Regarding further draw-down, the Defra Guide states that "*There are various rules of thumb quoted in the literature for the required rates of lowering of a reservoir, with significant variation between the alternate criteria, as summarised in Table 4.9*". Table 4.9 of the Defra Guide is reproduced as Table 1 below (omitting the columns giving Author and reference).

The Defra Guide also refers to draw-down criteria given in subsequent publication as British Waterways' standard (Brown D, 2009) that relates the required draw-down rates to both the overall consequence class of the dam (A to D as in ICE Floods and Reservoir Safety) and the frequency of surveillance (in recognition of the time that could occur between the incident occurring and being first noticed) (Table 2).

Table 1. Summary of rules of thumb for draw-down (From Defra, 2006)

Organisation	Criteria			No of days to reduce reservoir to percentage of initial height (load)[1]	
	Outlet capacity	Inflows	Initial Reservoir level	75% (50%)	50% (25%)
Overseas					
US Bureau of Reclamation, 1990	Varies with class of hazard and risk (9 classes in Table 4). Extremes shown in adjacent columns	Highest mean monthly inflow for duration of lowering	Spillway crest, exclude volume used for flood control	10-20 for high risk; 60-90 for low risk	30-40 for high risk; 90-120 for low risk
State of California, 1999	For reservoirs < 6.2Mm³: 50% of reservoir capacity in less than 7 days. For larger reservoirs 10% of reservoir depth in 7 to 10 days. (Logic is larger dams are more thoroughly designed and constructed). Excludes releases through power plants.	Nil (It is stated that in California this is true nine months of the year)	Not specified		See releases
French practice	Bottom outlets should be capable of reducing load on dam by 50% in 8 days			8	
United Kingdom					
Northumbria Water	Reduce reservoir contents to 25% of their storage over 28 days	Winter 28 day peak	Assumed at spillway crest	See releases	

[1] It is noted that the criteria could alternatively be based on dam crest level, as consideration of options for managing a piping incident developing during extreme floods.

Table 2. British Waterways' standards (Brown, 2009)

	Number of days to lower the reservoir to 50% of volume when full, with inflow of winter daily mean flow	
Overall Consequence Class	Surveillance once a week	Surveillance twice a week
A1	3	5
A2	5	7
B, C, D	7	9

Since issue of the Defra guide, a further criterion has been published by Anglian Water (Tam & Humphrey, 2012), set out in Table 3 below:

Table 3. Draw-down criterion of Anglian Water (Tam & Humphrey, 2012)

Organisation	Criterion	Inflows	Initial Reservoir level
Anglian Water	50% volume reduction in 10 days for impounding reservoirs (20 days for non-impounding reservoirs or reservoirs that are large compared with their catchment)	None	full

The target levels referred to are not necessarily the level at which danger posed by an emergency situation would be largely averted - this might be regarded as about half depth (quarter load), although it would depend on circumstances.

Overall, these criteria give a very wide range of times to achieve draw-down to depth. For drawdown to half-load (70% depth) or half volume, the range of times for a highest consequence dam (Category A1) ranges from a lowest of three days (Table 2 - British Waterways with weekly surveillance frequency) through to seven days (California smaller dams), eight days (France), 10 days (Anglian Water, published 2012), to "10 to 20 days" (USBR). It is recognised that the criteria are inevitably arbitrary although elements are based on rational principles.

The study found that for typical shape valley impoundments, the depths for half-load (70% depth) and half volume adopted in criteria were broadly similar in most cases, except at those reservoirs formed by raising of natural lochs.

Figure 1. Scour discharge (one of twin discharge pipes at Upper Glendevon Reservoir)

CRITERIA FOR DRAW-DOWN AND FACTORS INFLUENCING REQUIREMENTS

The starting point for deriving a set of criteria applicable to Scottish Water's stock of dams is the British Waterways' standard, that is the most onerous and most specifically risk-based of the criteria given in the Defra Guide. For weekly surveillance, this is a target of 3, 5 and 7 days for dams in categories A1, A2 and B-C-D respectively. Account is also taken of the Hinks' criterion for initial lowering.

Modifications are made to this basic set of criteria to take account of factors influencing the required draw-down rate. The Defra Guide states "*A theoretical assessment of the required draw-down capacity would take into account the factors [in Table 4.11], which would need to be established on a dam specific basis. ...*". The factors given are summarised below, with a note on how these have been taken into account to reflect the circumstances at SW's reservoirs.

The frequency of surveillance, i.e. how long could a problem develop before being noticed?
The frequency of surveillance is considered specifically in the British Waterways criteria. These are based on surveillance frequencies of once or twice weekly, with an increase in target draw-down times of two days allowed for the latter. This aspect has been incorporated in the proposed criteria by addition of two days in the drawdown requirements where surveillance visits are made twice or more per week.

How fast could the reservoir fail? And thus how fast does the reservoir load have to be reduced to avert failure?
These questions are fundamental to setting of criteria for draw-down along with the matter of frequency of surveillance. The British Waterways standard used as a starting point for this study is given in the series of three papers by Hinks, Brown & Brown (2009) that include assessment of rate of failure and associated risk. For this study specific factors, including type and age of dam that may relate to speed of failure, are addressed below under degree of intrinsic safety.

Inflows to the reservoir (base flows and flood flows) from both direct and indirect catchments, and whether direct or indirect inflows can be reduced, terminated or reversed.
Assessment of draw-down rate is made assuming a constant Q_{10} inflow from the catchment, and account is taken of any known installed means of diverting or turning out inflows from all or part of the direct or indirect catchment. Wet weather could be a factor in an incident requiring draw-down, although the Defra Guide recognises that "*at times of high inflow it*

may not be possible to lower the lake, or indeed hold it down if already lowered".

Various definitions of concurrent inflow for moderately wet conditions are referred to in the Defra Guide, but the assessments made for this study show that the choice of definition does not usually have much effect on the outcome.

Once the situation has been stabilized by lowering the reservoir, can inflows be controlled to keep the reservoir between defined draw-down levels?
This is not addressed specifically in the study beyond account being taken of constant Q_{10} inflow. It is particularly relevant to cases where outlets are limited or non-existent (Mann & Mackay, 2009) or draw only from upper levels, and needs to be considered on a case by case basis.

Figure 2. Emergency pumping at an incident at a privately owned reservoir

The consequences of failure defined by Overall Consequence Class.
Consequence class is considered specifically in the British Waterways' criteria that is the starting point for the proposed draft standards. The consequence class adopted in this study at each reservoir has been defined by taking the higher of that assigned by the previous Inspecting Engineer and that derived from a previous risk assessment carried out by SW to determine the inundated areas and the Consequence Score by the methodology in CIRIA C542.

For dams of Category C and D that are not essential for security of public water supply, a further relaxation is applied to the target period as the consequences are particularly small.

FURTHER FACTORS CONSIDERED
Further factors apparently affecting draw-down requirements have also been considered for this study. These are itemised below, again with a note on how these have been taken into account in deriving draw-down criteria

The degree of intrinsic safety (or likelihood of failure) of the dam.
As stated above, dam type and dam age are included as factors relevant to intrinsic safety, and corresponding relaxations are made to the draw-down targets.

The British Waterways'(BW) criteria reflect their portfolio of reservoirs (averaging almost a century older than the UK average age) dating from an era of less advanced understanding of principles of design and construction, and having had numerous problems and failures in the early years (Brown, 2009). Therefore the BW standard is applied in full to reservoirs built before 1840, as these are from an era when rational design methods were not well developed.

For reservoirs constructed after 1840, the BW standard is relaxed by one level – i.e. addition of two days to the BW target periods. For dams with weekly surveillance, this would require five and seven days for Category A1 and A2 respectively. Comparison with the other criteria shows that for dams at Category A2 level this is comparable with the California (Small dams) criterion and within all the others. For Category B and lower, this sets standards slightly lower than the Californian (Small dams) and the French criteria, and this appears appropriate in relation to the category.

For concrete dams founded on rock, and for earthfill dams built after rational methods of filter design (about 1960), the BW standard can be reduced by a further level. For dams with weekly surveillance, this would require seven and nine days for Category A1 and A2 respectively. Comparison with the published criteria shows that for dams at Category A1 level this is comparable with the California (Small dams) criterion and within all the others.

Design precedent at large capacity reservoirs.
Many large capacity reservoirs give draw-down times in the upper part of the range identified above. Their outlet works may have been designed to USBR or similar drawdown standards. Many of these are dams of modern design by rational methods that include a greater degree of intrinsic safety. (It is noted that the California criteria in Table 4.9 of the Guide apply a less strict criterion for larger impounded volumes exceeding an arbitrary threshold of 6.2Mm^3, on the logic that larger dams are more thoroughly designed and constructed).

Therefore such dams above a nominal threshold of 6Mm^3 are identified as a separate category where a lower "minimum" criterion can apply. This represents a significant concession to pragmatic considerations, and should only be adopted in situations where it appears to be justified on the ALARP principle (as Low as Reasonably Practicable) for risk.

The apparent potential for other emergency measures to avert failure
Alternative emergency intervention such as controlled breaching, strengthening or possibly coffer-damming appears to be more feasible options at low raisings of natural lochs. Draw-down provisions at such impoundments also tend to fall short of many or all of the range of criteria, as the impounded volume is high in relation to the low head and corresponding low discharge capacity of conventional outlet pipes. Therefore for reservoirs formed by raising a natural loch by less than about 3m, the criterion for emergency draw-down can be further relaxed and it appears appropriate to adopt the USBR criterion as a standard.

For raisings of very large natural lochs (Lochs Lomond and Katrine being particularly large) the circumstances are such that draw-down within the normal range of criteria is impracticable and a separate approach is justified.

To ensure that some fixed provision is available for immediate draw-down,
This is covered by the Hinks' criterion (2009) that excludes recourse to pumps or siphons for the initial draw-down rate, and also by adopting the recommendation in the Defra Guide that the permanent installation should never be less than 50% of the specified capacity. This ensures that draw-down can be achieved without the delay, cost and logistical drawbacks involved in deployment of pumps or siphons.

Hinks also states that *"where the reservoir has about 50% or more of the desirable capacity it is suggested that the deficit can sometimes be made up by temporary pumps subject to certain practical considerations"*. The main consideration for Scottish Water reservoirs would be how quickly and easily pumps could be deployed. This relaxation could apply for some reservoirs, but has not been applied for this study at this stage.

General and site-specific limitations on the capacity of supplementary pumps or siphons, and the rate at which they can be deployed
Practical considerations of mobilising and operating supplementary pumping and their performance have featured in recent UK publications, and some direct experience has also been gained by AECOM and SW staff.

Pump capacities adopted for this review are those given in a paper Windsor D M (2012) from actual trials (Table 4). The paper notes that these rates can be assumed to apply to suction or submersible pumps.

Table 4. Pump capacities for a "typical" reservoir with a maximum total head of up to 6m (from Windsor, 2012)

Pump size	Capacity
150 mm (6 in) submersible pump (200 mm (8 in) suction)	0.055 m^3/s
200 mm (8 in) submersible pump (250 mm (10 in) suction)	0.097 m^3/s
250 mm (10 in) submersible pump (300 mm (12 in) suction)	0.250 m^3/s

In practice there is a limit to the number and size of imported pumps or siphons that can realistically be installed at any site. Experience shows substantial installations are required to achieve a total pumped capacity of 0.58m³/s using three 300mm submersible pumps at Birkenburn (WET News Sept 2010 p18) and 0.28m³/s using five 150 mm suction pumps at Pebley (Windsor's paper).

Figure 3. Siphon trial

For the initial study, a pump capacity equal to the capacity of the existing permanent installation has been adopted, up to a practical maximum installed of 1.5m³/s, and one day per 0.5m³/s capacity is allowed for mobilisation and installation.

DRAW-DOWN CRITERIA PROPOSED FOR ADOPTION

The proposed draft criteria for emergency draw-down require both an appropriate "Initial" rate using fixed provisions only, and "Further" draw-down (allowing supplementary means) within an appropriate period to a target level.

The "Initial" rate adopted for lowering in the first 24hrs is the Hinks' criterion as a minimum, although this may not be applicable at low raisings of natural lochs, and a higher initial rate of up to 1m may be appropriate as a "best standard" at high category dams.

The "Further" draw-down target adopted is based on the most onerous and most specifically risk-based of the criteria, that is the British Waterways' standard, extended by considering relaxations with regard to specific aspects in terms of age and type of dam and impoundment as described above, to derive appropriate adjusted targets. The proposed draft criteria for emergency draw-down are set out in Table 5.

Table 5: Proposed draft criteria for Scottish Water emergency draw-down

Overall Conse-quence Class	"Initial" lowering in first 24 hrs under Q_{10} inflow (Not necessarily applicable to raising of natural lochs by less than about 3m)		"Further" draw-down to depth - Number of days to lower the reservoir to 50% of volume or 70% depth (half load)^, with Q_{10} inflow Weekly surveillance *				
	Minimum standard $	Best standard to apply, where feasible	Embankment pre-1840	Embankment post-1840	Embankment post-1960; masonry or concrete on rock	Raising of natural loch by less than c. 3m	Minimum standard for dams with volume > 6Mm³ $
A1	Hinks@	0.6 to 1.0m	3	5	7	10-20	14
A2	Hinks@	0.6 to 1.0m	5	7	9	10-20	14
B (all), C, D (essential for supply)	Hinks@	Hinks@	7	9	11	20-30	21
C, D not essential for supply	n/a	150mm per day	9	11	14	60-90	60-90 (probably n/a)

Notes:
@ Hinks' formula is for initial lowering of 300 plus 5H mm in the first 24 hours under Q_{10} inflow, without pumps (where H is the height of the dam in m).
^ The target should ideally be to achieve both half volume and half load. If one but not the other is met, this should be regarded as marginal.
* For surveillance frequency increased to twice weekly or more, the number of days can be increased by two except in the case of the Minimum standard.
$ Minimum standard allowable post 1960 or known high standard where precedent and practical considerations are significant.

If upgrades are being made to outlets, it may be appropriate to provide more capacity than meets the criteria, where the additional cost is not excessive.

These draft criteria are the outcome of a study to derive consistent, informed and risk-based targets from the wide range of standards published. Nevertheless, the desirable draw-down capacity at any dam is the responsibility of the appointed engineer under the Act to determine taking account of all relevant factors, in discussion with the Undertaker where appropriate.

INITIAL ASSESSMENT OF EXISTING DRAW-DOWN CAPABILITY
A simplified initial review of existing draw-down capability has been made, based on limited data. Following this initial review, a further assessment of draw-down capability is in hand to understand the extent of upgrading required to meet the proposed criteria.

ACKNOWLEDGEMENTS
The authors acknowledge the support given by Scottish Water (including detailed review of the study report by Gamini Karunaratne, Reservoirs Engineer, SW) and Aecom in preparing this paper.

REFERENCES
ICE (1996). *Floods and Reservoir Safety 3rd Edition.* Thomas Telford Ltd, London

Defra (2002). *Reservoirs Act 1975: Safety of embankment dams.* Letter Ref WS 194/24/10 dated 4 Dec 2002 to Managing Directors of water companies in England

Defra/Jacobs Babtie (2006) *Engineering Guide to Emergency Planning for UK Reservoirs*, Defra, London

Hinks J, Brown A and Brown D (2009) Low level outlets (papers 1 to 3). *Dams & Reservoirs Vol 19 No.1,* Thomas Telford Ltd, London

Mann R and Mackay A, Maich dam overtopping emergency, Renfrewshire – engineering and emergency response aspects, *Dams & Reservoirs Vol 19 No.1.* Thomas Telford Ltd, London

ITT Water & Wastewater (Sept 2010). *Float the pumps, boys.* WET News, Sept 2010, p 18

Tam & Humphrey (2012). Improving Anglian Water's emergency response for reservoir safety. *Proceedings of the 17th Biennial BDS Conference BDS, Leeds, pp 474-485.* ICE Publishing, London

Windsor D M (2012). Pebley Reservoir (Derbyshire) Emergency Drawdown Exercise. *Proceedings of the 17th Biennial BDS Conference BDS, Leeds, pp 474-485.* ICE Publishing, London

Maintaining the Safety of our Dams and Reservoirs
ISBN 978-0-7277-6034-0

ICE Publishing: All rights reserved
doi: 10.1680/mdam.60340.283

Public Safety at Dams – A Canadian Perspective

G J SAUNDERS, Revelstoke Design Services Limited, Winnipeg, Manitoba Canada

SYNOPSIS   Following several highly publicised events that resulted in loss of life at dams, the Canadian Dam Association, in 1995, embarked on an ambitious effort to develop a set of guidelines to assist dam owners and operators in understanding the principles of dam safety and the process and management of these principles to protect the public, not only from a dam failure, but from injury and/or death due to the public's close proximity to these facilities.

This paper will focus on the implementation of a public safety program at several of BC Hydro's dams in western Canada in 2013 and efforts undertaken to minimize risk to the public of hazards upstream and in the tailrace and spillway areas of a dam. Working together with BC Hydro, Revelstoke Design Services Limited conducted detailed site assessments followed by recommendations for the installation of public safety control measures in critically hazardous areas.

INTRODUCTION
Canada is abundant in natural resources having over 14,000 dams, of which in excess of 900 are considered large dams. Unlike other countries Canada does not have a federal regulatory agency guiding the development of safe management of these dams; this responsibility is one of the many regulatory functions given to the individual provinces.

The "Canadian Dam Association" (the Association) was formed in the 1980s by volunteers to provide owners and operators a forum to discuss the safe operation and management of dams. In the early 1990s several highly publicised deaths by drowning occurred in the hazardous waters created upstream and downstream of these dams.

This paper presents a brief history of the efforts and influence the Association has made to assist in reducing the public's risk around dams; a historical review of Canadian drowning statistics, and how Revelstoke Design Services Limited (the consultant) relied on these guidelines, specifically the Association's "Booms and Buoys for Public Safety Around

Dams" to reduce the public risk around several high profile dams owned by BC Hydro in British Columbia, Canada.

BRIEF HISTORY OF THE ASSOCIATION

In its infancy the Association saw a need for the development of a set of guidelines that could be used by provincial regulators and dam owners to assist in the operations and management of these facilities to protect the public from their hazards. In 1999 the Association published a "Dam Safety Guideline" which provided owners guidance into the standards and practices among professional engineers intended to cover the majority of the requirements for dam safety. These guidelines were revised and republished in 2007.

There has been three recorded dam failures in Canadian history resulting in the loss of life; the first in British Columbia in 1912 with the loss of one life; the second in New Brunswick in 1923 with the loss of one life, and the third in Quebec in 1966 with the loss of three lives. There have been other failures but none involving loss of life. In Canada more people have died in accidents around dam sites that are not associated with their structural failure. This was and continues to be a matter of concern to dam owners.

The Association has focussed more recently on the risks of the public's legislated right of recreation around dams which resulted in the 2011 publication "Guidelines for Public Safety Around Dams" and three technical bulletins which include: "Audible and Visual Signals for Public Safety Around Dams"; "Signage for Public Safety Around Dams"; and "Booms and Buoys for Public Safety Around Dams".

HISTORICAL DROWNING STATISTICS IN CANADA

The average number of water related drowning deaths in Canada averages approximately 550 per year. Between 1990 and 2010 the number of yearly drowning deaths has been on a downward trend from 683 to 483. Approximately 60% of these preventable drowning deaths result from recreational activities such as boating, fishing, and swimming on Canadian lakes, ponds, rivers, and streams.

In 2005 the National Lifesaving Society of Canada published statistics of "Drowning Trends from Artificial Aquatic Sources" which reported that between 1995 and 2005, there were 1,057 drowning deaths resulting from artificial aquatic sources (bathtubs, pools, canals, quarries etc.) of which 27 resulted from accidents around dams.

More recently a 2013 review of Canadian media reports of public safety incidents around dams found five fatalities, two of which occurred at low head dams/weirs, and five incidents which required emergency medical service rescue.

It is obvious, although there is a general downward trend statistically in the number of drowning deaths, 2013 events illustrate the importance of continuing to be proactive as owner/operators with respect to public safety around their facilities.

It is worth noting that although large dams command respect, drowning statistics in Canada show low head and small diversion dams can be equally or even more hazardous to the public.

GUIDELINES FOR PUBLIC SAFETY AROUND DAMS (2011)

The Association in 2011 published a set of guidelines that address the hazardous risks the public can encounter during recreational activities around these dams. The guidelines outline a "Management System" and the elements it should include to provide public safety at dam sites. The main elements of a successful management system include:

- The setting of policies, objectives, and requirements
- Assessment of the risks and plan corrective measures
- Implementation of a Public Safety Plan for each site
- Monitoring and evaluation of the plan
- Constantly audit, review, and act as necessary

To make the "Public Safety Management System" a success the policies, objectives, and requirements of the system must be developed and supported by the most senior levels of management in the company. It is important to understand and to incorporate into these policies any legal and legislative requirements the company may have. For example, the Navigable Waters Act of Canada legislates that it is illegal for any company to restrict the public's water access to the site without sufficient justification and approval even if the company owns the shores surrounding the site.

Senior management should develop the risk tolerance criteria to allow the categorisation of hazards at each site according to what risk the company is willing to accept. For example the risk tolerance of a dam in an urban setting would be significantly lower than a dam located 400 miles from any urban setting with limited access, requiring a much more stringent public safety programme.

The Public Safety Management System should be written and detailed to allow the implementation of a Public Safety Plan for every site. Site risks should be reduced or eliminated through the implementation/installation of control measures at the site. Any safety plan should include for training and ways to promote public awareness of the hazards.

The management system requires continuous monitoring, review, and evaluation once it has been implemented to ensure its effectiveness. If corrective or preventive measures are required this allows for a proactive approach.

Attached to the Association's "Guidelines for Public Safety Around Dams (2011)" are three technical bulletins: "Booms and Buoys for Public Safety Around Dams"; Audible and Visual Signal for Public Safety Around Dams"; and "Signage for Public Safety Around Dams".

These guidelines and the technical bulletin "Booms and Buoys for Public Safety Around Dams" were used by the consultant in its recommendation and design of public safety booms at ten of the owner's sites.

The publishing of the guidelines has created a current Canadian industry "standard of care" around dams that owners as a minimum are expected to follow. Historical evidence suggests owner liability will increase if the guidelines have not been implemented.

APPLICATION OF GUIDELINES
Early in 2013 the owner, as part of its Public Safety Management System and through a risk tolerance review, monitoring, and evaluation targeted ten sites that required remedial work to reduce the hazards in order to improve the public's safety. All sites were in close proximity to populated centres and used routinely by members of the public for recreational purposes including hiking, fishing, boating, kayaking, canoeing, surfing, and swimming. The consultant was contracted to work alongside the owner to survey and evaluate the sites, provide recommendations, and finally design the required remedial work at the sites to ensure conformance to the owner's management system and Association guidelines

The consultant's task was specific to water hazards created by the dam but required coordination with and involvement with other control measures such as "Signage For Public Safety Around Dams".

The contract was implemented by the consultant with reference to the Guidelines for Public Safety Around Dams 2011. The remainder of the paper describes in detail the methodology followed.

Set Policies, Objectives, and Requirements
The owner had previously completed this phase of their Public Safety Management System and had established and implemented a Public Safety Plan for each site visited. During the "Monitoring and Evaluation" stage by the owner ten sites were determined not to conform to the 2011 Association guidelines particularly in terms of signage, buoys, and booms.

Assessment of Risks
In order to assess the risks at each site, each location was visited and surveyed with the aid of a risk assessment worksheet developed by the

Association. Upstream and downstream hazards were identified; existing control measures and their effectiveness were recorded including location of fencing, signs, buoys, and booms; reservoir, tailrace, and spillway flow characteristics were observed; and potential public access points were located including boat launches, portages, parks, campgrounds, and swim areas. The non-existence or lack of control measures was also noted and recorded.

Also noted or requested from the owner/operator of the facility during the site visit were local environmental data including water flow, wind, wave, debris, ice, and water depth, minimum, and maximum elevation data. This information is required for any boom or buoy design.

The observation and recording of water flow patterns and current velocities was found to be most important. This information was critical during the design of the control measures to ensure proper results. For instance, a boom placed in an area of high velocity can become as much a hazard as the hazard itself. Figure 1 illustrates the salvage of a workboat capsized while stranded against an upstream fish guidance boom shown in the image. The water velocity at the boom was estimated at 1.37m/s.

Figure 1. Retrieval of capsized boat from fish guidance boom

In 2011 a young kayaker was stranded against the safety boom shown in Figure 2. The boom was perpendicular to the flow, so not conducive to self-rescue and located in a narrowing with the water velocity greater than 1.83m/s. The kayaker was unable to hold on to the boom due to the water force and entered the spillway.

Figure 2. Boom in reservoir narrowing subject to high water velocity

On completion of the site visit a drawing of the dam, and its upstream and downstream areas was created identifying the hazards, and existing control measures including signs, fencing, buoys, and booms. Figure 3 summarises the results of one of the ten site assessments.

The assessment found this site to be sufficiently fenced to prevent public access with "No Trespassing" signs posted at least every third fence section. The fencing was in good repair with locked vehicular access gates for owner access. Both upstream and downstream areas were fenced in, and a fence on the crown of the dam prevented hikers or disembarked boaters from accessing the downward slope of the dam and spillway gate structure. Signage was limited and not in conformance to the guidelines. There was no signage upstream or downstream.

A timber boom in poor repair was situated immediately in front of the spillway and gate structure and a smaller one in front of the low level outlet. The close proximity of the safety boom to the weir was a concern. The booms did not conform to the guideline's criteria for a public safety boom.

From the assessment it was determined a number of hazards above the owner's acceptable level of risk for the site remained and required the installation of a number of control measures. During low water the public could access the spillway structure around the end of the fence and the earth dam by crossing the log boom on foot or by swimming. Boaters and swimmers could also access the low level outlet tower to climb up and dive from. Motocross bikers and skate boarders were able to and had been accessing the concrete "half pipe" spillway structure from the Provincial

Park by climbing and jumping the weir. Figures 4 and 5 are pictures of the weir and low level outlet tower respectively.

Figure 3. Site assessment hazard map

Figure 4. Timber log boom in front of weir

Figure 5. Low level outlet tower and timber log boom

Recommendations made to the owner specific to this site included an increase in signage more consistent with a high use recreational area; the installation of a fence on the weir to prevent spillway access by motocross bikes and skate boarders; and the replacement of the existing timber booms with booms meeting the requirements of the guidelines.

Implementation of Recommendations

To conform to the Guidelines for Public Safety around Dams the owner was required to install additional signage at this site. Samples of appropriate signage are illustrated in the Technical Bulletin on Signs which is available from the Canadian Dam Association. Detailed signage requirements are not covered in this paper.

The consultant provided recommendations to replace the existing timber booms with a boom of higher visibility which included embedded wording to warn of the dangers associated with the spillway weir and gate, and the low level outlet tower. Safety booms and buoys when properly installed are an effective measure of identifying hazards and discouraging access by the public to the hazardous areas around dams. If coordinated with shoreline fencing and signage their effectiveness is increased.

The guidelines state that a safety boom is to provide a visual warning of a dangerous waterway; a physical barrier to catch and stop stranded boaters or swimmers preventing them from entering the hazardous waters; and a physical entity to assist in self-rescue. Safety booms should provide assistance to a stranded boater or swimmer in getting to safety due to their orientation to flow.

Safety boom requirements to conform to the guidelines include:

- Yellow in colour as regulated by Transport Canada
- Resistant to ultraviolet light
- Floats to have at least 305mm of freeboard in still water conditions
- Floats to be a minimum of 600mm long
- Floats to be foam-filled to prevent sinking if punctured
- The distance between floats a maximum of three metres to aid in preventing the passage of a stranded boat
- Situated if possible in calmer water at sufficient distance from the structure for safety and to prevent interference with the structure's operations should it fail
- Placed so the ends of the boom coincide with the location of the large "Dangerous Water" signs on shore to better delineate the area.
- Be in a configuration to promote self-rescue

The two most common boom shapes which promote self-rescue are shown in Figure 6 below.

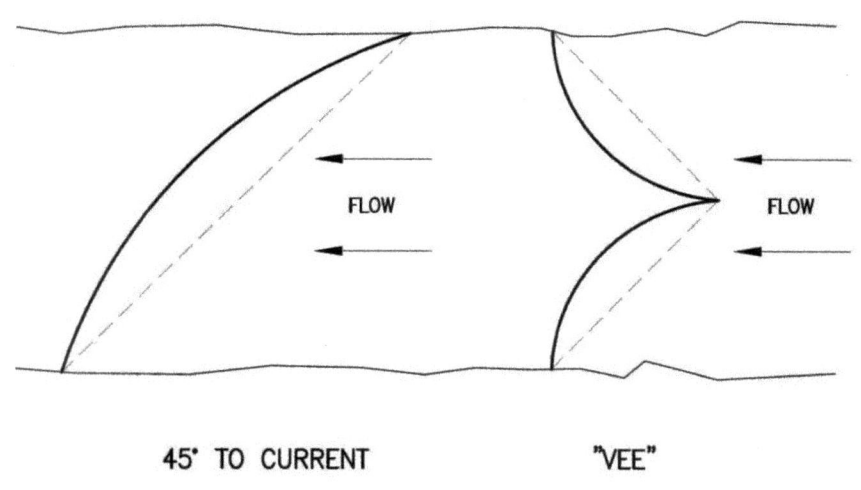

Figure 6. Boom configurations with good self-rescue characteristics

A safety boom design and specification was developed for the site. It was recommended to relocate the spillway boom further away from the weir to coincide with the shoreline fence in order to provide increased effectiveness. Locating the boom further from the weir would also lower water velocities at the boom and increase the likelihood of self-rescue.

To relocate the boom from its current position, because of the close proximity to the provincial park and the owner's property limits, required approvals from the province and Navigable Waters Canada. Approvals can be a time consuming process and due to the high level of public interaction at this site it was decided by the owner to install the boom at the existing location until approvals could be obtained. It was argued a boom with higher visibility would serve, even in the current location, to increase public awareness to the inherent dangers at the site.

The boom specified for the site was constructed of floats meeting the Association guidelines as illustrated in Figure 7.

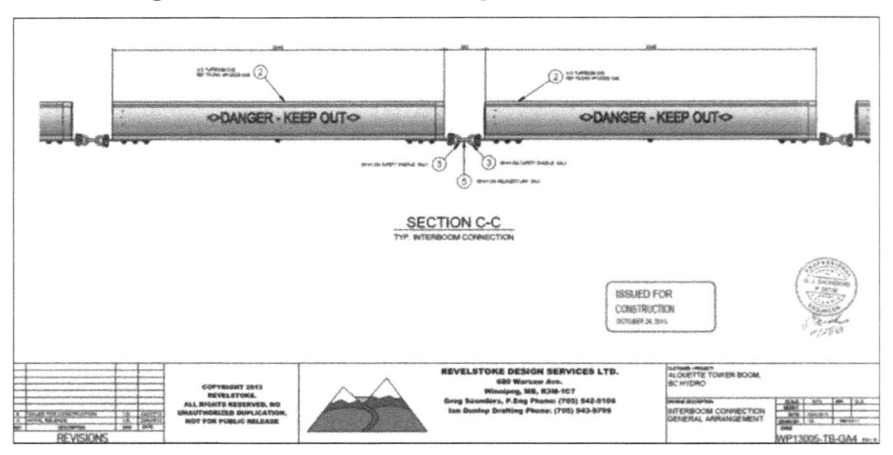

Figure 7. Boom specified to meet association guidelines

Figure 8. Replacement booms upstream of the spillway weir

The boom line in front of the spillway was configured in a "Vee" shape to promote self-rescue. The requirement to have a "Vee" around the low level outlet tower was determined unnecessary because of the extremely low water velocity. Figures 8 and 9 illustrate the safety booms at near completion. A number of individual booms sticks were to be removed to eliminate excess sag in the boom in order to further the boom's distance from the hazard.

Figure 9. Replacement booms around the low level outlet tower

A fence above the weir to prevent motocross bikes from accessing the spillway is also being installed. This and the increase in signage are planned for this spring.

Safety Plan Update
To complete the process, the consultant provided the owner with the material to update the site's safety plan including risk assessment worksheets, as-built drawings, maintenance requirements, design notes and parameters, and any further recommendations that would increase the publics' safety around the sites.

This procedure was completed at all ten sites the owner deemed a priority due to the sites risk tolerance.

CONCLUSION
Public awareness of the hazards around dam structures in Canada has increased significantly since the early 1990s. The Canadian Dam

Association has been an industry leader and significant force in providing guidance to owner/operators and consultants with respect to public safety around dams. The guidelines developed including the technical bulletins have become the industry standard and owners will be viewed in terms of these guidelines should an incident occur on their site.

Despite the efforts of the owners and the Association, members of the public continue to access these hazardous areas resulting in incidents and/or death. However, if these efforts save the life of even one individual, it is the author's opinion that they are well worth the emphasis, and the goal of eliminating these incidents seems attainable.

The consultant utilized the Association guidelines to make recommendations and design effective corrective measures at ten dams in British Columbia owned by BC Hydro. These sites today have an increased level of public safety to one year ago and upon completion of the remaining corrective measures this spring will conform to the current "Industry Standard" and latest set of guidelines concerning public safety around dams.

REFERENCES

Canadian Dam Association (1999). *Dam Safety Guidelines.* Toronto, Ontario, Canada

Canadian Dam Association (2011). *Guideline for Public Safety Around Dams.* Toronto, Ontario, Canada

Canadian Dam Association (2011). *Technical Bulletin, Signage for Public Safety Around Dams.* Toronto, Ontario, Canada

Canadian Dam Association (2011). *Technical Bulletin, Booms and Buoys for Public Safety Around Dams.* Toronto, Ontario, Canada

Maintaining the Safety of our Dams and Reservoirs
ISBN 978-0-7277-6034-0

ICE Publishing: All rights reserved
doi: 10.1680/mdam.60340.295

Enhancements in reservoir flood risk mapping: example application for Ulley

A D SMITH, University of Southampton, UK.
C A GOFF, HR Wallingford, UK.
M PANZERI, HR Wallingford, UK.

SYNOPSIS In July 2007, at Ulley Reservoir, South Yorkshire, a catastrophic dam failure was narrowly averted due to emergency preventative actions. During the event a number of homes were evacuated and roads were closed for precautionary measures. Within very close proximity of the reservoir lies the town of Rotherham, the busy M1 motorway and a trunk freight railway line. The incident highlights the need for detailed flood risk and hazard modelling to improve management of the risk and provide better incident planning.

Hazards and population vary in both time and space, but when traditionally modelling flood risk, the population is invariably located within the residential housing stock. This paper innovatively combines flood inundation and spatio-temporal population modelling for better estimates of the population potentially at risk. This is demonstrated though application to Ulley for the most probable worst case failure scenario should the preventative measures not have been undertaken and the dam have failed.

This paper proposes an enhanced flood risk assessment in three stages: (i) probabilistic modelling of a failure scenario using embankment breach models; (ii) hydrodynamic inundation modelling for assessment of flood water spreading, depths and velocities; (iii) spatio-temporal population modelling to assess the risk to the population likely to be present. The combination with spatio-temporal population outputs aims to demonstrate the enhancements achievable in reservoir flood risk mapping when vulnerable populations are concerned.

INTRODUCTION
The number of people potentially at risk during rapid onset flood events, such as dam failures, varies at a range of temporal scales. Traditionally decadal censuses alone are linked to residential housing datasets and therefore consider a static 'night-time' residential population estimate. Additional approaches are required to assess the impact on people. For

example, the same flood occurring in an urban centre during a weekday afternoon may have a greater effect on the temporarily present population (e.g. workforce, children concentrated at school sites) than in the evening. Dam failure events can occur with little warning and rapid onset times. This may result in devastating catastrophes in downstream areas (He et al, 2008). The risk from such events remains high in locations with significant potential for severe losses. In these events human susceptibility and key infrastructural assets heighten vulnerability and the risk posed from sudden dam failures.

This paper aims to demonstrate one example of an enhanced method to assess the impact of a rapid onset dam breach event in the United Kingdom (UK). The approach adopted concerns the combination of numerical modelling of an embankment scenario, resultant flood inundation extent and exposure to any population likely to be present at the time of event. It is acknowledged that flood hazard and population vary in both time and space at a range of scales (Aubrecht et al, 2012). The example application provided combines flood spreading and dam breach models developed at HR Wallingford with the 'SurfaceBuilder24/7' spatio-temporal modelling tool to estimate gridded population densities at a range of times. This is demonstrated on an evaluation of a scenario for the failure at Ulley Reservoir in South Yorkshire, a nineteenth century clay-earth embankment dam that is believed to have been close to failure during severe UK-wide flooding of June 2007.

BACKGROUND
Ulley reservoir is located three miles south-east of Rotherham and five miles east of Sheffield, Yorkshire, UK. It is presently a country park, owned by Rotherham Metropolitan Borough Council (RMBC). Construction of the earth embankment dam was completed in 1873. The supply of drinking water from Ulley ceased in 1986 when it was taken over by RMBC as a recreational facility. During exceptional widespread flooding experienced in the UK during summer 2007 the dam was destabilized. Although the modern spillway coped admirably with the high overflows, the older masonry spillway along the left mitre of the embankment suffered from out-of-bank flows and deterioration of the channel. This in turn led to a large erosion hole developing in the downstream face of the embankment, putting the stability of the entire dam at risk.

Historical context
This region surrounding Ulley is not immune from unprecedented, catastrophic dam failures. The collapse of Dale Dyke Dam in 1864 (13 miles west of Ulley) caused the Great Sheffield Flood, resulting in considerable downstream destruction and 244 fatalities. The dam collapsed

under severe storm conditions while being filled for the first time. The breach resulted in the discharge of c.3Mm³ of water (Amey 1974) into the narrow catchment below. The embankment was of the same earth/clay construction type as Ulley, which was to be constructed less than ten years later.

Increasing industrialisation and population growth within the Yorkshire region during the nineteenth century increased the demand for an adequate and clean water supply. This was driven by the increase in the cotton and steel industries and concerns over healthcare and access to safe drinking water. Poor health and intermittent water supply caused by shortages prompted the construction of the reservoir at Ulley to alleviate these concerns. The dam was constructed by Messrs, Lawson and Mausergh of Westminster between 1871 and 1874. It consists of an earth embankment and puddle clay core.

Ulley June 2007

On 25 June a slow moving depression bought prolonged heavy rainfall to northern and central England, with more than 90 mm of rain falling in 18 hours (Environment Agency 2007b; Met Office 2011). In 2007 June was the wettest for England and Wales since 1860 (Marsh and Hannaford 2007). Intense slow moving frontal rainfall on the 25 June fell on saturated ground with some rivers already exceeding capacity and reservoir levels high. It is estimated that the rainfall levels that led to this event had an annual probability of occurrence of 1% (Warren and Stewart 2008).

Flooding on the River Don at Sheffield nearby was also at its worst extent since the 1864 collapse of Dale Dyke dam (Environment Agency 2007a). The prolonged rainfall had already caused widespread flooding in this region. A potential collapse of the Ulley embankment would have been exacerbated by significant volumes of standing flood water already immediately downstream due to the excessive rainfall.

Spillway failure

The mechanics of the events leading to the risk of destabilisation at Ulley have been well documented (e.g. Hinks *et al,* 2008; Mason and Hinks 2009, 2008) and are therefore not reproduced in detail for this paper. Despite a larger concrete spillway constructed in 1943, flood water reverted to the original masonry stepped spillway in the left mitre of the main earth embankment. The flow exceeded the channel capacity and the masonry channel collapsed, resulting in a large erosion hole in the embankment (Warren and Stewart 2008). During the flood, peak flow on the failed spillway was estimated at 6.1m³/s with 4m³/s down the newer concrete spillway (Horrocks 2010). RMBC was advised to take immediate emergency action to prevent major flooding downstream (Environment Agency 2007b).

Population at risk and response
Approximately 1000 people were evacuated by the emergency services in downstream areas of the dam from the villages of Catcliffe, Whiston and Treeton. This was based on rough estimates of possible water levels should the dam have failed. The M1 motorway was closed northbound between junctions 32 and 34, and southbound between junctions 34 and 36 (Sturcke et al, 2007) for 40 hours at an estimated cost of £2.3M (Environment Agency 2007b). In addition to the population exposure there was also a substantial risk to critical national infrastructure in the form of two high pressure gas mains and a regional electrical substation. Other assets at risk during this period included telecommunication towers, highways, sewage treatment works and the M1 motorway.

Emergency work to stabilise the dam and reduce water levels continued before the motorway was reopened. The initial remedial action resulted in packing the scour hole with 2,500T of 150mm (maximum) crushed limestone. Water was pumped from the reservoir to the newer spillway channel to lower the reservoir level in addition to the discharge from the existing scour pipe. Repair of the dam cost £3.8M and resulted in the construction of improved scour pipe capacity and a new reinforced concrete spillway in the centre of the dam. The new scour pipe has twice the capacity of the previous one and can drain 40,000m^3/day, enough to lower the reservoir water level by one metre per day (Horrocks 2010).Following this incident Sir Michael Pitt's (2008) review recommended evacuation plans should be drawn up for all registered reservoirs. These were completed by the end of 2008.

MODELLING METHOD AND DATA
The modelling method consists of three main components which are discussed in turn; embankment breach, hydraulic flood spreading, and spatio-temporal population modelling. Two modelling tools developed by HR Wallingford were used for breach analysis and checking; the EMBREA (EMbankment BREch Assessment) complex model and AREBA (A Rapid Embankment Breach Assessment) simplified model (Van Damme et al, 2014). These models were used to simulate the failure mechanism of the dam and derive the resultant outflow hydrograph. The initial water level was set to 0.52m above TWL, and the failure mechanism was assumed to be erosion of the downstream toe causing a major slip on the downstream face leading to overflowing of the embankment and subsequent external erosion. Initial slope stability analysis suggests that the supporting embankment material is liable to slippage following erosion of the toe material and exposure of the core. The process was initiated during the 2007 flood event

but was fortunately prevented from worsening following emergency remedial work. Should the breach have continued, preliminary core stability calculations suggest that the core would have failed under these conditions. After block failure of part of the core due to the initial slip of supporting embankment material, breach flow causes further removal of embankment material supporting the core. With increasing exposure of the core, stresses in the core increase and subsequently give rise to two further block failures indicated by the 2nd and 3rd peak in the breach hydrograph (Figure 2a). Due to the block failures, the head driving the flow suddenly increases leading to high breach flows.

The extent of a potential inundation following a breach at Ulley was modelled using the open source TELEMAC-2D hydraulic model. The breach hydrograph was used for the reservoir discharge parameter. A two-metre LiDAR digital elevation model (DEM) (Environment Agency 2013) was used to generate a mesh for the model input. Depths for the existing downstream flood extent were estimated from aerial photographs taken during the emergency response and evaluated using Ordnance Survey spot heights. This was accounted for in the modelled scenario to simulate realistic conditions should a breach have occurred within the already inundated catchment. Culverts through notable barriers downstream such as the railway and motorway embankments were accounted for in the DEM. The modelled outputs for the spreading for a breach event concerning water depth and velocity at a 15 metre resolution were analysed using a Geographic Information System (GIS).

Finally, spatio-temporal population modelling was undertaken for an 8 x 10 km study area centred on Ulley and Rotherham using the 'SurfaceBuilder24/7' tool (Martin 2011). This employs a variable kernel density estimation technique with a distance decay function. This facilitates the spatial redistribution of population datasets in space and time based on centroid locations and ancillary datasets. A population centroid is a georeferenced point with an associated population count. The model utilises 'origin' centroids taken from the UK census and georeferenced residential postcode locations. The model redistributes population from these to 'destination' centroids (such as schools and places of work) based on their location and site capacities informed by administrative datasets (e.g. school census, census workplace data). The proportion of the available capacity occupied at a destination centroid varies by time of day and is governed by a site-specific time profile. A destination example would be location and number of pupils on the roll at a school (informed by the school census) who are present during school hours on a term-time weekday. The school aged population is then drawn from the surrounding origin centroids within the school's catchment area to fulfil the destination's expected capacity.

Mid-year population estimates for 2007 were used as the baseline residential population for the creation of the data library for this paper.

A background mask is also utilised to constrain population allocation to habitable locations (e.g. excluding water bodies) and to represent the road transport network. The population on the road network also varies by time of day and is informed by the distribution of vehicle count data and capacities from the Department for Transport's (DfT) National Transport Model. The modelled population output is in the form of a rasterised regular grid at 100m resolution, based on the current resolution of available input data. The output is disaggregated and adjusted for a 15m resolution to match the output from TELEMAC-2D. The gridded results are analysed at the output resolution comparing water depths, velocity and population for each cell. This has been used to calculate a hazard rating and fatality estimate for each cell based on the method outlined by Penning-Rowsell *et al*, (2005).

RESULTS

Water depth and velocity results derived from TELEMAC-2D for the post-breach inundation extent for Ulley Reservoir are shown in Figure 1. The output extent recognises the antecedent flood conditions. The greatest depths occur in river channels, while increased velocity occurs from the initial breach and through culverts.

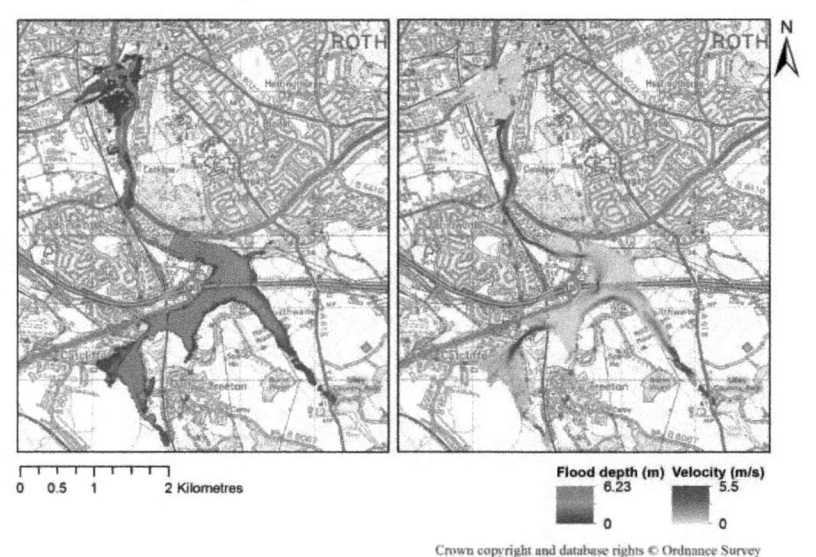

Figure 1. Flood inundation results for water depth (left) and velocity (right)

An example reservoir breach hydrograph used in the model identifies three distinct peaks (Figure 2a) representing initial overtopping followed by downstream undercutting and core failure.

depth normalised to height above ordnance datum (AOD).

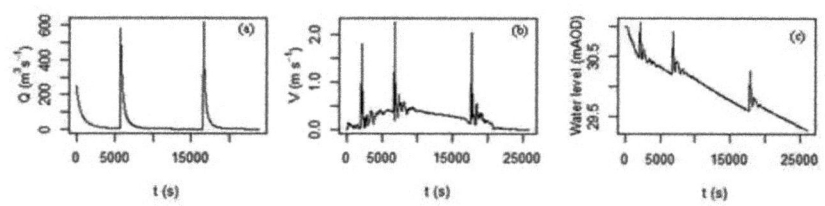

Figure 2. (a) AREBA reservoir breach discharge hydrograph, (b) velocity profile at motorway embankment, (c) flood level at motorway embankment.

A velocity time-series was taken at the motorway embankment immediately downstream from the reservoir (Figure 2b) with a profile closely aligned to the initial hydrograph. Figure 2c represents another location on the motorway embankment at its lowest elevation (at the River Rother culvert) and a time-series for water

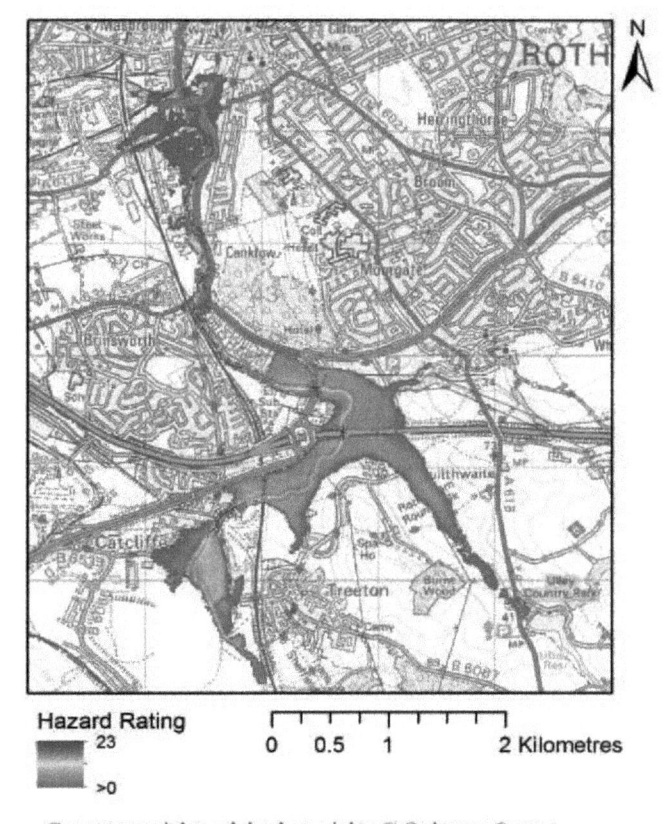

Figure 3. Hazard rating per cell for an Ulley flood scenario

The hazard rating (Figure 3) primarily identifies the locations of greatest depth and velocity. The highest rating occurs within the original channels. The population exposed also varies by time of day. Figure 4 displays the estimated population for three different times of day (00:00, 08:30 and 12:00) for a represented weekday.

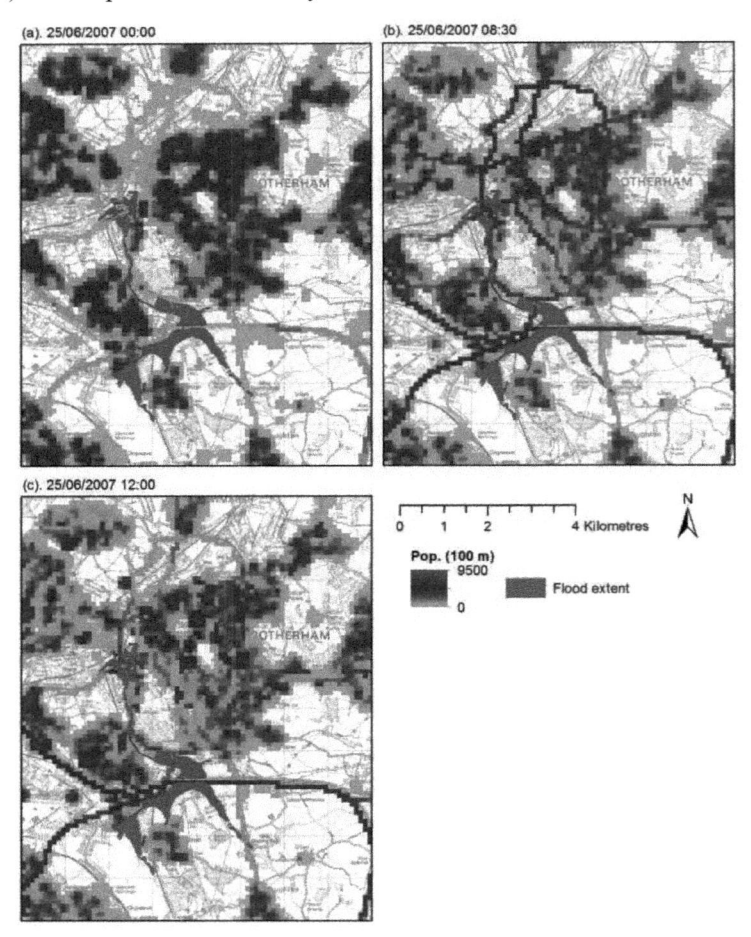

Figure 4. Gridded (100m) spatio-temporal population estimate for the Ulley region.

DISCUSSION
There are an exponential number of possible permutations for the temporal variation in the population modelled outputs (e.g. time of day, day of week, season) when combined with scenarios for flood extent, parameters and time step. Therefore this paper only attempts to demonstrate a few of the

possibilities while striving to demonstrate the enhancements in reservoir flood risk to people when considering spatio-temporal population data.

The comparison of the potential water depths adjacent to the motorway and the lowest elevation of the motorway surface (31.7mOD) according to the LiDAR data suggest that the motorway embankment may not have been overtopped. However this is based on the assumption that the culverts (or road underpasses acting as temporary culverts) are unobstructed. Nevertheless, the level of the maximum water depth for the scenario modelled indicates that the water level could have come within 0.70m of overtopping the carriageway. The depth time-series at the embankment also indicates that the breach events could add up to a metre to pre-existing flood water levels. This is consistent with the same estimation made by Mason and Hinks (2009).

The TELEMAC-2D simulations show that the leading edge of a breach flood wave could reach the motorway embankment in less than 30 minutes. If destabilisation of the embankment at Ulley had continued unchecked or rapidly deteriorated then the model simulations, and historic events, indicate there would have been no time for an effective warning. Flow velocities are highest immediately downstream, but rapidly dissipate when entering the existing standing water. The channelization effect of the culverts creates localised intensification in velocity (Figure 1).

A velocity time-series was taken adjacent to the southern edge of the motorway embankment closest to the reservoir (Figure 2b). Although flood flows have slowed significantly at this point velocities around 2m/s are still estimated. This is within the velocity threshold for masonry and concrete significant for the onset of structural damage (Priest et al. 2007). Therefore it is possible that the integrity of the motorway structure could be comprised, particularly at the locations of culverts and bridges. In turn the flood hazard rating (Figure 3) adjacent to the motorway embankment ranges from 1-1.5, high enough in places to pose a significant hazard with a danger to most people (Priest et al. 2007). It is possible that the resultant risk to the saturated earth motorway embankment (due to preceding flooding conditions) could have been unacceptably high.

Table 1. Population exposure

Time	Flood extent	Study area
00:00	633	118,937
08:30	2,155	127,714
12:00	1,608	121,995

Spatio-temporal population estimations (Figure 4) highlight the variability throughout the weekday of the incident had the onset occurred at different times. Table 1 gives the population estimates for the study area contained within Figure 4.

On the representative weekday in questions there is a large notable shift of population to the potential flood extent at 08:30 and 12:00 compared to the 00:00 'night-time' population. The peak at 08:30 is attributed to commuter flows on the road transport network. The section of motorway (M1) passing through the flood extent has an average annual daily flow of 116,000 vehicles with occupancy typically ranging from 1 to 13 (DfT 2013, 2005). The 12:00 representation (Figure 4c) illustrates the localised concentration of the temporally present population at 'day-time' destination locations.

A hazard rating was calculated (Figure 3) based on the method proposed by Penning-Rowsell et al, (2005). This was used to estimate the number of fatalities based on the area and people vulnerability. The initial result suggests that there could have been up to 137 fatalities within Catcliffe, the downstream village that was evacuated, had the event occurred with no warning and action taken. Further analysis is required assess how this fluctuates with cyclical population change. The very nature of modelling an unknown variable make this value difficult to validate and therefore it is provisional. However, this would account for 14% of the population actually evacuated had they remained behind.

The number of possible variations within a flood scenario has already been noted. This paper does not attempt to provide a single definitive answer or to even identify all possible combinations. It attempts to demonstrate one possible scenario based on an embankment failure at Ulley reservoir and provide an enhancement to considering risk to people in flood mapping. There are a number of uncertainties and external factors to consider.

CONCLUSION

This paper highlights the potential advantages with integrating spatio-temporal population estimations with established flood modelling techniques. The population component does not attempt to predict individual moments in human behaviour but rather represents predictable trends based on range of available datasets. Results suggest that the closure of the motorway and evacuation of residents was necessary and proportionate, and would have undoubtedly prevented fatalities had the dam failed. The possible impact of a failure at Ulley has been analysed and subsequent remedial work on the dam's embankment justified, within this context in potentially preventing a future catastrophic flood. Modelled assessment for a typical weekday shows that the worst time for this dam to fail is likely to have been during the morning peak commute under standard conditions. When considering worst case, but possible, failure scenario the money spent by the relevant authorities to reduce future flood risk potentially could have prevented in excess of 100 fatalities.

ACKNOWLEDGMENTS

The authors are grateful for the data and technical information kindly provided by Rotherham Metropolitan Borough Council and the Environment Agency, as well as comments on this manuscript provided by the reviewers Funded as part of the EPSRC Impact Acceleration Account through the University of Southampton.

REFERENCES

Amey, G (1974). *The Collapse of Dale Dyke Dam 1864*. London: Cassell.

Aubrecht, C, Özceylan, D, Steinnocher, K, & Freire, S (2012). Multi-level geospatial modeling of human exposure patterns and vulnerability indicators. *Natural Hazards*, 1-17 (DOI: 10.1007/s11069-11012-10389-11069), doi:10.1007/s11069-012-0389-9.

DfT (2005). National Transport Model: FORGE The Road Capacity & Costs Model. http://webarchive.nationalarchives.gov.uk/20110202223628/http://www.dft.gov.uk/pgr/economics/ntm/etheroadcapacityandcosts3031.pdf. Accessed 18/03/ 2014.

DfT (2013). Great Britain Road Traffic Survey. http://www.dft.gov.uk/traffic-counts/. Accessed 27/03/ 2013.

Environment Agency (2007a). 2007 summer floods - Environment Agency - A table showing the likelihood of the 2007 summer floods occurring at places where we measure river flows and levels. http://www.environment-agency.gov.uk/static/documents/Research/returnperiods_1918541.pdf. Accessed 13/11/ 2013.

Environment Agency (2007b). Case study 2007 summer floods: reservoir safety – learning from Ulley. http://www.environment-agency.gov.uk/static/documents/Research/reservoirscasestudy_1917484.pdf. Accessed 03/04/ 2013.

Environment Agency (2013). 2m LiDAR Data Acquisition. http://www.environment-agency.gov.uk

He, X Y, Wang, Z Y, & Huang, J C (2008). Temporal and spatial distribution of dam failure events in China. *International Journal of Sediment Research*, 23(4), 398-405, doi:http://dx.doi.org/10.1016/S1001-6279(09)60010-X.

Hinks, J L, Mason, P J, & Claydon, J R (2008). Ulley Reservoir and high velocity spillway flows In H. Hewlett (Ed.), *Ensuring reservoir safety into the future*. London: Thomas Telford.

Horrocks, J (2010). Rebuilding Ulley Dam. *Hydro Review Worldwide, September 2010*, 22-24.

Marsh, T J, & Hannaford, J (2007). *The summer 2007 fl oods in England and Wales – a hydrological appraisal.* Wallingford: Centre for Ecology & Hydrology.

Martin, D (2011). SurfaceBuilder247: User Guide. http://www.esrc.ac.uk/my-esrc/grants/RES-062-23-1811/outputs/Read/ece508b5-6438-4e96-99eb-8fd68f1d3b99. Accessed 30/01/ 2014.

Mason, P J, & Hinks, J L (2008). Security of stepped masonry spillways: lessons from Ulley dam. *Dams and Reservoirs, 18*, 5-8

Mason, P J, & Hinks, J L (2009). Conclusions from the post-incident review for Ulley dam. *Dams and Reservoirs, 19*, 43-44

Met Office (2011). Flooding — Summer 2007. http://www.metoffice.gov.uk/about-us/who/how/case-studies/summer-2007. Accessed 03/04/ 2013.

Penning-Rowsell, E, Floyd, P, Ramsbottom, D, & Surendran, S (2005). Estimating Injury and Loss of Life in Floods: A Deterministic Framework. *Natural Hazards, 36*(1-2), 43-64, doi:10.1007/s11069-004-4538-7.

Pitt, M. (2008). *The Pitt review: Learning lessons from the 2007 floods.* London: Cabinet Office.

Priest, S, Wilson, T, Tapsell, S, Penning-Rowsell, E, Viavattene, C, & Fernandez-Bilbao, A (2007). Building a model to estimate Risk to Life for European flood events - Final Report. HR Wallingford.

More evacuations as floods threaten to burst dam. (2007, *The Guardian, London, 26 June 2007*.

Van Damme, M, Morris, M W, Borthwick, A G L, & Hassan, M A A M (2014). A rapid method for predicting embankment breach hydrographs In F Klijn, & T Schweckendiek (Eds.), *Comprehensive Flood Risk Management.* London: Taylor & Francis.

Warren, A L, & Stewart, E J (2008). The Implications of the 2007 Summer Storms for UK Reservoir Safety In H. Hewlett (Ed.), *Ensuring Reservoir Safety into the Future.* London: Thomas Telford.

SECTION 5:
DESIGN AND CONSTRUCTION CASE STUDIES

Maintaining the Safety of our Dams and Reservoirs
ISBN 978-0-7277-6034-0

ICE Publishing: All rights reserved
doi: 10.1680/mdam.60340.309

Planning a new Water Resource Development - Cheddar Reservoir Two

P KELHAM, Arup
R GROSFILS, Arup
M BROWN, Arup
M ATYEO, Bristol Water

SYNOPSIS   In its 2009 Water Resources Management Plan, Bristol Water identified the need for an additional raw water storage reservoir at Cheddar to meet their forecast resource deficit. Arup has been working with Bristol Water to identify the best reservoir location, develop the scheme design and prepare a planning application covering the reservoir and its associated infrastructure, environmental, social and recreational facilities. The search for a suitable reservoir site and the requirement to produce a scheme design have been supported by an extensive ground investigation and geotechnical analysis with the aim of minimising the need to import construction materials by winning as much suitable material as possible on site.

This paper describes the development of the scheme, from initial site selection to submission of the planning application in December 2013 and highlights some of the engineering, environmental and stakeholder challenges faced in promoting the reservoir.

INTRODUCTION
In its 2009 Water Resources Management Plan, Bristol Water identified the provision of an additional raw water storage reservoir at Cheddar as a key scheme in satisfying their forecast resource deficit. The draft 2014 Water Resources Management Plan sets out the company's water resource strategy for the next 25 years and confirms that Cheddar Reservoir Two remains a priority scheme.

The existing Cheddar reservoir was constructed between 1933 and 1937 to the west of Cheddar village. R.W. Hall (1936) provides a good record of the construction process, and there are also a number of high quality and informative photographs in the archives.

It has a total capacity of 6,140Ml and a top water level of 18.288mAOD. The approximately circular reservoir is formed by a continuous 3,600m long earth embankment ranging in height from 1m to 14m above the natural ground level. The earth embankment has a puddle clay core with shoulders of head material and weathered Mercia Mudstone. The inner and outer slopes are respectively 1:3 and 1:2.5. A typical section through the higher, western embankment is shown in Figure 1 below.

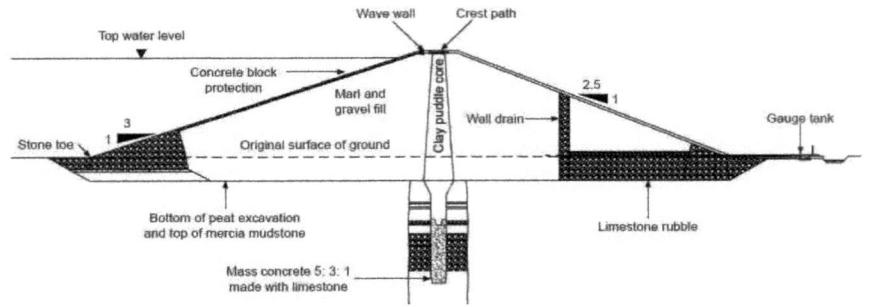

Figure 1. Section through existing Cheddar reservoir embankment.

The embankment has generally performed well over time, though recent investigations have been carried out to assess one area of suspected leakage at the south-west corner of the reservoir. These investigations were managed by the reservoir's Inspecting Engineer, who is also a member of the Independent Reservoir Review Panel established for Cheddar Reservoir Two. It was concluded from the investigations that there are no substantial leakage paths.

SITE SELECTION AND ASSESSMENT
A rigorous site selection study was undertaken between August 2010 and January 2012 to determine the most appropriate location for a new raw water storage reservoir fed from the Cheddar springs water source. It was considered essential that the assessment methodology adopted should provide a robust and transparent audit trail of decisions made, based on an appropriate range of evaluation criteria encompassing engineering, social and environmental considerations. The two stage site selection methodology adopted can be broadly summarised as follows.

First Stage– identifying a shortlist
The process started with the identification of an initial 53 potential locations within a defined search area. These were subsequently refined to a long list of 36 possible sites with adequate space, which also avoided absolute constraints such as environmental designations, housing and major infrastructure.

A desk study assessment of each long list site was undertaken, comparing them against a range of operability, sustainability and planning criteria. A number of weighted assessment matrices were used to assess the relative merits of the long list sites, and to reduce these down to a short list of six.

Second Stage – identifying the preferred location
One of the key factors influencing constructability and cost was the depth of superficial deposits deemed unsuitable for founding an embankment dam. The estuarine alluvial clays present over much of the Somerset Levels are logged up to 30m thick in places (Green & Welch, 1965). Intrusive ground investigation was undertaken at each of the shortlisted sites, with the principal aim of determining the thickness of these superficial deposits.

The result of these investigations informed comparative construction costs, and as a result four of the shortlisted sites were discounted at this point. Outline schemes for the remaining two sites were developed and assessed in greater detail, with further consideration of the criteria set out below.

- Whole life costs
- Archaeology and cultural heritage
- Ecology and biodiversity
- Ground conditions/geology
- Risk

- Planning process
- Wider economic and social benefit
- Sustainability
- Community opposition
- Proximity to source and integration to network

With the exception of costs these were assessed in a qualitative way.

Preferred location
The preferred site for the new reservoir was confirmed in January 2012 and is shown in Figure 2. It is a fairly constrained site, with Cheddar village to the east, the existing Cheddar reservoir to the north, and the River Cheddar Yeo to the south and west. Nevertheless, there were a number of key factors that weighed significantly in its favour:

- Proximity to the water source and Bristol Water's trunk main network, reducing the need for long connecting pipelines and pumped transfers;
- Suitable ground conditions to minimise material imports;
- Operational efficiencies and lower operational costs;
- A single planning authority to evaluate the planning application, and limited local opposition anticipated;
- Greater potential economic and social benefits to the local area.

Figure 2. Aerial view showing existing and proposed reservoirs

RESERVOIR DESIGN
With the site location confirmed, and after a competitive procurement process, Arup was appointed to assist Bristol Water to prepare and submit a Planning Application. This necessitated a significant degree of scheme development and a robust design basis. Key elements of this are discussed in subsequent sections.

Ground investigation
Given the importance of understanding the ground conditions, a more comprehensive intrusive investigation was undertaken between October 2012 and January 2013 with key objectives including:

- To better understand the geology and ground water conditions;
- To ascertain permeability of the in-situ and re-worked materials for construction of the embankment;
- To ascertain levels and quantities of the various materials, to inform cut/fill modelling, and to confirm the feasibility of construction;
- To determine characteristic strength properties of both the in-situ and re-worked materials;
- To ascertain workability of the material, particularly relating to optimum compaction and the likely groundwater control requirements.

This investigation has provided the bulk of the geotechnical data that will be required, though additional targeted ground investigation works will be required during subsequent phases of the project, as the design develops.

Geology

The investigations broadly confirmed the British Geological Survey's published records. The solid geology comprises the Triassic Mercia Mudstone, which is partly overlain by variable thicknesses of Head, Higher Alluvium, or Estuarine Alluvium at various locations. The Head was typically described as a poorly sorted gravel of Carboniferous Limestone and Dolomitic Conglomerate, mixed with the top of the weathered Mercia Mudstone to create a reddish clayey gravel or gravelly clay. The Alluvium was logged as a soft pale brown, blue or grey slightly sandy silty clay. Under much of the proposed embankment footprint, the thicknesses of these superficial materials do not exceed 2m-3m, enabling easy removal.

The Mercia Mudstone was logged by weathering grade I to IV in accordance with the classification provided by Chandler & Forster (2001), and are hereafter termed MM1 to MM4. It was often described as a red brown firm to stiff silty clay near its surface, quickly becoming a weak thinly bedded mudstone with depth.

Embankment stability assessment

In order to develop the embankment design, a stability assessment was required to ensure that the proposed side slopes maintained a sufficient factor of safety in various design cases. Shear strength parameters were derived from a number of direct and empirical correlations, and compared well with the literature (Hobbs, *et al*, 2002, Chandler & Forster, 2001).

Table 1 Stability calculation summary

Case	Embankment side	Factor of safety	Pore water pressure condition
At end of construction	Both	1.3	Un-drained in all materials
During operation	Downstream	1.5	$r_u=0.15$ in the embankment $r_u=0.5$ in underlying geology
During rapid drawdown	Upstream	1.2	$r_u=0.5$ in all material
Seismic loading during operation	Downstream	1.0	$r_u=0.15$ in the embankment $r_u=0.5$ in underlying geology

Stability analyses were undertaken to satisfy the requirements set out in Table 1 above. These resulted in a 1:3 downstream (outer) slope gradient, and a 1:4 upstream (inner) slope gradient.

Water tightness and hydrogeology
A fundamental engineering consideration was to provide confidence in the water tightness of the proposed reservoir, together with appropriate drainage provision. Blower and Jarvis (2012) provide a comprehensive review of the performance of earth filled dams constructed on Mercia Mudstone, concluding that whilst it is possible to construct reliable and economic embankment dams using this material, leakage must be considered.

A suite of in-situ and laboratory permeability tests were undertaken on both undisturbed and re-worked samples of the mudstone. The results of these tests are aggregated in Figure 3.

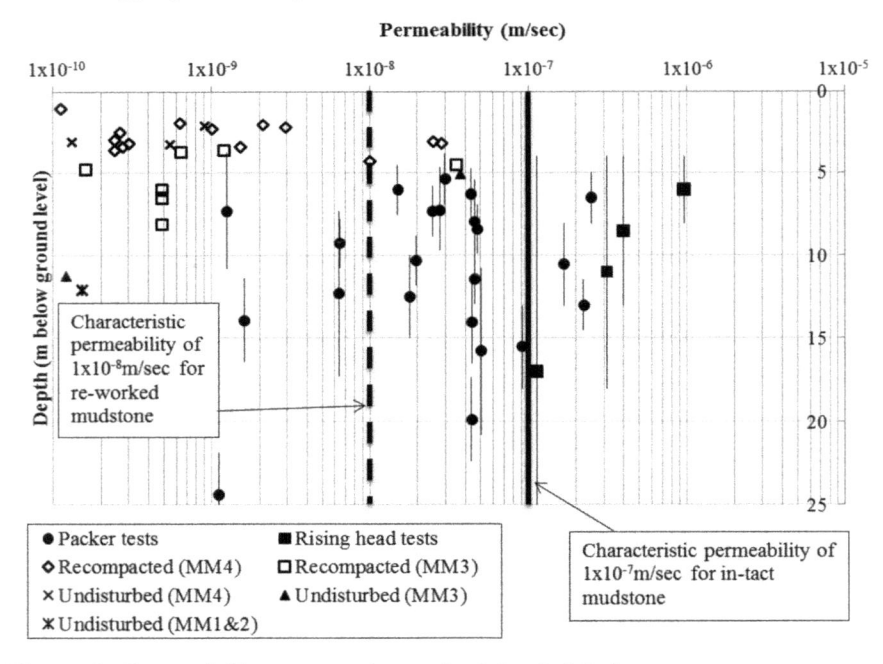

Figure 3. Permeability test results on the Mercia Mudstone.

Note that in Figure 3 the vertical bars denote the packer or rising head test section length, with the test result centred in each test section.

Characteristic permeability values of 1×10^{-7} m/s and 1×10^{-8} m/sec were derived for the in-situ and re-worked mudstone respectively for the purpose of seepage analyses. The higher measured permeability of the in-situ material is attributed to the existing fissuring and evaporate deposits within the bulk material, which were destroyed in the re-working laboratory process. No variation in permeability was observed for either weathering grade or depth. The derived permeability values are broadly comparative to published values (Hobbs *et al*, 2002).

The geometry of the packer testing equipment precluded assessment of the in-situ near surface material (within circa 5m of existing ground level). It was originally proposed to undertake a number of soakaway tests in trial pits, but high groundwater levels prevented meaningful measurement.

To mitigate this unknown, provision for a cut-off penetrating into the MM3 was included in the planning application, though its extent and depth will be subject to detailed design. Not only will this minimise leakage from the reservoir through the foundation material, but if installed early in the construction process it could reduce groundwater inflow into the borrow pit excavations. The most likely form of cut-off will be either a cement bentonite slurry trench, or a 'key' of re-worked Mercia Mudstone (being of lower permeability than the in-situ mudstone).

Preliminary seepage calculations have been undertaken using the 2D finite element software Plaxis 2D v2012, with groundwater conditions calculated using the steady state groundwater flow facility. A number of cases have been analysed, with a range of possible foundation and cut-off types considered. The general magnitude of seepage losses calculated is considered acceptable to the scheme, and will not be significant from the overall reservoir resource perspective.

Embankment Design
Figure 4 shows a preliminary cross section for the embankment dam which makes maximum use of site-won material from the borrow pit within the reservoir footprint. Other dam cross sections using a conventional rolled clay core and flanking chimney drains are still under consideration and will be assessed further during the detailed design stage, which will follow on from the proposed trial excavation and construction of a trial embankment.

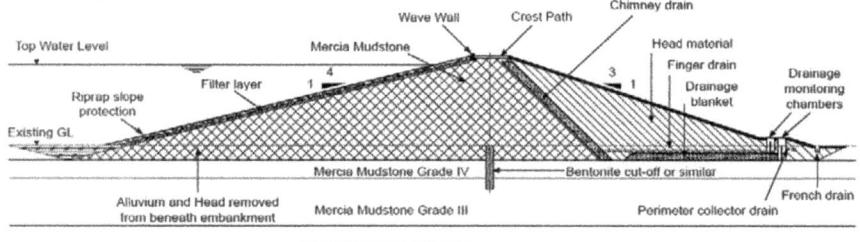

Figure 4. Preliminary embankment cross section

The embankment will contain a chimney drain and a drainage blanket, to control seepage flow through the embankment and to prevent internal erosion developing. Both of these will be designed in accordance with filter rules and be formed of graded granular materials that will normally comprise a coarser filter layer within a finer filter sandwich. It is anticipated that the majority of the drainage materials will be imported, however further investigations will be undertaken to assess whether some of the granular material may be site won.

The chimney drain will act as an interface between the Mercia Mudstone and Head material sections of the embankment. It will discharge into finger drains running through the outer shoulder to V-notch drainage monitoring chambers. These chambers will then connect to a collector drain running along the perimeter of the embankment toe before discharging into the local 'rhyne' drainage system at appropriate locations.

The drainage blanket will be constructed on the foundation for the outer shoulder of the embankment dam. Flows through the drainage blanket will be fed into a separate series of monitoring chambers before discharging to the same perimeter collector drain.

Embankment Construction
The embankment construction methodology is to excavate superficial materials (Alluvium and Head deposits), and found the embankment on in-situ Mercia Mudstone. The Mercia Mudstone will also be used to form the majority of the embankment, with Head material potentially placed in the downstream shoulder. Unsuitable material excavated from the embankment foundation or reservoir area will be placed into the worked out borrow pit as construction proceeds. The only earthworks material that is likely to be taken off site will be topsoil that is surplus to landscaping requirements. As this will have a high organic content it is not considered suitable for placing back into the borrow pit.

Wave protection
To protect the inner slope from erosion by wave action and weathering, the following options were evaluated:

- Riprap
- Tied concrete block mats
- Reinforced concrete slabs
- Open stone asphalt (OSA)
- Pre-cast concrete blocks
- Hand placed stone pitching
- Elastocoast® system

These options were compared across a number of factors, including ease of placement, effectiveness in wave run up reduction, durability, flexibility, maintenance, cost and aesthetics.

The most favourable options that resulted from this assessment were dumped riprap and OSA. Dumped riprap was eventually recommended as the most suitable for the following reasons:

- More effective at reducing wave run-up
- Easily placed without the need for specialist plant and labour
- Uneven surface should deter the public from attempting to access the reservoir
- Reduced embankment height and fill volume as a result of reduced freeboard
- Material can be obtained from local Mendip quarries
- Considered better aesthetically

A pre-cast concrete wave wall is proposed, to reduce embankment fill volume, provide wave freeboard and prevent accidental access to the inner slope by vehicles and pedestrians. The wall will be similar to the wave wall on the existing reservoir, and will typically project some 300mm above the crest track. However it is anticipated that the wave wall will incorporate pedestrian-friendly features by locally raising areas to 'break up' the regular appearance around the top of the reservoir, and to provide seating for those enjoying a stroll around the reservoir.

Reservoir and associated infrastructure

The proposed reservoir will have a total capacity of 9,400Ml which will provide an additional dry year yield of 16Ml/d. It will have a top water level of 19.288mAOD, which is one metre higher than that of the existing reservoir. It will be formed by an encircling 3.6km long earth embankment, ranging in height between 7m and 15m above existing ground levels. Figure 5 below shows the proposed infrastructure necessary to integrate the new reservoir with the client's existing water supply network.

The reservoir will be fed from the same sources that currently feed the existing reservoir, with the principal source continuing to be from Cheddar Springs, located at the mouth of Cheddar Gorge. The existing transfer mains will be retained, and in order to increase the potential transfer from this source, a 250Ml/d low lift transfer pumping station will be built within Bristol Water's Cheddar WTW. A new 1400mm diameter inlet main will run from here to the new reservoir. Water can also be abstracted from the River Axe at Brinscombe, to the west of the reservoirs. This can feed the existing reservoir and will also be capable of feeding the new one.

The reservoir will have inlet and draw-off structures to the east and west of the reservoir respectively. The proposed draw-off configuration will have the ability to draw water from three levels through separate passively screened intakes. The draw-off from the reservoir will be connected to the existing raw water pumping station at Axbridge for transfer either to

Cheddar WTW for local supply, or northwards to feed south Bristol. A 60Ml/d low lift draw off pumping station will be constructed to maintain the supply when reservoir levels are low.

A combined overflow and emergency draw down main will be provided as part of the draw-off tower and culvert arrangement. This will be sized for a maximum overflow of 10.9m^3/s, based on probable maximum rainfall plus unrestricted inflow, and an emergency draw-down flow of 9.0m^3/s based on a draw down rate of 1m per day. This will discharge via a stilling basin to a wide landscaped channel which will discharge to the River Cheddar Yeo.

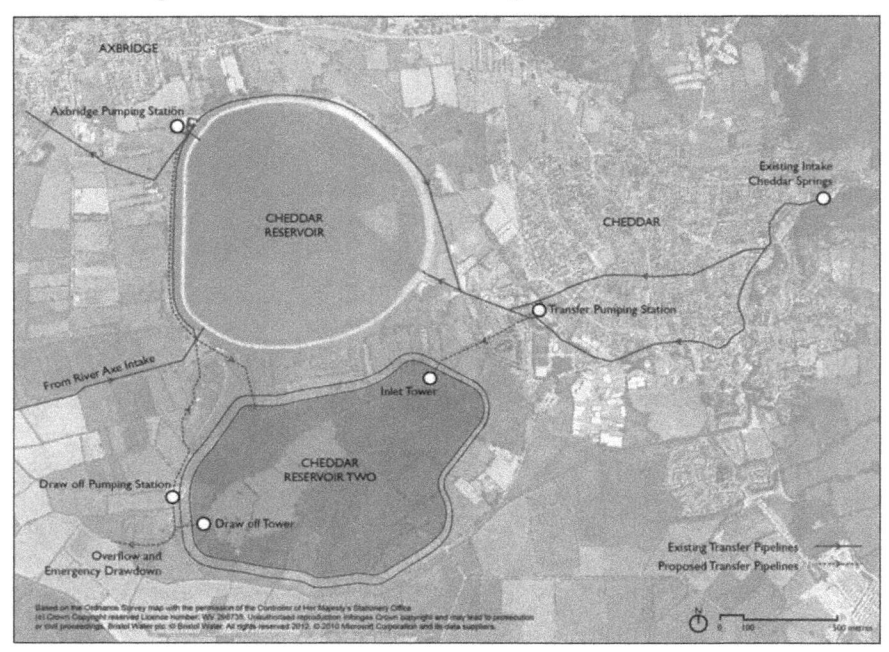

Figure 5. Proposed infrastructure

MASTER PLANNING

Bristol Water's vision is to create a leading 21st century reservoir that sets a quality benchmark for future water infrastructure design. Whilst the primary purpose for the project is the construction of a new water resource, it provides an excellent opportunity to include a range of environmental and recreational enhancements to benefit the local communities of Cheddar and Axbridge. The design of an innovative piece of enhanced green infrastructure will also provide stronger links between Cheddar village and the Somerset levels.

An extensive master planning exercise has sought to integrate the reservoir and associated infrastructure with their surroundings, whilst making the most of the enhancement opportunities. Extensive ecological and environmental surveys and studies have been undertaken and an Environmental Impact Assessment produced to support the planning application.

Engagement with statutory consultees, stakeholders and the public has played a key role in developing the scheme. A two stage consultation process included three-day public exhibitions at each stage. This was positively received, enabled public concerns to be discussed first hand, and allowed the feedback to inform the on-going engineering design and master planning.

A number of significant factors have influenced the final scheme masterplan and are discussed in subsequent sections.

Flood compensation
The reservoir lies partially within the Environment Agency's flood zone 3, and the amount of existing flood storage volume that will be lost by its construction will be some 95,000m^3. The compensatory flood storage volume will be provided by lowering fields to the west of the reservoir, but preserving as far as possible the existing rhynes and hedgerows that demark field boundaries. A range of habitats will be developed in this area including grazing marsh, wetland and wet woodland, and the area will also have boardwalks, viewing platforms and bird hides to encourage public interaction with nature.

Utility and watercourse diversions
Overhead 33kV cables, a sewage rising main and the Helliers Stream will all be diverted through the Middle Moor Lane corridor between the two reservoirs. This corridor was widened following the first public consultation exercise, to preserve a greater number of mature trees and hedgerows, and to retain existing bat foraging routes.

Pedestrian, cycle and bridleways
A network of foot, cycle and bridle routes is proposed, integrating the existing and proposed reservoirs with each other, and linking with the wider local network.

Archaeology
The area has a rich cultural heritage and an extensive archaeological investigation programme is underway. Geophysical surveys and a programme of archaeological trial trenching have been carried out, to help determine the extent, nature, age and state of preservation of any potential remains, and to enable planning of further targeted investigations.

To the south west of the reservoir is a Duck Decoy dating back to the 17th century. This was a pond formed to attract ducks that could then be captured. Although this is now in filled, the buried form still remains and it is proposed to restore this as part of the landscape design proposals.

Recreation and amenity
A visitor centre will form part of the scheme, and is expected to include a café, toilets, community meeting areas and space for exhibitions, arts & crafts etc. An adjacent play area, picnic areas and a community orchard are also included.

THE FUTURE
The planning application was submitted to Sedgemoor District Council as the Local Planning Authority (LPA) in December 2013. At the time of writing this paper the LPA were evaluating the proposals, with a decision expected in spring 2014.

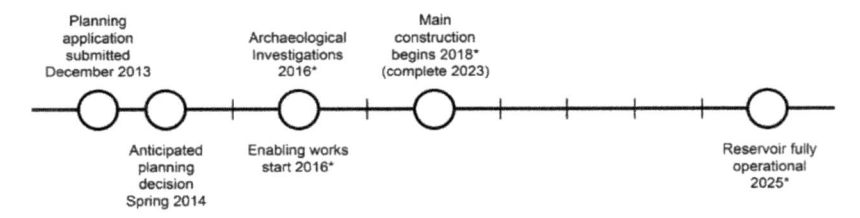

Figure 6. Anticipated project timeline

Funding for the scheme is also being sought. Bristol Water has included provision in its Business Plan, which is currently being reviewed by the water industry regulator Ofwat. Assuming positive responses from the LPA and Ofwat, the anticipated timeline for the scheme development is shown in Figure 6, with the reservoir operational from 2025.

The authors consider that the scheme submitted for planning is appropriate for a 21st century reservoir. It is not just the design of a major piece of infrastructure but the creation of a new place that happens to contain a major piece of infrastructure.

ACKNOWLEDGEMENTS
The authors are very grateful to Bristol Water for permission to publish the work undertaken to date, and the input provided by the Independent Reservoir Review Panel (Alan Warren and Tim Blower). Furthermore, recognition is given to our many colleagues who have input to the scheme's development, and have shaped it to become what it is.

REFERENCES

A L Warren. (June 2012). *Cheddar Reservoir: Report of the Result of a Periodic Inspection under Section 10 of the Reservoirs Act 1975.*

Arup. (2013). *Cheddar Reservoir Two Planning Application.*

Blower, T., & Jarvis, L. F. (2012). Some aspects of embankment dams constructed on the Mercia Mudstone. *Dams: Engineering in a Social and Environmental Context*, (pp. 71-82). ICE Publishing, London

Bristol Water. (December 2009). *Final Water Resources Management Plan.*

Bristol Water. (March 2013). *Draft Water Resources Management Plan.*

Bristol Water. (n.d.). *Mendip Resources System - Description and Operation.*

Bristol Water. (November 2003). *O&M Manual No. OM215.*

British Geological Survey. (1984). *Geological Survey of Great Britain (Sheet 280).*

Chandler, R. J., & Forster, A. (2001). *Engineering in Mercia Mudstone, CIRIA C570.*

Green, G. W., & Welch, F. B. (1965). *Geology of the Country around Wells and Cheddar (Geological Memoirs & Sheet Explanations).* British Geological Survey.

Halcrow. (March 1999). *Cheddar Reservoir Report on Dambreak Analysis (Draft).*

Hobbs, P. R., Hallam, J. R., Forster, A., Entwisle, D. C., Jones, L. D., Cripps, A. C., *et al* (2002). *Engineering geology of British rocks and soils - Mudstones of the Mercia Mudstone Group Research Report, RR/01/02.* British Geological Survey.

Institution of Civil Engineers. (1996). *Floods and Reservoir Safety, 3rd Edition.*

R.W. Hall. (1936). The Construction of Cheddar Reservoir. *Water and Water Engineering.*

Maintaining the Safety of our Dams and Reservoirs
ISBN 978-0-7277-6034-0

ICE Publishing: All rights reserved
doi: 10.1680/mdam.60340.322

Construction challenges at a roller compacted concrete dam in Vietnam

A M KIRBY, Mott MacDonald Ltd
T BLOWER, Mott MacDonald Ltd

SYNOPSIS   This paper describes the challenges faced during construction of a 110m high roller compacted concrete (RCC) dam in central Vietnam. The paper details the main features of the dam's design and construction including the method of river diversion; the mix design, trials, production, placement and testing of the RCC and the programme of curtain grouting. Geological conditions at the left abutment proved much worse than were foreseen.   The paper sets out how the geological conditions were investigated and tested and what practical measures were put in place to overcome the problems.

INTRODUCTION
The Song Bung 4 Hydropower Project is a 156MW hydropower scheme under construction in the mountains of Quang Nam province in central Vietnam. It comprises a 110m high by 345 m long roller compacted concrete (RCC) gravity dam; an intake in a side valley next to the dam; a 3.2km headrace tunnel of 7.2m diameter and a powerhouse housing two Francis turbines.

Figure 1. View of Song Bung 4 dam from upstream

Table 1. Key facts

Catchment area: 1,448km²

Reservoir area: 15.65km² at FSL

Total/active reservoir volume: 510.8Mm³ /234.0Mm³

Installed capacity: 156MW

Mean annual energy: 514.1 x 10^6 kWh

Construction cost (dam): £44M*; total scheme cost: £200M*

*costs estimated at 2014 prices, total scheme cost includes financing, resettlement, land clearance, compensation and contingency

The scheme owner is Electricity Vietnam (EVN), which set up a management organisation, Song Bung 4 Hydropower Management Board (SB4HPMB), to manage the delivery of the project. The designer is Power Engineering Consulting Company Nr 1 (PECC1). Sinohydro Corporation Ltd from China is the contractor for the two civil works contracts, comprising the dam and the headrace tunnel/powerhouse. Procurement support and construction supervision is being provided by Mott MacDonald Ltd and Power Engineering Consulting Company Nr 4 (PECC4).

Dam foundation excavation commenced in October 2010, river diversion was achieved in January 2012. Placement of RCC commenced in June 2012 and was completed at the end of March 2014. Spillway construction started in September 2013. Impoundment is planned for August 2014 and the reservoir should be full less than two months after that. Power is expected to be produced from the end of 2014.

GEOLOGY

The dam site, intake, draw-off tunnel and power house are all constructed in or on the Song Bung Formation, comprising mainly sandstones and siltstones. Jointing is a prominent feature of rock at the site, with two joint sets dominating. The main set is sub-vertical, dipping at typically more than 70° towards 030° to 090°. A secondary joint set is virtually vertical with a dip direction of 125°/305°.

Table 2. Rock weathering scheme used on Song Bung 4 Dam

Zone	Rock	Description
IB	Weathered	Joint surfaces are substantially altered, but centre of rock blocks not altered. Durability markedly reduced.
IIA	Fractured	Rock is hard, jointed, fairly fresh, joint surfaces slightly oxidised, but rock colour almost unchanged
IIB	Intact	Rock is fresh, hard and mineral components of rock not altered; joints are tight with low permeability.

The other major feature controlling the ground conditions at the site is weathering, the most relevant categories being described in Table 2.

The project is located in a low seismicity area and this is reflected in the low ground accelerations developed of 0.04g for the Operating Basis Earthquake (OBE), based on a return period of 145 years and 0.126g for the Maximum Design Earthquake (MDE), with a return period of 2,000 years.

DAM LAYOUT

The RCC gravity dam has a vertical upstream face except for the lower 40m, which is at 1V:0.4H, and the downstream face is stepped with 1.2m high steps at a slope of 1V:0.8H. The upstream and downstream faces of the dam are formed in grout enriched RCC (GERCC).

The spillway is located off-centre on the dam with a total crest length of 72m. It comprises six ogee weirs with flow controlled by 12m wide by 12m high radial gates. Energy dissipation is provided by a flip bucket and plunge pool. The Safety Check Flood is 10,798m^3/s (a 0.02% AEP event). A 0.6m diameter outlet pipe is provided for flow compensation; however there is no scour arrangement or bottom outlet. River diversion was achieved using a twin box culvert, each barrel 5m wide by 9m high, located at the right bank. This will be closed with a gate and then plugged with concrete on impoundment.

FOUNDATIONS

Original design

Ground Investigations (GI)
The principal components of the investigations at the dam site included 49 no. rotary drilled boreholes up to 110m deep, numerous in-situ tests (e.g. 90 no. Lugeon tests) and a range of laboratory tests. In addition, a 60m long exploratory adit was excavated on the dam axis in right abutment, allowing various in-situ rock tests to be performed, including direct shear tests and rock compression tests. Geophysical surveys were also performed.

Rock Shear Strength
The shear strength parameters for the rock mass were derived by two means. Firstly by direct shear tests in the trial adit, and secondly by calculation from the GI data, using the method of Hoek & Brown (2002). There was reasonable correlation between the two methods, and it was also notable that the calculated strengths for Unit 2 were lower than for Unit 1.

Following on from the investigations, parameters were derived for the design of the dam. The table below gives the required shear strength parameters for the foundation rock, as set out in the PECC1 Technical Design. The Mohr-Coulomb parameters c' and ϕ' represent the best fit straight line to the curved shear strength envelope given by the Hoek & Brown method.

Table 3. Rock Mass Shear Strength Properties Required to Support Design

Rock Grade / Location	UCS (MPa)	GSI	mi	σ_3 (MPa)	c' (MPa)	φ' (Deg.)
IB Dam	30	30	14	2.0	0.33	28.5
IIA Dam	60	55	14	2.7	0.94	43.3
IIB Dam	65	57	14	2.7	1.08	46.0

The original dam design required the dam to be founded primarily on Zone IIA rock or better, except at the ends of the abutments, where steep slopes were to be cut through Zone IB material.

Excavation of the Left Abutment

Excavation Stability
Excavation of the left abutment began in earnest in late 2011. Early excavations suffered with slope failures. In March 2012 a series of wedge failures occurred in the upstream slope of the left abutment foundation slot. These were controlled by the bedding planes dipping out of the slope. Similar problems occurred on the right abutment later the same year. Around about the same time cracks appeared in the sprayed concrete surface just above the downstream slope of the left abutment foundation slot where some significant rockfalls had occurred. The reason was believed to be stress relief and a reduction in support due to such a deep excavation. It was thus clear that some remedial slope works would be required to support the rock on both sides of the foundation slot.

Left Abutment Foundation Conditions and Investigation
As excavation proceeded it also became clear that the nature of the rock at the design foundation level was not Grade IIA rock, but was Grade IB or worse, and could be excavated by tracked excavator. Accordingly, the contractor carried out some exploratory drilling to examine the depth to Grade IIA rock in the foundation area. A total of 6 no. boreholes were drilled in the left abutment foundation slot, and the depth to Grade IIA rock was found to be between 11m and 23m below the design level.

Implications for Dam Foundations
It was clear that further excavation was required to find a suitable founding level. However, simply excavating right down to the Grade IIA rock would have had a number of significant impacts, as follows:

- A deeper, wider foundation and thus a larger dam footprint;
- Increased slope heights around the foundation "slot" up to 70m;

- Substantially increased slope stabilisation works;
- Additional soil and rock excavation, disposal and RCC placement quantities;
- Additional cost and time delay.

Trial adit and re-analysis of rock strength
In the light of these severe consequences it was decided to re-assess the requirements for rock strength for the stability of the dam. In July 2012 it was recommended that a trial adit should be driven into the left abutment to allow the rock to be physically examined, and to allow direct shear tests to be carried out to get an accurate assessment of the rock mass strength. The adit was completed in August 2012 and the rock testing undertaken in September. The rock over much of the length of the adit was of Grade IB.

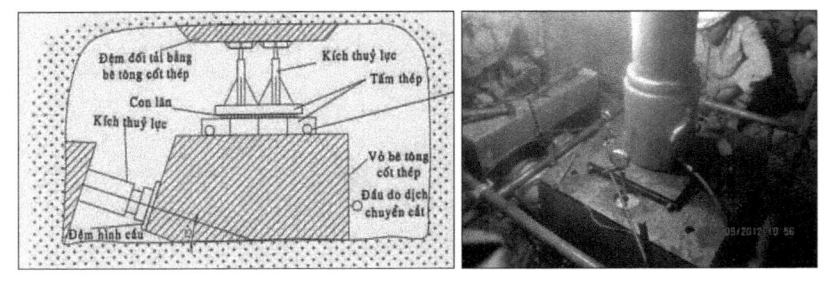

Figure 2. Direct Shear tests in left abutment adit

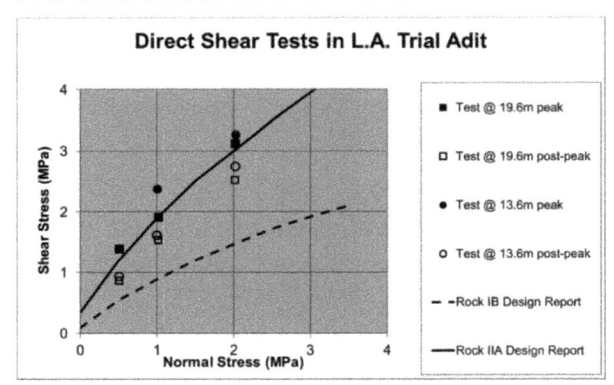

Figure 3. Direct Shear test results

Various tests were undertaken, but the most significant were the direct shear tests. The set-up is shown in Figure 2. Figure 3 shows the results of the direct shear tests on a normal vs shear stress plot. On the same plot is added the shear strength envelopes for Grades IB and IIA rock. All of the peak and post-peak data points are well above the strength envelope for Grade IB rock. The peak data points are also above the envelope for Grade IIA rock. It would thus appear that notwithstanding the description of the rock tested in the adit as "Grade IB", it may nevertheless have sufficient strength to be equivalent to Grade IIA, at least in relation to its peak strength. The post-peak results are a little below this envelope, and the movements to get to these values of strength are modest (<15mm).

Thus to close out the design review, a sliding check was undertaken. This was done and the design considered suitable if the dam were to be placed on rock deemed to be "Grade IB rock with strength equivalent to Grade IIA".

Amended left abutment design
Thus the left abutment foundation design was amended to accept the dam being founded on Grade IB (equivalent to IIA) rock. This was to substantially reduce the amount of additional excavation over that which had originally been feared. However, it did require the development of criteria for the on-site acceptance of Grade IB2 rock as a suitable foundation for the left abutment. These acceptance criteria were determined in terms of visual assessment, GSI value, Fracture Index and UCS values as determined by Schmidt hammer rebound values.

Although the foundation issue had been resolved by these amendments, the revised design still required substantial additional excavation, and thus substantial additional support. This was provided in the form of additional slope anchors of up to 30m length, some of which were pre-stressed. During the implementation of these works, regular monitoring was undertaken of the cracks and slope movements on the downstream and the upstream slopes in this area.

Consequential effects
The significant difficulties encountered on the left abutment had a number of additional consequences. The most significant of these was a substantial delay in the progress of the excavation of the left abutment whilst the various investigations, the adit, the in-situ tests and the redesign were carried out. During 2012 this started to threaten the progress of the whole RCC placement in the dam, so to avoid this the placement sequence was reviewed and amended. It was decided to allow the left side of the dam to lag behind the remainder of the dam, so that bulk RCC placement could continue uninterrupted. This was a fundamental shift in the sequencing of the dam construction.

Other effects included the additional rock anchors to support the greatly increased slopes around the left abutment excavation, and additional consolidation and curtain grouting, as discussed in more detail below.

RCC DESIGN AND CONSTRUCTION

Materials

The RCC mix adopted was a medium-high paste mix and is detailed in Table 4.

Table 4. RCC mix proportions per cubic metre

	RCC	Bedding mortar	GERCC grout
Cement PC40	80kg	517kg	891kg
Pozzolanic mineral admixture	120kg		
Sand	783kg	1437kg	
50mm-25mm Aggregate	526kg		
25mm-12.5mm Aggregate	387kg		
12.5mm-5mm Aggregate	470kg		
Water	135 l	310 l	710 l
Admix	1.5 l	2.07kg	2.67kg

The PC40 cement was obtained from the Kim Dinh Cement factory about a three hour drive from the site. The mineral admixture used was a natural pozzolan from Gia Qui in southern Vietnam near Ho Chi Minh City. A Sika TM25 admixture was used as a retarder.

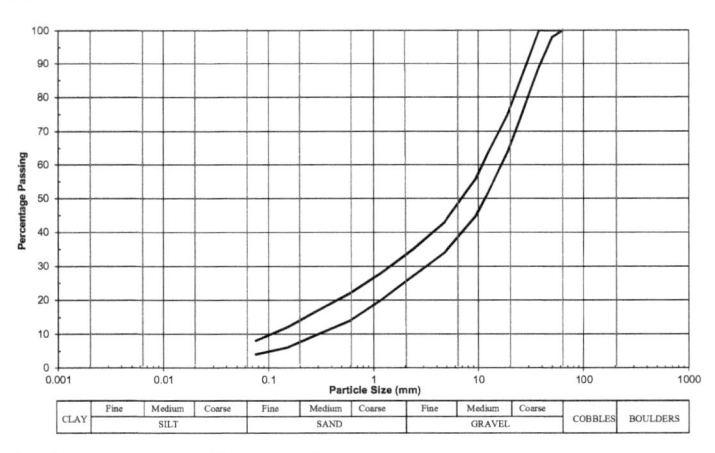

Figure 4. Aggregate grading envelope

Sand and aggregate were exploited by developing a quarry and crushing plant at a granite outcrop about 10km from the dam. The aggregate was processed through conventional jaw crushers. Vertical shaft impact crushers produced the sand/fine aggregate. The maximum size of aggregate was limited to 55mm, due to angularity of the granite. Three aggregate sizes were targeted: 50mm-25mm, 25mm-12.5mm and 12.5mm-4.75mm. These combined with the sand fraction in the proportions in Table 4 resulted in the combined grading envelope shown in Figure 4.

Laboratory trials and trial embankment
The designer carried out extensive laboratory and full scale trials to develop RCC mixes prior to construction, including trial embankments at another hydropower scheme Song Tranh 2, which used a similar granite aggregate. Leading from these initial trials, two alternative mixes based on different sources of pozzolanic material, a processed flyash (Pha Lai) at 140kg with 60kg of cement and the alternative of a natural pozzolan (Gia Qui) at 120kg with 80kg of cement, were indicated in the Specification.

A further trial embankment was carried out at the start of construction with these two alternative mixes and the contractor elected to use the mix with the natural pozzolan, Gui Qui, due to problems of availability and transport in the case of the processed flyash.

Production and transportation
The total volume of RCC in the dam is 870,000m³ placed over a period of 22 months. An RCC continuous batching plant with a capacity of 360m³/h was erected at the top of the right abutment.

Ambient temperature conditions are fairly typical of a tropical climate without major variations throughout the year with mean monthly temperatures in the range 20-28°C. This relatively low variation in ambient temperature enables RCC placing throughout the year and this was necessary to meet the overall construction programme. As part of the design studies, a comprehensive FE thermal analysis was carried out based on the specified allowable rate of rise of the dam wall and this established a maximum allowable placing temperature of 23°C, in conjunction with a typical transverse joint spacing of 20m.

A temperature rise of 3-4°C during transport could be expected, with an additional 1°C allowance as a safety margin meant that a maximum temperature of 18°C was required for RCC leaving the batching plant and this was closely monitored. The cooling of the RCC was achieved by firstly, in the case of all four different aggregates, good shading of the immediate stockpiles and then a major plant facility for air-cooling of coarse aggregates. Air-cooling of coarse aggregates was continuous, in addition ice was added to the mix during the hotter periods of the year (April-September). These cooling requirements reduced the overall production rate for the RCC to 220m³/h.

In practice the placement temperatures ranged from as low as 15°C to 23°C. The design studies had indicated a maximum permissible post-placement temperature after heat of hydration of 42.1°C to limit thermal stresses and these temperatures have been carefully monitored with an array of temporary and permanent temperature gauges. Figure 5 shows typical temperature readings.

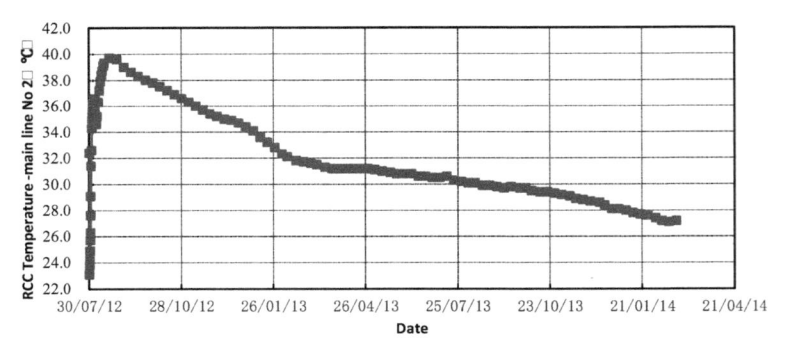

Figure 5. Typical temperature development over time

RCC transportation was for the most part by dump trucks to the point of placement. The contractor had intended making more extensive use of conveyors but a change in the placement sequencing of the RCC caused by the problems at the left abutment prevented this. In any case the conveyors that were used proved problematic with difficulties preventing segregation at the inlet hopper, mechanical breakdown and problems re-aligning the conveyors after each placement.

Placing and compaction
RCC was placed in layers of 0.3m compacted thickness spread using 14t dozers and compacted with 12t vibratory smooth drummed rollers with 7.5t dozers used in more confined areas. Placements were typically continuous with hot joints only for heights of 1.2m to 2.0m and sometimes as higher as 3m, the height primarily being constrained by the height of RCC that could be supported by the type of formwork used. In the lower sections layers were placed horizontally but the contractor then changed to the slope layer method for large placement areas. This continuous placing of the RCC in layers was encouraged to significantly reduce the number of cold and warm lift joints. Reasonably efficient RCC delivery and placements meant that cold joints were generally only formed at the end of each 1.2m to 3m placement.

Lift joints were classified as either hot, warm or cold joints with different treatments:

- **Hot joint** (less than 14h until the next layer placed): only removal of loose materials and water from the surface between lifts;
- **Warm joint** (14h-24h): clean loose materials and water, and lay a thin layer of mortar between lifts;

- **Cold joint** (more than 24h): greencut with compressed air jet, clean loose materials and water and lay a 10-15mm thick mortar layer.

GERCC was used to form the upstream and downstream dam faces and at any interfaces where rollers could not reach: the rock face, around waterstops, galleries and other structures (e.g. pendulum and lift shafts). It was also used to infill hollows and clefts in the foundation rather than using dental concrete. The GERCC at the upstream and downstream faces was nominally 0.6m wide, though was thicker to suit conditions, e.g. where formwork tie bars extended further into the RCC. The optimum arrangement of constructing the GERCC by adding cement grout to the RCC and then vibration using conventional pokers was first established at the trial embankment and proved very successful, with minimum honeycombing evident after stripping the forms.

High quality placement requires a carefully orchestrated sequence of plant movements, a workforce well-trained in RCC and effective foremen in the placement area. In general placements were efficient; however problems did occur in the following areas:

- Segregation of RCC when tipped - re-mixing of larger aggregate back into the RCC heaps was required before spreading by dozer;
- Vehicles tracking over freshly compacted RCC – where the sequence of plant movements had not been well thought-through trucks would be found reversing and turning on fresh RCC, churning up the surface;
- Over-wetting RCC surfaces – periodically workers seemed to have a compulsion to wet up the RCC surfaces by spraying with water even though this was not needed and would weaken the lift joint;
- 'Feather edges' were not always treated correctly: at the end of each slope using the slope-layer method is a narrow triangular wedge of material - a 'feather edge' which needs rolling in and trimming back to avoid a weak area.

Placements continued throughout the year including during the rainy season (September-December) with only minor disruption for rainfall. When rain was forecast any uncompacted RCC would be rolled off in advance of the rainfall and the surface protected with tarpaulin. RCC was placed 24 hours a day with two 12 hours shifts. The total quantity of RCC placed was 870,000m^3. Placement rates peaked at 75,000m^3 per month and 4,200m^3/day. Figure 6 compares planned and actual placement rates.

Transverse Joints
The transverse joints at 20 m spacing were formed using a vibrating kerf cutter directly before any hardening of the RCC, and this also punches in a textile to act as a bond breaker. Watertightness was provided by two

500mm wide PVC waterstops with hollow central bulbs located near the upstream face, with a drainage hole behind. The waterstops are located within an area of GERCC that is contiguous with the upstream GERCC facing.

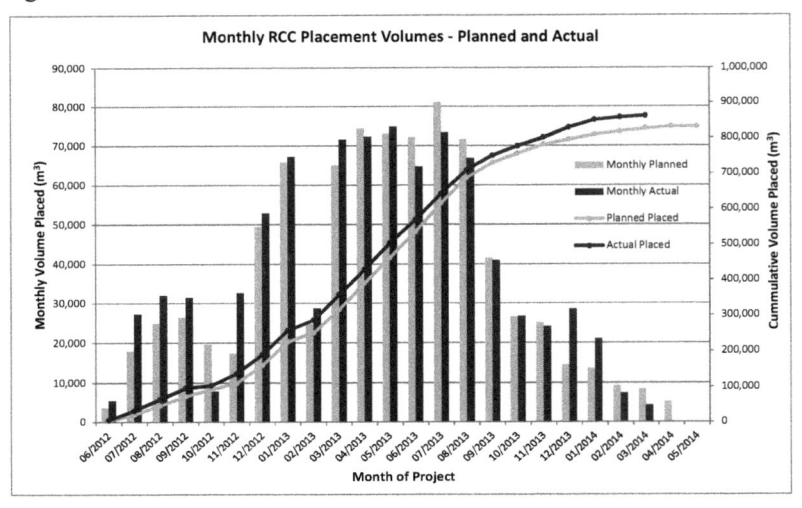

Figure 6. RCC placement: planned versus actual

Testing

The testing of the RCC has two main components; firstly directly during construction and then post-placing; and after at least 90 days, core drilling in the RCC and the laboratory testing of cores.

Testing during actual placing has involved Vebe tests, effectively checking the required workability of the RCC which falls within close limits of 8-12 seconds, the taking of cylinders for strength testing and then nuclear densometer tests on the RCC directly after compaction. The latter showed values consistently above the required 98% of the average wet density at the batching plant. In terms of compressive strength, Figure 7 shows cylinder strength gain against age and compares the results achieved on site with those of the RCC trial embankment and those carried out for the design. The required compressive strengths at 365 days for the RCC were 17MPa for cylinders and 15MPa for cores.

The programme of testing of cylinders has included elastic modulus and also tensile strength testing, covering 90 days and up to 365 days, the latter to meet the specification requirements with respect to strength.

The main check on the quality of the placed RCC is by 150mm diameter core drilling and an extensive programme has commenced, with one deep drillhole for each 20m wide block. To date the cores have generally shown a dense RCC throughout and very little evidence of hot joints, as would be expected. There has been some shearing at cold joints, but typically these have also remained intact during coring. The emphasis of the testing of the cores will be 365 day results, covering compressive strength and tensile strength of unjointed RCC, but more importantly the strength at lift joints. This will be examined by tensile strength testing targeting joints supplemented by some shear strength tests.

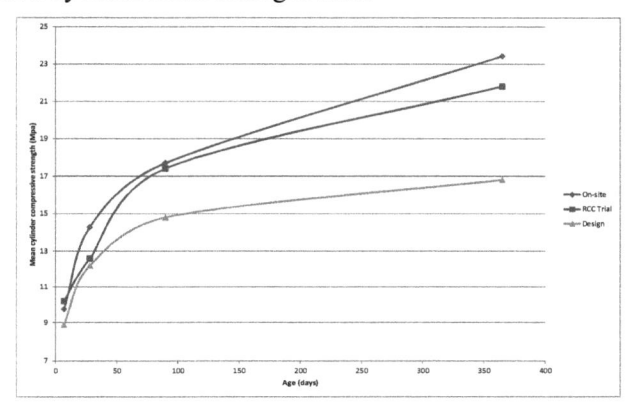

Figure 7. Mean cylinder compressive strength development over time

GROUTING

Consolidation grouting
Consolidation grouting was carried out over the dam foundation to a depth of 5m at 3m centres. Where the dam was founded on IB2 rock consolidation grouting was increased to 10m depth with the aim of improving the stiffness of the IB2 rock. The original design had a 0.5m thick CVC slab at foundation level from which to carry out consolidation grouting. Placing of this CVC slab as well as being slow to construct would have created additional interfaces between the RCC and the rock foundation. The slab was therefore omitted and instead consolidation grouting was carried out from the RCC, between RCC placements, area by area as the RCC rose up the abutments. Average grout takes were 127kg/m in areas of IIA rock and 211kg/m in areas of IB2 rock. Water tests were carried out pre and post grouting with Lugeon values reducing from an average of about 16Lu down to about 1.2Lu.

Curtain grouting
The original design had a main grout curtain row at 3m spacing to a depth of 0.5H where H is the retained head of water, with a minimum depth of 15m. A short auxiliary row was planned but only to a depth of 10m. The curtain grouting arrangement was amended in the light of the poorer quality rock at the left abutment.

The depth of the main row was increased from a minimum depth of 0.5H up to a depth of up to 0.75H in holes where the Lugeon value in three consecutive 5m stages was greater than 3Lu. Tertiary holes were also introduced between the primary and secondary holes to give a hole spacing of 1.5m. These were to a minimum depth of 10m and continued deeper in locations where adjacent primary or secondary holes had shown high grout takes of over 50kg/m. The auxiliary row was also extended in areas of IB2 rock to be the same depth as the main row.

Curtain grouting for both main and auxiliary rows was carried out from the lower gallery. The first 10m of each hole was drilled, water tested and grouted in descending 5m stages. Below 10m depth water tests were carried out in descending stages and grouting was carried out in ascending stages. Grouting was started with a water:cement ratio of 5:1 and then thickened to 3:1, 2:1, 1:1 and 0.6:1. Grouting was carried out at pressures from 0.5MPa for the first 5m stage up to 2.0MPa for depths exceeding 20m.

Table 5 summarises the mean grout takes in different areas of the foundation. As expected the lower grout takes were in the central section (at the river bed), being significantly higher at the left abutment. The improvements in water tightness from the grouting are indicated by the reducing grout takes from primary to secondary to tertiary. Grout takes even in the tertiary holes in the left abutment were still quite high, being >100kg/m in a number of locations, providing confirmation that adding the tertiary holes had been worthwhile.

Table 5. Mean grout takes in main grout curtain row (kg/m)

Location	Left abutment	Centre (river bed)	Right abutment
Primary	308	198	221
Secondary	117	135	163
Tertiary	62	n/a	28

Figure 8 compares grout take with depth at the left abutment and in the central river bed section. The central section shows a marked reduction of grout take with depth, which is not seen at the left abutment. Counter-intuitively mean grout takes at shallow depth in the central river bed section tend to be higher than at the left abutment.

After grouting inclined check holes were drilled along the grout curtain at 20m spacing and water tests carried out. A Lugeon value of less than 3Lu was the test criterion adopted. To date, water tests have been well within this limit.

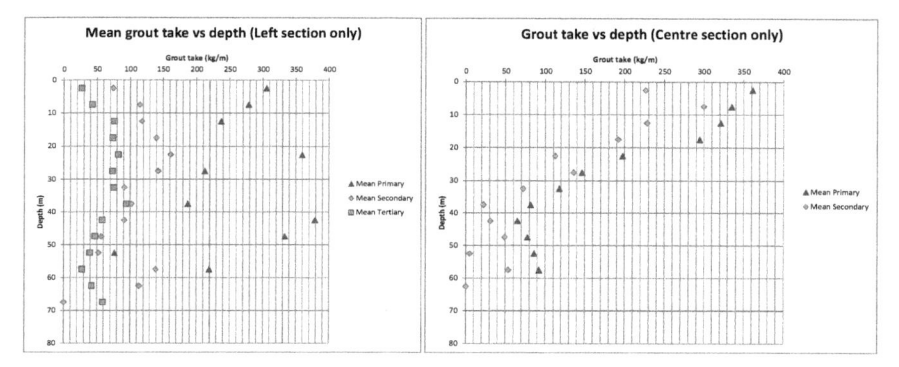

Figure 8. Comparison of grout takes between left abutment and river sections

CONCLUSIONS

The paper has described some of the important construction issues and approaches used at Song Bung 4 that is hopefully of interest and relevance for RCC dam construction elsewhere.

Construction of the dam has progressed well. In spite of the unforeseen geological conditions at the left abutment the planned completion date of the dam has not been significantly affected. This required a pragmatic and collaborative approach from all involved. It also required some imaginative re-scheduling by the contractor to limit its impact and a willingness to ramp up resources to maximise RCC productivity.

The geological problems at the left abutment highlighted the need for good interpretation of site investigation data, the benefits of additional investigations in advance of foundation excavation and the need for a pragmatic approach during construction to deal with unexpected conditions.

ACKNOWLEDGEMENTS

The authors gratefully acknowledge the permission granted by Mr Truong Thiet Hung, Director of Song Bung 4 Hydropower Management Board and Vice President of GENCO 2, the new Owner, to publish this paper. We would like to thank in particular Mr Le Minh Toai, Manager, PECC4 and his staff who provided a lot of the background data used in the paper. We would also like to thank Mr Li Xuejiang, Project Director of Sinohydro Corporation Ltd, Mr Richard Ramsden, Mr Ed Davies, Mr Jack Meldrum and Mr Chris Wagner for reviewing the draft paper.

Maintaining the Safety of our Dams and Reservoirs
ISBN 978-0-7277-6034-0

ICE Publishing: All rights reserved
doi: 10.1680/mdam.60340.336

The design and construction of an enlargement scheme for Black Esk reservoir (Scotland)

J C ACKERS, Black & Veatch, Redhill
D A GETHIN, Black & Veatch, Glasgow
G KARUNARATNE, Scottish Water
S A PRYCE, Black & Veatch, Chester
T A SCOTT, Black & Veatch, Glasgow
J TUDHOPE, Scottish Water
M WHEELER, Black & Veatch, Chester

SYNOPSIS   Work started in early 2013 on a scheme for enlarging Black Esk reservoir in the Scottish Borders, and was substantially completed within twelve months. The overflow level was raised by about 2.5m, increasing the original storage volume of 2200 Ml by about 40%.

The original reservoir, built around 1962, was impounded by an embankment dam that was about 20m high, with a rolled clay core and the overflow works comprising a bellmouth shaft and tunnel spillway.

The overflow level was raised by the innovative adoption of precast piano-key (PK) weirs around the rim of the bellmouth – believed to be the first PK weir in the UK and a world first for the adoption of PK weirs at a bellmouth spillway. The new PK weirs are designed to pass the design flood of 183 m^3/s with a flood surcharge of just under 1m, saving about 0.7m from the amount of dam raising that would have been required in conjunction with simple raising of the circular weir around the bellmouth rim.

This paper covers the design and construction of the dam raising, using material won from the original boulder clay shoulder to heighten the core, then replacing the shoulder excavation with relatively free-draining material, the development of the PK weir design and its adaptation to the circular rim of the bellmouth, together with the use of precast construction, and the raising and modification of the valve shaft.

INTRODUCTION
Scottish Water, in the planning period to 2031–32, predicted a maximum deficit in the Black Esk water resource zone of 4.4 Ml/day. A number or projects were developed to address this deficit, of which the enlargement

project for Black Esk reservoir was the largest. The increase in the storage volume by about 40% to 3150 Ml increased the estimated yield to 19 Ml/d, which fully resolves the deficit and creates an estimated surplus of 1.1 Ml/d.

BLACK ESK RESERVOIR

Black Esk reservoir, which is situated in Dumfries & Galloway Region about 75km south of Edinburgh, was completed in about 1962 to designs by Robert H. Cuthbertson & Associates. The reservoir had an original surface area of about 33 hectares and a storage capacity of about 2200 Ml, being impounded by an embankment dam, whose crest was about 20m above the downstream valley floor and whose crest length was about 250m.

The original dam features a rolled clay core formed of glacial till, and the shoulders of the dam are also built with similar material, but include a chimney drain at the downstream face of the core and approximately horizontal drainage layers (Figure 1).

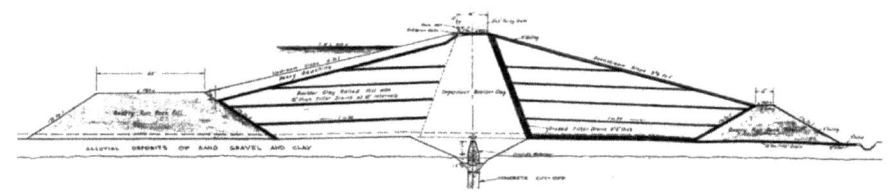

Figure 1. Indicative cross section of original Black Esk dam (from record drawing)

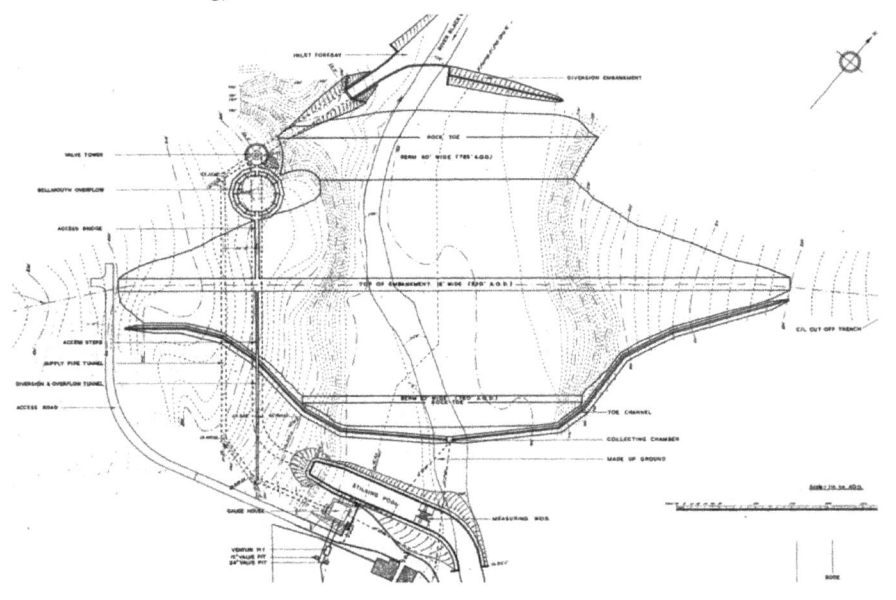

Figure 2. Layout of original Black Esk dam (from record drawing)

Figure 2 gives the overall layout of the original embankment dam, with the spillway and drawoff works located at its southwest or right abutment, and Figure 3 contains a longitudinal section of the bellmouth and spillway tunnel.

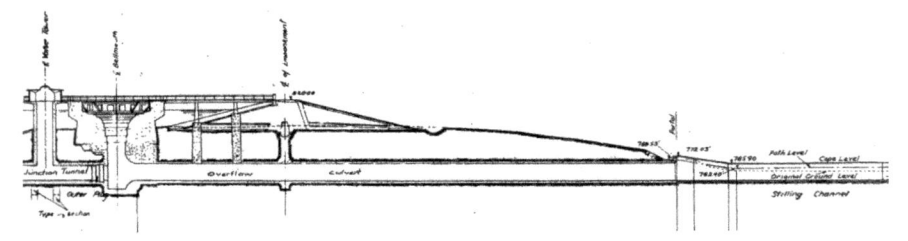

Figure 3. Long section of original bellmouth shaft and spillway tunnel (from record drawing)

BLACK ESK DAM RAISING
The scheme has increased the storage capacity of the reservoir by raising the full supply level, corresponding to the overflow weir crest level, by about 2.5m. An earthworks solution was recognised as providing a cost-effective means of heightening the embankment. Earthworks would also have the benefit of allowing for further enlargement of the reservoir, if ever needed, in line with the original design for the dam that foresaw the potential for an increase in embankment height of 9.1m (30 feet).

The original proposal had considered increasing the embankment height symmetrically, raising the upstream and downstream shoulders by similar amounts about the centreline of the embankment. But the importance to Scottish Water of maintaining water supply from Black Esk during this scheme precluded drawing down the reservoir fully to allow construction of a higher embankment on this principle. Through coordinated design with the unique spillway arrangement, it was found that the amount of raising required to accommodate the PMF (probable maximum flood) and wave freeboard could be limited to 2.5m, dispensing with the need to include a wavewall on the raised dam.

Earthworks design
The original embankment section comprised what was in effect an homogeneous rolled clay embankment, but zoned by defining a central core zone delineated by shoulders provided with horizontal drainage blankets at regular intervals, together with a chimney filter drain to the downstream side of the core (Figure 1). The upstream slope of the embankment was 1 on 4 and the downstream slope 1 on 3.75. The upstream slope was protected by riprap over a bedding layer, and the downstream slope grassed. A wavewall was provided on the upstream side of the crest.

In order to minimise drawdown of the water level of the reservoir for construction an earthworks solution was developed that had most of the construction on the downstream side of the embankment, and reduced interference with the upstream side of the crest, thereby maintaining reservoir safety during construction. The construction detail added earthfill to the crest and downstream shoulder, extending the central core zone of the dam by adding an inclined core to provide continuity of the water-retaining element of the dam. Recognising the importance of filters in embankment dam safety, a filter zone was included on the downstream side of the inclined core connecting into the existing filter layer (Figure 4).

The clay used in the original construction of the embankment was won from local borrow pits in the glacial till of the valley sides. In order to carry out the raising, additional clay would be needed. Although opening up a new borrow pit could potentially provide a similar material for use in extending the core, this faced a number of practical problems. Locating a source of clay elsewhere and importing it was also possible. However, recognising that the embankment was essentially constructed of the same clay till material throughout, the unconventional approach of borrowing material from the embankment shoulder was identified as an opportunity to use the same rolled clay material as in the existing core. This unusual approach required some confidence in the stability of the existing fill materials in winning the clay from excavations into the downstream shoulder. The temporary borrow pit thus formed had to be filled, and to do this, imported granular fill of quarry scalpings from a local quarry was used.

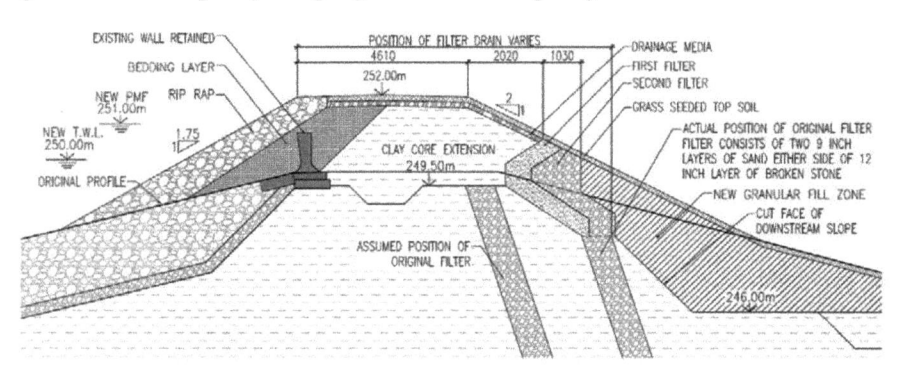

Figure 4. Cross section of raising of dam crest

The granular backfill had advantages in that it was readily available, unlike suitable clay for rolled clay fill, was less costly than clay to import, was easier to place than clay, and was sustainable. It also had the added benefit of having a greater friction angle than the clay, allowing an increase in the gradient and further reducing the volume of earthworks needed to provide the heightened cross section. A highly sustainable solution was thus developed that facilitated construction.

Figure 4 shows the original crest detail, together with the crest raising detail developed to allow raising of the embankment in the future if this were to be required. The top section of the downstream slope was increased from 1 on 3.75 to 1 on 2, while the top section of the upstream face was increased from 1 on 4 to 1 on 1.75. The triangular wedge indicates the area from which clay was abstracted.

The original wavewall was buried within the bedding layer as upstream filling progressed. This was beneficial to reservoir safety during construction, as the wavewall continued to protect the crest, as well as saving time, simplifying construction and lessening the amount of drawdown needed, minimising the temporary impact of the works on the security of water supply for the region. The new crest arrangement does not need a wavewall, as sufficient freeboard was built into the embankment crest level, saving in the duration and complexity of construction.

Abutments
The original construction at the left and right abutments comprised a concrete cutoff wall, about 1.8m wide, constructed through the glacial till and weathered rock into the top of the rock with a grout curtain down into the bedrock below. Based on the record drawings, the existing foundation concrete cutoff trench and grout curtain were thought to extend up each abutment, but the exact level and alignment were not clear on the drawings and had to be established during the construction works.

The top of the original cutoff wall at both the left and right abutments was below the new top water level. As part of the dam raising works, the contractor excavated down to the top of the concrete cut off and backfilled the excavation to provide a simple clay-filled trench connected to the concrete and up to the new PMF level, thereby ensuring the continuity of the cutoff.

The original dam was constructed with an intended proposal to allow for future raising of the dam by up to 9.1m (30ft) showing a level of confidence in the general watertightness of the reservoir basin up to that level.

Construction
Although the earthworks were on a relatively small scale, an end-product specification was the approach preferred for control of placement and compaction of fill. More emphasis was therefore placed on field trials prior to dam earthworks commencing and conformance testing during the works.

The original filter layers were located during the excavation of the downstream face. Exposure of the top of the filters adjacent to the central clay core has shown that the original filter comprised a layer of gravel sandwiched between two layers of sand which appears to match the description on the completion drawing of two 9in layers of sand on either side of a 12in layer of broken stone. The filter seen during construction did not correspond to the exact dimensions shown on the completion drawings being slightly narrower as placed than the total thickness 2ft6in (760mm) expected.

The position of the filter was not as shown on the original record drawings, being found further downstream than had been assumed for the crest raising design (Figure 4). The distance from the wall to the edge of the filter drain increased from around 6m at the right end (looking downstream) of the dam to 8m at the left end of the dam. However, the earthworks design was sufficiently flexible to accommodate this by some realignment of the details without compromising the integrity of the construction or unduly increasing its complexity.

Figure 5. Cut slope in downstream shoulder after excavation of glacial till

It had been expected that the original filter would terminate against the left and right abutments, where the dam meets the original ground. However, this was found not to be the case, with the original filter terminating around 40m short of the abutments. Given the importance of the filter in preventing internal soil movement and controlling seepage, where the filter was not present a new filter was therefore added against the core zone from the cut level up to original crest level. This new filter detail was then connected to the filter being provided as part of the crest raising detail.

As the position of the filter was further downstream than had been assumed, this meant that the estimated quantity of clay won from the downstream face would be less than assumed. The design, however, had included an option to over-dig and remove clay to a greater depth on the downstream shoulder if required. This proved to be a fortunate precaution and an area of over-dig was utilized to win additional clay fill, with excavation taken down a further 1m to a level of 245m OD over a 100m section. The additional extraction was subject to the condition that excavation from the over-dig zone was promptly followed by backfilling and compaction of the replacement granular material in order to restore the area to 246m OD, with the length of additional open excavation no more than 15m at any time.

Figure 6. Sheepsfoot roller in use on the embankment crest

Artesian pressures at left abutment
During the course of construction a wet spot was found on the dam crest at the left abutment, close to the end of the wavewall. After stripping of topsoil in this area water could be seen bubbling up to the surface. At the time, the ground level here was 3.5m above the water level in the reservoir, which was being held down for construction works.

Historical records showed a borehole sunk in this area, as part of investigation for the original construction of the dam, had encountered water as a 'spring tapped' in the bedrock found beneath about 10m of till. Artesian groundwater was also encountered in a borehole drilled in 2011 about 50m from the wet spot. This borehole encountered artesian water from highly fractured Greywacke bedrock at 8m below ground level. The borehole was sealed and provided with a pressure gauge that has recorded a constant 1.65 bar above ground level.

The wet area was investigated by careful excavation to try to identify the source of the water. The top of the concrete cut off was found at 1.8m depth; water could be seen as a small flow emanating from a transverse crack in the castellated top of the concrete and entering the troughs in its top surface (Figure 7). The flow was clear and contained no sediment when observed over a period of a few days.

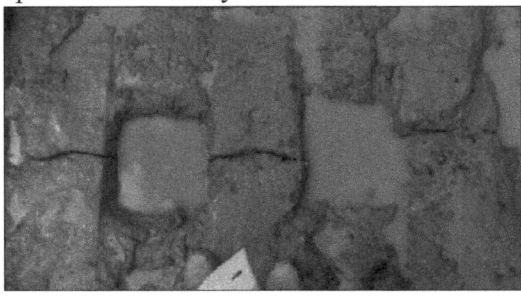

Figure 7. Crack seen at the top of the original concrete cutoff.

Rather than attempt to seal the crack, the decision was taken to capture and control the flow and direct it to a point where it could be monitored for change. Using filters and drainage proved an effective measure in managing the flow and allowing the clay core to be reinstated and effectively compacted, thereby allowing the earthworks to be completed safely. The flow is monitored and there has been no change observed.

Monitoring during and following construction
Standpipe piezometers had been installed in site investigation holes done in 2011. These piezometers were monitored regularly before and during the construction works so far as practicable. Those on the crest and downstream face were protected during the works and were extended through the new fill with new headworks provided to complete construction. Permanent settlement marker pins were installed on the dam crest in November 2013. V-notch weir plates were fitted for flow measurement in the chambers that collect water from the wet spot and from the mitre drains. As monitoring of embankment behaviour is significant to safety evaluation, the monitored performance of the embankment was compared with readings taken prior to the works and also with those predicted from SEEP/W mathematical modelling. The results of the monitoring are summarised as follows:

- Groundwater levels in the boreholes show that groundwater levels in the rock and glacial till of the foundation are not directly connected to the reservoir, appearing to fluctuate seasonally or with rainfall.
- The piezometric level in the embankment fill remains fairly constant and compares well with that predicted by SEEP/W modelling.
- The groundwater level in the siltstone rock on the right side of the dam is different from that in the Greywacke on the left side that is under artesian pressure.
- Flows into the collection chamber at the base of the mitre drains are dominated by rainfall with no sign of a seepage contribution.
- Flow into the collection chamber from the 'wet spot' on the left abutment remains constant.

It has been concluded that the earthworks and modifications to the embankment have had no detrimental effect on its performance.

SPILLWAY RAISING
The 9.1m raising scheme envisaged at the time of the original construction of the reservoir required the building of a new bellmouth profile directly above the original, thus heightening the shaft. The same concept was adopted in feasibility studies that preceded the recent enlargement scheme.

In order to provide a more economical solution, the design-and-build contractor was keen to investigate the potential for reducing the flood surcharge and thereby reducing the amount of dam heightening required.

Hydraulic design development of PK weirs
Conventional labyrinth spillways of course have a large footprint, so are not well suited to implementation around the rim of a bellmouth. But recent research into piano-key (PK) weirs (Lempérière, 2009; Erpicum et al, 2011), in which the upstream and downstream ends of the zigzags are cantilevered beyond the foundation, led to the idea of adapting the concept to retrofitting PK weirs around the rim of a bellmouth.

PK weirs have been developed over little more than a decade. At an early stage of design, in November 2011, a study tour was arranged to enable Scottish Water and the Black & Veatch design and construction team to see PK weirs that had been retrofitted to augment spillway capacity at two hydropower dams (St Marc and l'Étroit) owned and operated by Electricité de France and another on a water supply reservoir near Limoges (Figure 8).

Figure 8. Retrofitted PK weirs at St Marc dam (left) and Étang de Gouillet

The original ogee weir around the rim of the bellmouth shaft at Black Esk comprised 12 straight segments, with angles of 30° between them. In order to suit this, the initial concept for the PK weir was based on accommodating a single weir cycle into each of the 12 segments, as illustrated in Figure 9 (left), with each cycle of the PK weir positioned so that its outward-projecting bay (inlet key) was centred within the segment. The inward-projecting bays (outlet keys) straddled the 30° angles between the segments and were consequently truncated. Initial rating calculations were performed using empirical equations developed by various researchers, but were then modified in the light of the results of analyses by computational fluid dynamics (CFD).

It was subsequently decided that the external overhang of this weir design would be excessive, so attention turned to a range of 24-cycle weirs, arranged in pairs with 30° angles between them. These designs offered about the same aggregate weir length as the 12-cycle design, but with a smaller external overhang. Unfortunately, they gave slightly poorer hydraulic performance than the 12-cycle design, but were nevertheless favoured because they would be small enough to allow precasting of individual units. It was then realised that further simplification could be made by avoiding the handed pairs of units, making all 24 units identical and adjusting the concrete shoe arrangements around the original bellmouth rim to suit a 24-segment PK weir, as shown in Figure 9 (right).

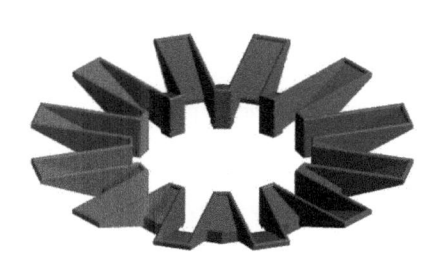

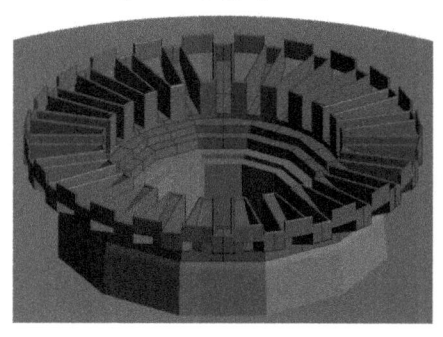

Figure 9. Initial 12-cycle design (left) and adopted 24-cycle design with debris booms across inlet keys (right)

The CFD model of the PK weir comprised a 15° sector between the centres of successive inlet keys. The CFD modelling did not include the shaft and tunnel, largely because it would not be possible to achieve a sufficiently fine mesh in such a model. In addition, it was considered preferable to obtain an undrowned weir rating using CFD, then apply an empirical treatment of the tunnel hydraulics (which was supported by the results of the original model tests) and of the ultimate drowning of the PK weir after the transition from weir crest to tunnel control.

For further details of the hydraulic design, CFD analyses and flood performance of the PK weirs, see the paper by Ackers *et al* (2013).

Designing the PK weirs for construction
The design of the PK weirs has developed with ease of construction a significant consideration, in particular to facilitate the use of precast concrete units, thus minimising working directly over water by ensuring that most of the on-site construction activities occurred on the original bellmouth rim. A crane platform about 10m from the outside of the bellmouth facilitated the extensive lifting required for removing the original debris boom and walkway, installation of the PK weir units and also the raising of the original valve shaft adjacent to the bellmouth.

In order to maximise the developed weir length, the weir units were pushed to the outside edge of the bellmouth. With the arrangement chosen, this sets the minimum inlet key width, which was chosen to allow a small section of *insitu* wall and base to be cast to stitch individual precast units together. The upstream overhang was limited to approximately half of the whole upstream–downstream length, to ensure the stability of each precast unit while being placed and aligned, before being secured to the foundation, and to ensure stability in transit and standing on site.

Manufacturing the units in precast concrete allows smaller bay widths than would be feasible for *insitu* construction, as no scaffolding is required within the outlet keys. Casting the units inverted allows the casting of the rounded crest to be achieved with greater control than for insitu methods and also allows thinner sidewall sections to be used.

A mixture of waterproofing barriers was employed depending on the connection detail. Between precast and *insitu* wall sections, waterstops have been designed to be the primary barrier. Between the precast units and the foundation concrete, hydrophilic strips were used in conjunction with pressure grouting to provide a double barrier.

Figure 10 shows the completed PK weir structure on top of the original bellmouth spillway and the first overspill in December 2013.

Figure 10. Completed PK weir structure and first overspill, December 2013

VALVE SHAFT RAISING AND MODIFICATIONS

The new overflow level for the reservoir is about coincident with the level of the original floor of the valve tower. The hydraulic control panel for the main intake and scour valves and the manual control for the flow regulation valve which provided the compensation flow were located at this floor level and housed within a steel and concrete superstructure.

The floor level was raised by extending the thick concrete walls of the valve tower by 1.7m to a level of 251.6m OD, so that it would be above the new PMF level. At an early stage the decision was taken not to provide a new superstructure building. Instead a new hydraulic control panel was located within a small kiosk and access to the valves was provided through a large access cover within the steel flooring. The concrete panel of the floor supporting the kiosk was built as a precast unit, so that, should the need arise in the future to remove any of the larger valves, the kiosk could be removed by crane along with the concrete panel.

The crane platform that was used during installation of the PK weir units was also used for craneage work on the valve shaft. The platform was left in place (below the new overflow level) for any possible future lifting requirements.

The original valve shaft had a concrete platform cantilevered out from the main shaft. This was partly removed to create a regular circle, but was otherwise left in place to act as a secure platform from which to erect the valve shaft extension.

New penstocks were provided on the outside of the upper two intake pipes to act as guard valves. These replaced the previous isolation system which involved lowering a large rubber coated copper ball through the cantilevered platform, to be held on the intake bellmouth by hydrostatic pressure.

Access to the valve tower was originally via a concrete bridge which provided access to the bellmouth from the embankment and then followed the circumference of the bellmouth before continuing onto the valve tower. After being satisfied that a good view of the submerged elements of the PK weirs was possible, it was agreed to provide a bridge straight out to the valve tower, spanning the open bellmouth.

Two piers that had supported the original bridge were extended and new slender columns were positioned in an area having the least hydraulic effect within the bellmouth area. This suited the use of a lightweight steel bridge manufactured in five main spans, which were lifted into place by crane and secured. A steel door and barrier were also added to prevent public access to the section of the bridge over the bellmouth.

REFERENCES

Ackers J C, Bennett F C J, Scott T A and Karunaratne G. (2013). Raising the bellmouth spillway at Black Esk reservoir using PK weirs, in Erpicum S *et al* (editors) (2013) *2nd International Workshop on Labyrinth and Piano Key Weirs – PKW 2013*, 20–22 November 2013, Paris, Taylor & Francis.

Erpicum S *et al* (editors) (2011). *Labyrinth and piano key weirs –* PKW 2011, Proceedings of the International Conference on Labyrinth and Piano Key Weirs, 9–11 February 2011, Liege, Balkema.

Lempérière F (2009). *New labyrinth weirs triple the spillways discharge – Data for an easy design of P K weir*, www.hydrocoop.org.

Eller Beck Flood Storage Reservoir – the challenges of low impact flood storage design

P BRINDED, Arup, Leeds, UK
R GILBERT, Arup, Leeds, UK
P KELHAM, Arup, Leeds, UK
A PETERS, Arup, Leeds, UK

SYNOPSIS   The proposed Skipton Flood Alleviation Scheme comprises two flood storage reservoirs and in-town flood defence works. The scheme is to provide a 1 in 100 year standard of protection to Skipton, North Yorkshire.

One flood storage reservoir, on Eller Beck, is located within a golf course and a national park. To ensure that the 14m high earth embankment dam is sympathetic to its surrounding landscape the preferred spillway solution for the Category A dam was a reinforced grass channel.

However, the A65 road embankment immediately downstream of the dam has the potential to create a range of tailwater levels, the greatest of which is only 1m from the spillway crest. This could result in a hydraulic jump forming anywhere on the downstream face of the embankment. Consequently, the design team investigated the performance and restrictions in the use of reinforced grass systems under hydraulic jump conditions.

The proposed site is constrained by transport infrastructure on both abutments as well as significant geological features, which have required detailed investigation and analysis to confirm their likely impact on the earth embankment dam.

This paper explores the decision-making process which led the design team to the adopted solution and discusses the design challenges encountered.

INTRODUCTION
Skipton is an historic market town situated in the upper Aire Valley on the southern edge of the Yorkshire Dales National Park. The town is regarded as the gateway to the Yorkshire Dales and the majority of the town centre is designated as a conservation area.

The watercourses which pass through the town are critical to its character; these include Eller Beck, Waller Hill Beck and the Leeds-Liverpool Canal.

The town is surrounded by steep hills apart from to the south, which opens out into the Aire Valley.

The town of Skipton has experienced a significant degree of flooding with major events occurring in 1908, 1979, 1982, 2000, 2004 and 2007. There are currently 378 residential and 165 non-residential properties at risk of flooding during a 1 in 100 year flood event.

A Flood Alleviation Scheme is being developed for Skipton by the Environment Agency to tackle the flooding issues. This scheme will result in Skipton having a 1 in 100 year standard of flood protection. The scheme comprises:

- Eller Beck Flood Storage Reservoir
- Waller Hill Beck Flood Storage Reservoir
- In-town flood defences at three distinct locations

The sites of the dams were pre-determined and their development is crucial to the success of the overall scheme. This paper is concerned primarily with the design of Eller Beck Flood Storage Reservoir and the issues faced on a challenging site.

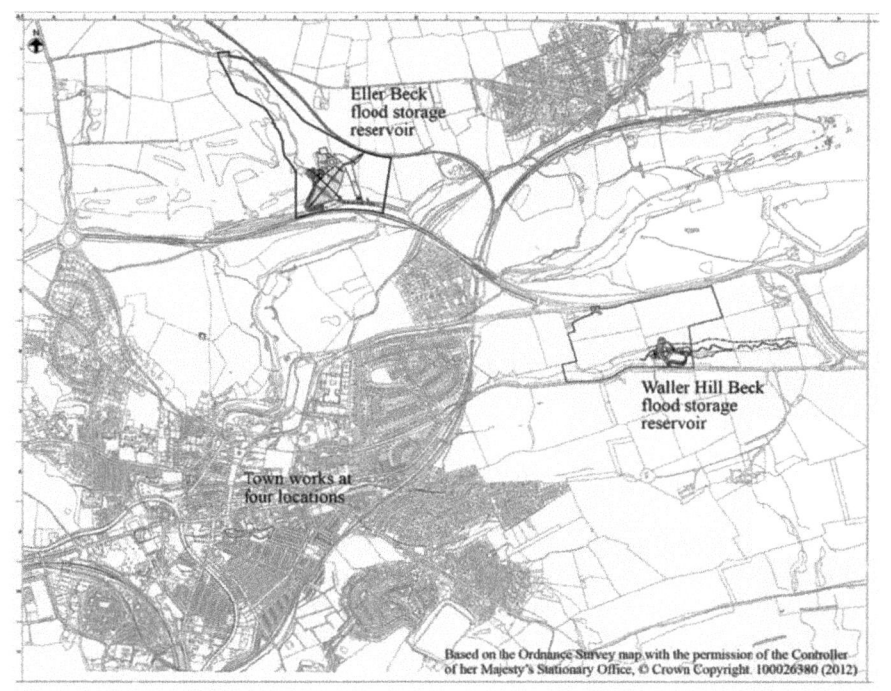

Figure 1. Plan of Skipton FAS

ELLER BECK FLOOD STORAGE RESERVOIR

To the north of Skipton, Eller Beck flows through farmland before passing through land operated by Skipton Golf Club. It is joined by Embsay Beck immediately upstream of the A65 road embankment before flowing through Skipton Wood and eventually through Skipton itself.

Eller Beck and Embsay Beck combine to form a direct contributor to flooding in the town. To ensure the defence heights in the town are kept to a manageable level and the historic nature of the town is not adversely affected, flows above 17m³/s need to be restricted and stored upstream. During the 1 in 100 year flood event (before the spillway is in operation) the reservoir across Eller Beck will store 433,000m³ of water. This results in an embankment crest 13.63m above the valley floor.

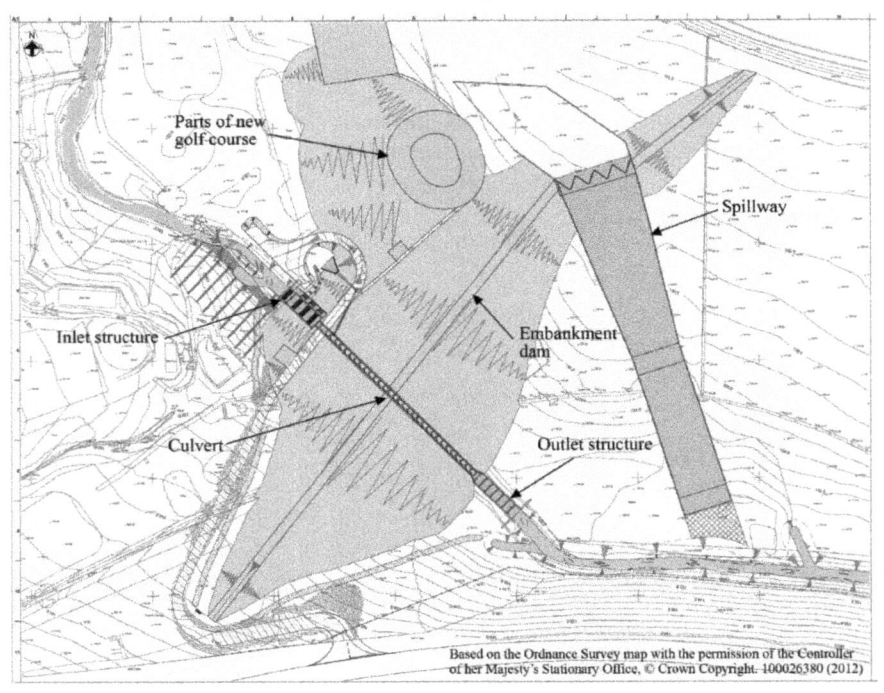

Figure 2. General Arrangement of Eller Beck dam

Eller Beck site constraints
The topography of the valley operated by Skipton Golf Course led to the site being selected as the most feasible location for a flood storage reservoir, upstream of the A65 road embankment. The valley upstream of this location is much wider and flatter and therefore a similar storage solution would not be feasible. Embsay Beck is much more developed, including an existing reservoir and a sewage treatment works. Early in the project development it was realised that the only feasible solution would be to

provide all storage provision solely on Eller Beck, albeit the most appropriate location, the site selected for the Eller Beck Flood Storage Reservoir, has a number of challenging constraints. These include catchment characteristics, geology, site context and existing infrastructure. This paper explores a few of these constraints in more detail and how they have been overcome through the design process.

SITE CONTEXT

The proposed dam site is located at the southern end of Skipton Golf Course and the plan area of the dam conflicts with a number of holes within the course. Consideration was made during development of a design solution to locate the dam embankment to ensure minimal permanent disruption to the golf course; however some permanent disruption was inevitable. To compensate for this, parts of the golf course (currently one green and two tees) are incorporated into the upstream shoulder of the dam, through construction of berms within the upstream face and careful planning of access tracks to ensure inclusion of these spaces for both maintenance vehicles and users of the golf course.

Existing infrastructure

The proposed dam site is bounded at each abutment by two critical routes of transport infrastructure. The left abutment ties into the valley side close to a Network Rail track, which services commercial trains to a nearby quarry. The right abutment ties into the natural valley side immediately upstream of the A65 road embankment.

Both of these existing infrastructure routes have an impact on the dam design - both in the height of the embankments and the location of the dam. The embankment height and peak reservoir water level have been carefully engineered to ensure they do not impact adversely on these stakeholders. As the design progressed, the presence of the A65 road embankment has resulted in relocating the dam further upstream than originally envisaged. This was due to concerns that the significant dam structure could impact on the stability of the A65 road embankment, as well as the risk of induced settlement due to the local geology.

GEOLOGICAL CONSTRAINTS

A number of geological constraints were identified through the feasibility stage desk study and ground investigations, although a number of the constraints were not fully realised until the detailed design stage ground investigation. The methods of ground investigation had to consider the sensitivity of working on the Golf Course, as well as obtaining the quality data from the varied ground conditions. A summary of the site geology is provided below:

- Alluvial Deposits - Soft low strength Alluvial deposits in the base of the valley under the proposed embankment footprint.
- Head Deposits – Predominantly firm clay which provides a good formation for the embankment, however zones of granular content are present.
- Glaciolacustrine Clay – Soft thinly laminated low strength clay up to 4 metres thick on the valley sides. Deposited with glacial lake formed due to blockage of the valley by ice sheet movement. Subsequently eroded in the base of the valley by Eller Beck.
- Glacial Till Deposits – Predominantly stiff clay which provides an impermeable base to the reservoir, however granular zones are present at shallow depth within the embankment footprint.
- Bedrock – Steeply dipping interbedded Limestone, Mudstone and Siltstone of the Clitheroe Formation. Generally encountered at over 10 metres depth, however at the downstream toe of the embankment limestone outcrops in the river bed. A normal fault also shown on the geological maps in the north of the site.
- Groundwater – Artesian groundwater was encountered within bedrock during drilling operations. Perched groundwater is also anticipated to be contained within granular glacial layers.
- Historical Landslip – The geomorphology of the site indicates movement has occurred within the Glaciolacustrine deposits in the reservoir basin.

Geotechnical Engineering Considerations
The challenging ground conditions and site constraints resulted in a number of geotechnical engineering solutions being incorporated into the design of the dam embankment. Construction over the large thickness of soft Glaciolacustrine Clay has required careful consideration. The material has governed embankment stability design with slope gradients limited at 1 in 4. Piezometer monitoring and a limit on the rate of construction are also proposed. Significant settlements are also anticipated. To remove the impact of settlement on the spillway weir and its associated retaining walls, localised excavation and replacement is proposed.

The historical landslip within the Glaciolacustrine Clay has been analysed for stability during operation of the reservoir. Past contributions to movement are considered to be river erosion at its toe, and elevated pore pressures within the hillside from the underlying granular Glacial Till layers. Further slope movement was considered a risk to the embankment inlet structure. A section upstream of the embankment is therefore proposed to be remediated with deep shear trenches, which increase the mass shear strength of the slip surface and promote drainage of the underlying Glacial Till.

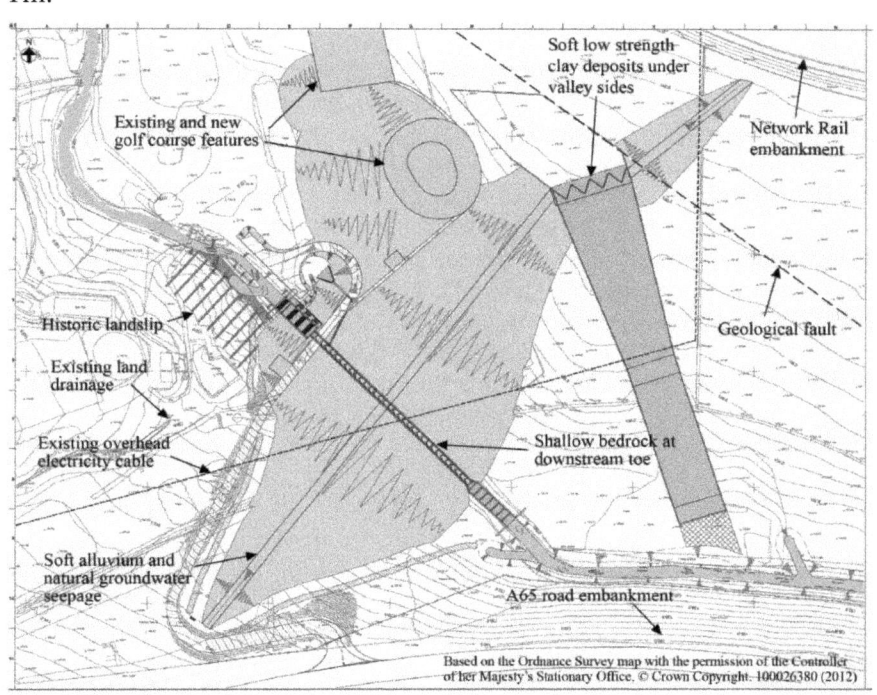

Figure 3. Plan of major site constraints

The shallow bedrock encountered in the downstream toe presents a risk to the embankment integrity through differential settlement. A geophysical survey was commissioned which included seismic refraction and electrical resistivity with the aim of mapping the depth of bedrock. A 3D foundation settlement model was created which found the induced differential settlement due to the bedrock topography was limited. Design of the normal-flow culvert through the embankment, however, required localised over excavation of the shallow bedrock to allow a gradual settlement profile to be achieved. The culvert is designed with a pre-cambered invert to account for larger settlement under the embankment crest.

The site groundwater conditions are considered to be beneficial to reservoir performance. Artesian groundwater was encountered within bedrock at depth which indicates the Glacial Till provides an impermeable base to the reservoir. The interbedded nature of the bedrock and artesian strikes encountered also show that down-valley flow of groundwater through bedrock is low. Seepage modelling has proven this to be the case, however a key trench through near-surface granular zones is required, and a drainage blanket is required over shallow rock at the downstream toe.

HYDRAULIC CONSTRAINTS

The catchment area for the proposed reservoir is in the region of 12.4km². This leads to a reservoir outflow of 159.5m³/s during the PMF event.

Eller Beck joins with Embsay Beck at a confluence immediately upstream of the A65 road embankment. The combined watercourse flows through two 2.85m diameter Armco-type culverts under the A65 road embankment. The watercourses have separate catchments, and during a storm event these catchments may respond in different ways. This has an impact on tailwater levels, as the tailwater may be generated by one or both watercourses as shown on Figure 4 below. The combined effect of the two watercourses result in predicted tailwater levels ranging between a relatively low level, up to potentially 1m below the spillway crest during the combined PMF flow from the Eller Beck FSR spillway and a PMF flow on Embsay Beck. This has resulted in a more considered approach to spillway design, ensuring the solution would be suitable for all potential tailwater conditions.

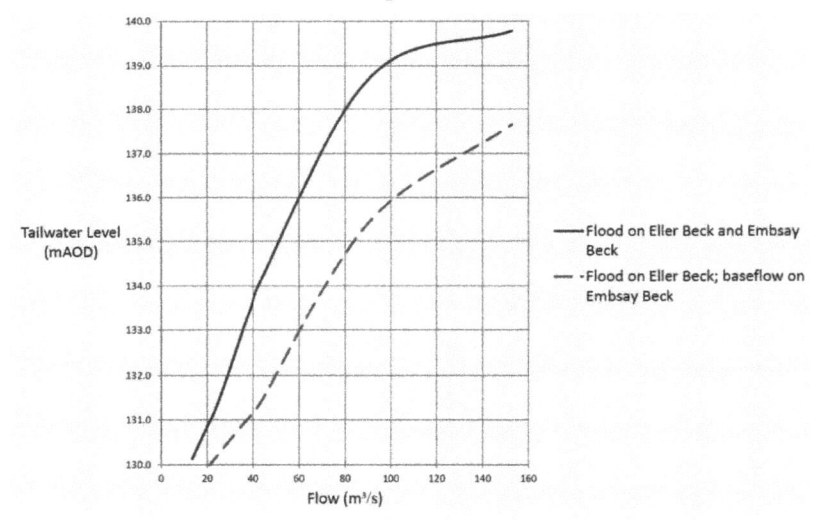

Figure 4. Graph to show anticipated tailwater levels. Note that the spillway weir level is 140.80mAOD.

Spillway Optioneering

Once the dam location, the maximum allowable height of the structure and the flood rise were confirmed, the most appropriate spillway solution could be investigated and developed. Any spillway solution must convey the PMF outflow but also be able to withstand the potential differential settlement in the embankment due to the underlying geology. It must also be visually sympathetic to the local environment whilst minimising peak reservoir water level so that the adjacent infrastructure is not adversely affected. This led to a preference for a reinforced grass spillway over the central section of the dam.

The outline design was undertaken following guidance provided in CIRIA 116 (Hewlett et al, 1987). Following an iterative design process a weir length of 100m was adopted to ensure favourable hydraulic conditions on the downstream face of the embankment. The predicted velocities on the spillway during a PMF event were in the region of 7.3m/s with a Froude Number of 5.2 (assuming the flow reached terminal velocity) for an anticipated duration of 10 hours.

These parameters suggest that the reinforced grass spillway should comprise a concrete system. However, with the varying tailwater level discussed above, the reinforced grass spillway system must also be capable of withstanding the fluctuating pressures, velocities and erosive forces of a hydraulic jump, as supercritical flow from the spillway meets the deeper tailwater.

The CIRIA design guide states that it may be advisable to provide heavier armour in the region of the hydraulic jump. This could be in the form of a concrete scour apron. However, on the Eller Beck embankment the hydraulic jump could occur at any point on the downstream face. This suggests that a concrete scour apron is required over the entire downstream face of the embankment. To avoid the need for this visually intrusive and expensive solution the performance of reinforced grassed systems under hydraulic jumps was investigated further.

Three reinforced grass systems were considered: cable-tied concrete blocks, in-situ concrete surface and open stone asphalt. All three materials are proven in reservoir spillway applications, therefore the investigation focussed on their performance under hydraulic jump conditions.

Cable-tied concrete blocks
Cable-tied concrete block revetment systems are designed to provide protection against embankment erosion above and below water level whilst ensuring the flexibility to cope with variations in embankment profile.

During the CIRIA 116 field trials, a cable-tied concrete block system withstood velocities of 8m/s when laid with a geotextile under-layer. The under-layer is of fundamental importance to the stability of the structure as a whole.

A cable-tied concrete block system can withstand the anticipated maximum velocities and provides a solution sympathetic with the local landscape. However, under hydraulic jump conditions there are two concerns:

- Individual blocks be plucked out of the revetment system by negative pressures within the hydraulic jump, causing the revetment to systematically unravel;
- The under-layer may become eroded away, leading to local settlement of the revetment and increased turbulence of flow.

Due to the concerns detailed above and the lack of evidence of the product's performance under hydraulic jump conditions, the authors did not consider cable tied concrete blocks suitable for use on the Eller Beck spillway.

In-situ concrete surface (concrete grass)
In-situ concrete surfaces are cellular reinforced concrete slabs. The voids are infilled with topsoil and seeded. In-situ concrete surfaces are often used on reservoir spillways and flood defences. A geotextile is recommended as an under-layer to prevent sub-grade scour in the event of loss of infill to the individual voids. During the CIRIA 116 trials a product reached the maximum test velocity of 8m/s.

As an in-situ concrete surface is a monolithic construction material it is not envisaged that parts of the revetment would be plucked out by the uplift forces generated in a hydraulic jump. However, joint details are important in order to prevent differential settlement between panels. Differential settlement may lead to deformations in the surface which in turn may lead to unfavourable hydraulic conditions.

It is also feasible that the under-layer may become scoured. Being monolithic and therefore unable to articulate, scour holes and embankment settlement may remain unnoticed.

Once again, the authors found no information or guidance with regards to hydraulic jumps on in-situ concrete surfaces at the scale predicted on Eller Beck.

Open stone asphalt
Open Stone Asphalt (OSA) is discussed in detail in Bieberstein *et al* (2004). OSA is a gap-graded mixture of mastic asphalt and aggregate. OSA is a flexible material and therefore can follow surface deformations and settlement of new earthworks.

Sufficient voids remain in the revetment, ensuring it remains permeable to water and air and can accommodate vegetation roots. It is noted that the voids in OSA are small and irregular.

The authors concluded that the product can withstand the anticipated velocities at the Eller Beck flood storage reservoir. As with in-situ concrete surfaces, the product is monolithic therefore the plucking out of individual elements is not considered an issue. Although one known supplier has examples of OSA being placed in zones of anticipated hydraulic jumps in Europe, the examples are not on the same scale as Eller Beck, namely the size of embankment, anticipated velocities or duration of overtopping.

Conclusions

The aforementioned products are proven in their use in reservoir spillways. They can be a cost effective, visually acceptable solution. However, the design team concluded that none of these systems could demonstrate satisfactory evidence that they could withstand both the hydraulic and geotechnical constraints at the Eller Beck site. In the context of a Category A dam located upstream of a populous town it was concluded that a reinforced grass spillway was not appropriate for this site.

Spillway development

In order to safely convey the PMF flow over the dam a reinforced concrete spillway solution was adopted. Numerous alignments and weir types were considered, with the aforementioned constraints still applicable.

To minimise the visual impact of the structure as much as possible a labyrinth weir was chosen. A labyrinth weir provides the opportunity to house a relatively long weir in a relatively narrow structure. The maximum head over the weir was set to 1.5m in order to prevent any adverse impacts of the reservoir on the adjacent infrastructure.

Hand calculations were undertaken using the design guidance in Tullis *et al* (1995) and Falvey (2003). An approach for determining the optimum design coefficient of discharge was adopted using Ghare *et al* (2008). The resulting design housed a 51.4m long weir in a 31.8m wide structure. The initial weir was 2m high with a labyrinth angle of 35°.

The spillway channel depth was limited to 2m due to geotechnical constraints and concerns regarding the stability of the deposits on the valley side. The resulting channel remained at 31.8m wide to accommodate PMF flow depth and freeboard requirements. The channel had a slight curve, ensuring it remained normal to the valley contours, in order to minimise the cut/fill during construction. The channel was also divided in to two distinct gradients with a third transitional zone. This ensured that the channel remained within the topography of the hillside, also minimising the cut/fill during construction.

The spillway channel terminated in a Type III stilling basin designed for the 1000 year flow. Above the 1000 year flow the stilling basin is significantly submerged by the tailwater.

Physical hydraulic model

In order to confirm the hydraulic performance of the spillway a physical hydraulic model was commissioned. The model extents included the approach channel and upstream face of the embankment, the spillway structure and adjacent topography, including the A65 road embankment.

The physical hydraulic model confirmed the labyrinth weir worked effectively. The model proved that the weir height could be reduced, thereby reducing the volume of excavation required in construction.

Figure 5. Initial testing of labyrinth weir (at half PMF flow)

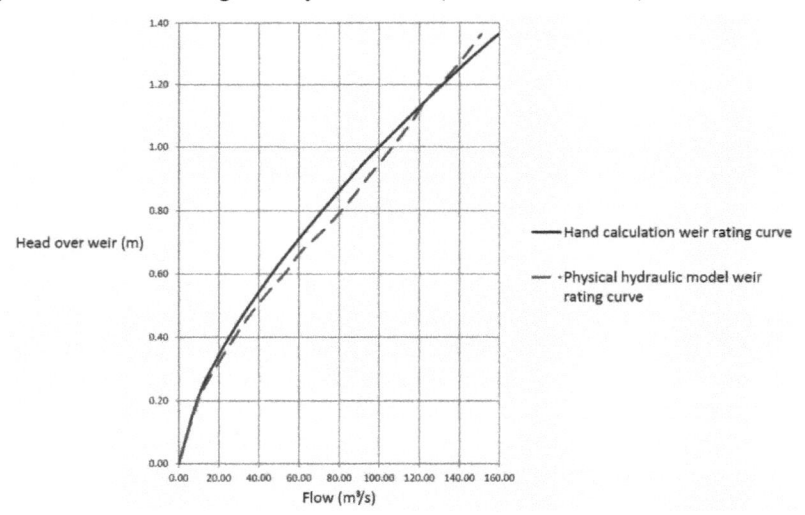

Figure 6. Comparison of weir rating curves

Figure 6 shows a comparison between the two weir rating curves (design and modelled). The data set from the physical model can be considered in two distinct parts. Up to 80m³/s the weir is more efficient than the hand calculation results showed, owing to a greater coefficient of discharge than the constant coefficient of discharge assumed in the hand calculations.

After 80m³/s the weir starts to lose efficiency. This is likely to be due to two factors:

- The weir starts to become hydraulically shortened as the apexes of the labyrinth become partially submerged by rising tailwater
- Jets from the upstream apexes start to strike each other causing flow disturbance (known as nappe interference)

Both sets of results suggest that the weir meets the design criteria of 1.5m maximum head for a discharge of 159.5m³/s.

Other factors which may impact the efficiency of the weir include:

- The weir is not fully perpendicular to the approach channel therefore approach losses may be present
- The weir shape for the physical model was chamfered for ease of construction, which may be slightly less efficient that the quarter-round crest shape assumed in the Tullis design method

Figure 7. Photograph of the final model with low tailwater.

Due to the curvature of the spillway channel, it was found that the bottom section was inefficient and there was a preferential flow path down one side of the channel. During higher flows, water was noted to jet over the spillway wall and cause significant eddying when coming into contact with the varying tailwater. Optioneering of the physical model minimised the risk of eddying and potential for scour of the valley side, as well as a reduction of the gradient of the channel and its width. This reduced the size of the structure and minimised the excavation required during construction.

The physical model also showed that the chute and baffle blocks of the Type III stilling basin were inefficient in low flows and ineffective once submerged with tailwater. The stilling basin was therefore omitted from the design, in favour of a 1m high end sill.

CONCLUSIONS

This paper describes the constraints and design development of the dam and spillway of the Eller Beck Flood Storage Reservoir.

The site was constrained by infrastructure at the abutments of the dams; complex geological issues and hydraulic constraints due to the presence of the A65 road embankment and widely varying tailwater levels.

The complex nature of the glacial geological features within the valley has required detailed geophysical and intrusive ground investigations. The findings of these investigations have required significant geotechnical engineering considerations be incorporated into the embankment design. Understanding of the local hydrogeological conditions has allowed efficiencies in the design of the reservoir. Verification of the stability of the reservoir basin during its operation has also required design of remedial works to a historical landslip.

Initially, the favoured solution was a reinforced grass spillway which would have resulted in a more environmentally sympathetic solution. However, the multitude of site constraints led the design team to conclude that a more traditional reinforced concrete spillway was required.

This paper summarises the technical issues that can be faced by a design team when the site selection process suggest only one feasible site. The paper aims to show that an engineering solution can be found to suit any number of technical issues, in this case the geological and hydraulic issues being paramount and that a fitting and sympathetic solution can still be achieved.

REFERENCES

Bieberstein A, Leguit A, Queisser J and Smith R (2004). Downstream Slope Protection with Open Stone Asphalt. In *Long Term Benefits and Performance of Dams; Proceedings of the 13th Biennial Conference of the British Dam Society, Canterbury*. Thomas Telford Ltd, London, UK.

Falvey H T (2003). *Hydraulic Design of Labyrinth Weirs*. ASCE.

Ghare A D, Mhaisalkar VA and Porey PD (2008). An Approach to Optimal Design of Trapezoidal Labyrinth Weirs. In *World Applied Science Journal 3 (6): 934-938*. IDOSI Publications.

Hewlett H W M, Boorman L A, and Bramley M E (1987). *Design of reinforced grass waterways CIRIA 116*. CIRIA. London, UK.

Tullis J P, Amanian N, and Waldron D (1995). Design of labyrinth spillways. *J. of Hyd. Engrg, ASCE, Vol 121, No. 3, pp. 247-255*

Shon Sheffrey Reservoir – Labyrinth overflow and replacement of masonry spillway

R J TERRELL, Dŵr Cymru Welsh Water,
D M PRISK, Black & Veatch
J C ACKERS, Black & Veatch

SYNOPSIS   Lack of adequate freeboard and concerns over ability of a masonry spillway to pass the design flood were resolved by providing a labyrinth weir and replacing the masonry spillway. The existing spillway was evaluated using computational fluid dynamics and the design was tested by physical modelling. The design rehabilitated the reservoir by providing improved freeboard and durability to withstand the design flood.

INTRODUCTION

Shon Sheffrey reservoir is impounded by an earthfill embankment dam 260m long and 15m high, with a puddle clay core, located 3km to the northwest of Tredegar in South Wales.

Investigations demonstrated improvements were required to ensure the overflow and spillway can safely convey the design flood without posing significant threat to the integrity of the dam. The dam was found to be at risk in three ways; high velocities may result in masonry being plucked from the spillway, the flow may overtop the spillway and there is insufficient freeboard to the dam.

Additional issues involved the poor condition of the drawoff and scour system; a risk of flooding at the adjacent water treatment works; clay ingress to the drawoff tunnel along with seepage past and voids below the overflow. These were resolved as part of the project but not discussed further here.

BACKGROUND

The dam was constructed in 1886 impounding the Sirhowy River to provide drinking and industrial water to the Tredegar area. The dam crest was raised by 1.8m in 1948 by providing a mass concrete crest block, raising the overflow weir and providing an auxiliary overflow weir.

Historically the dam has suffered seepage issues below the spillway and into the drawoff tunnel. Grouting operations were completed in 1984, including installing a row of sheet piles, with further grouting in 2013. Seepage adjacent to the spillway and into the tunnel has been reduced.

The masonry spillway has suffered damage through blocks being plucked from the surface during spill events. The spillway walls had moved inwards in places; thought to be due to maintenance loading (grass cutting) behind the walls. The spillway floor has been partly replaced with concrete slabs where previously damaged.

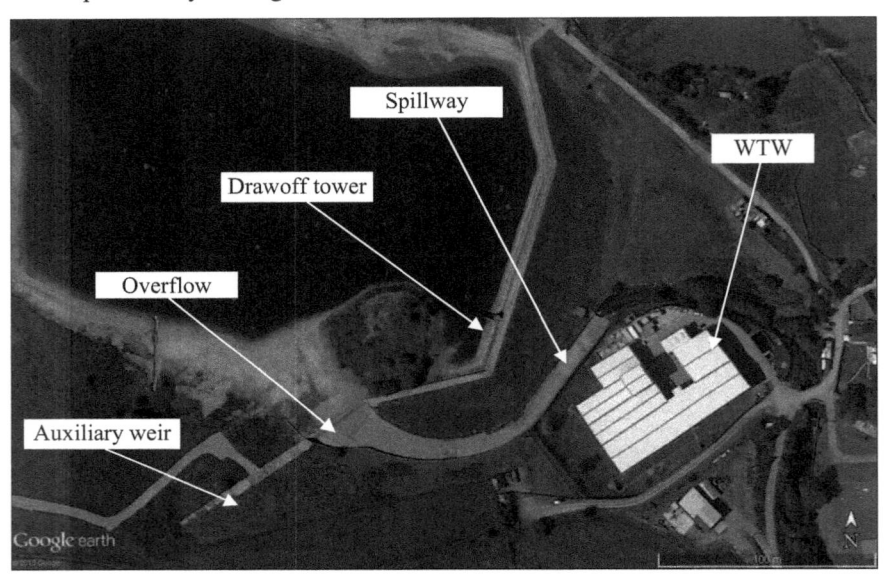

Figure 1. Shon Sheffrey reservoir

INVESTIGATIONS

The statutory inspection of 2010 required a number of investigations to be undertaken. Those relating to the overflow and spillway were to determine whether improvement works are required to ensure that the overflow and spillway can safely convey the design flood without posing a significant risk to the integrity of the dam. If these demonstrated concern then a feasibility study was to be carried out.

The three issues related to the spillway that raised concern are:

- The estimated PMF surcharge is close to the dam crest leaving no residual freeboard for wave action.
- Flow behaviour in the spillway during the extreme floods may result in significant quantities of high-velocity water overtopping the sidewalls and eroding the downstream shoulder of the dam.

- Performance of the masonry spillways at Boltby in 2005 and Ulley in 2007 has raised concerns generally over the durability of old masonry spillways at reservoirs in the UK.

The 2010 inspection also required short term rehabilitation works to the spillway and a separate investigation into clay ingress into the drawoff tunnel. The spillway works involved rebuilding short sections of side wall and local repointing to the floor and walls. The clay ingress investigation works resulted in the grouting contract of 2013. These two items are not discussed further in this paper.

A programme of investigations was undertaken described in Table 1.

Table 1. Investigation works

Problem	Investigation	Outcome
Unreliable record drawings	Topographic survey	DCWW held a good quantity of record information but substantial elements were various historic proposals for modifications. The survey set a baseline to allow further study.
The design flows and available freeboard were to be confirmed	Review of the flood study and determine the overflow stage rating curve	The PMF is 139m³/s with 95m³/s passing over the main overflow and 44m³/s over the auxiliary weir. Remaining freeboard at PMF is 100mm.
Flow behaviour of the existing weir	Computational fluid dynamics (CFD)	The flow behaviour was determined for the 150, 1000, 10,000-year and PMF flows with risk areas identified to focus an outline design. The maximum velocities at bed level reached 15m/s. Locations of out of channel flow were identified.
Unknown existing spillway construction details	Cores through the spillway and into the foundation	The masonry block thickness is 250mm to 300mm laid either directly onto natural clay formation or onto a gravel bed. The gravel increased 'plucking' risks. Cores did not find voids below the spillway.
Durability of existing materials	Materials testing	Testing of materials to establish their strength and physical characteristics.
Voids upstream of the overflow	Rotary coring	The rotary cored drillholes identified significant voids upstream of the overflow. These were grouted as part of the investigation.

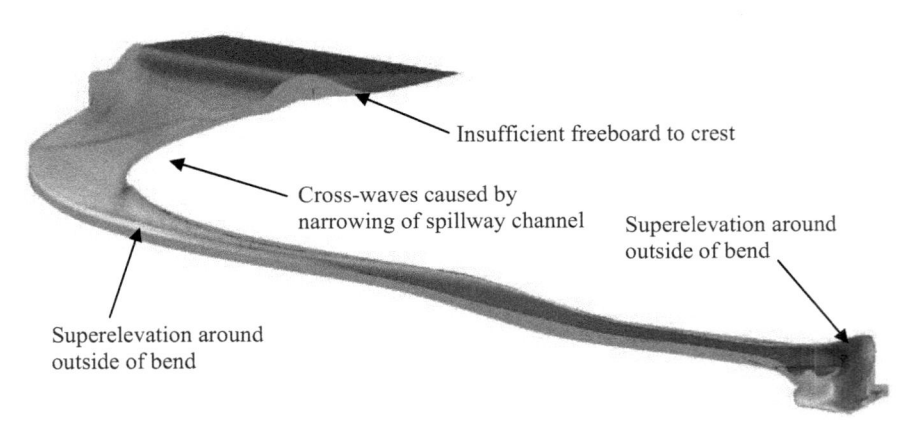

Figure 2. CFD analysis and significant issues

REFURBISHMENT SCOPE

The investigation works identified the deficiencies in the current overflow and provided sufficient information to complete a feasibility study. After considering a number of options DCWW agreed the scope for detailed design. The scope of the refurbishment works included:

- A labyrinth weir to reduce the flood surcharge and thereby leave sufficient residual freeboard for wave action
- Provide a new spillway to ensure PMF flows are fully contained
- Replace components of the drawoff system to facilitate the works

The existing spillway is of partly masonry and partly concrete slab construction with low masonry sidewalls. CFD modelling at the investigation stage demonstrated the high velocity risks to the masonry and that side walls would be overtopped at spills in excess of a 150-year event.

A labyrinth weir was proposed to lower the water level during the design flood without the need to lower the overflow level, to reduce the storage volume, or to heighten the dam. The shape of the spillway approach allowed a large four spoke labyrinth to be adopted, providing an efficient labyrinth design, which was also demonstrated through physical modelling.

In order to construct the new overflow and spillway the reservoir had to be lowered for a prolonged period of time. The lower existing drawoff valve was found to be jammed partially open and required replacement. It was decided to replace the full drawoff stack as part of these works; provide an intake screen to prevent future blockage and replace the downstream isolation valve.

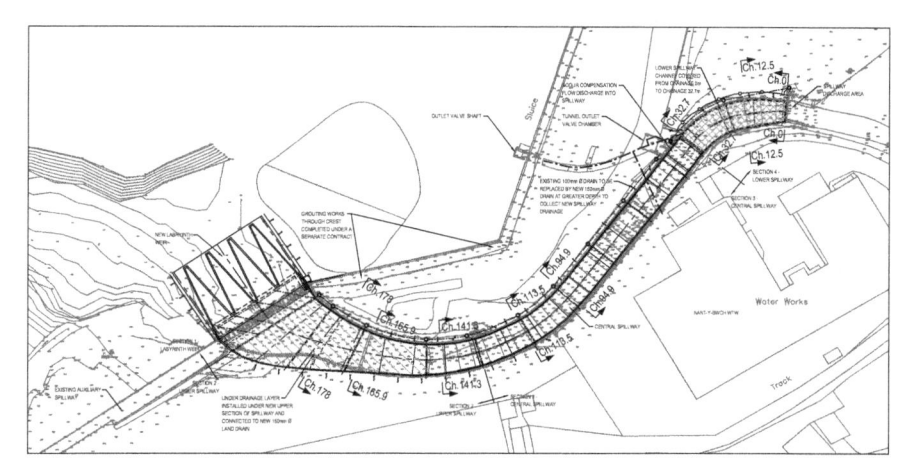

Figure 3. The proposed solution of labyrinth and new spillway

LABYRINTH WEIR

A number of labyrinth weirs have been constructed recently to provide additional freeboard to reservoirs. The labyrinth provides an extended weir length to reduce flow per unit structure width and therefore a lower upstream water level. The method has inefficiencies where the upstream corners of the labyrinth are drowned or flow is starved from the downstream corners. The most efficient design is a single V arrangement. Multiple V arrangements are used to permit installing long and low head weirs in confined places, for example where the upstream shoulder drops steeply.

A four segment labyrinth was possible at Shon Sheffrey because of the local topography. DCWW had carried out a bathymetric survey previously which identified a large and shallow shelf at the approach to the existing overflow. This allowed an approximately square arrangement to be installed. Both a three-cycle and a two-cycle labyrinth were also buildable in the space available, but from a high level cost/benefit analysis the four-cycle option was considered the most efficient arrangement for this particular location.

Labyrinth weirs must be designed for an unusual flotation risk. The structures are normally built on the wet side of the original overflow and dam cutoff, where it is presumed the water pressures exerted below the base slab will equate to the weir level. This pressure is balanced in the flooded upstream V-shaped bays but not within the empty downstream bays. This results in an overturning flotation risk with the structure rotating back about the upstream heel into the reservoir. A base slab of 900mm thickness is required at Shon Sheffrey to restrain this overturning flotation risk. The base slab is 26m long by 33m wide. Constructing the structure downstream of the existing overflow was considered to avoid this issue and allow a much thinner base slab, but this was not taken forward for two reasons. The ground had not been preloaded previously and differential settlement may occur along the 180m long weir (previously 33.5m long). Also, the structure would be more visually imposing and may have caused planning objections.

By providing a more efficient main overflow, the main spillway attracts flow away from the auxiliary weir. To provide the required freeboard the PMF reservoir level also needed to be below the auxiliary weir crest. Due to these two issues the main spillway flow at PMF increased from 95m^3/s to 139m^3/s. This did provide an added benefit. The water treatment works commissioned in 1996 had been built across the auxiliary spillway and would be flooded at events in excess of the 1000-year event. This risk is removed through carrying all the spillway flows through the controlled spillway.

Figure 4. Physical model demonstrating the labyrinth at PMF flows and 1000 year flows

SPILLWAY REPLACEMENT
To overcome the risks of the very high velocities during a spill event causing the masonry blocks to be 'plucked' from the bed, the entire spillway was to be replaced in reinforced concrete. A number of issues needed to be resolved:

- Erosion of the spillway floor during construction
- Construction work causing blockages resulting in out-of-channel flow
- Unforeseen voids below the spillway
- Energy dissipation at the discharge point

To achieve an efficient labyrinth weir design the upper spillway needed to be excavated to below its current formation level. This posed the risk of eroding the channel bed if a storm event spilled during early construction. The spillway is very wide at the upstream end; 32m wide reducing to 8m. This enabled the upper section to be constructed in two halves, always retaining a minimum of 8m wide flow channel and retaining the existing protection to the dam.

Where possible the spillway was to be constructed within the footprint of the existing spillway to avoid both excavation costs and to avoid the spillway formation being exposed to erosion during construction. The construction sequence was restricted to building in a downstream direction only. This ensured that if there was a spill during construction it passed down the new spillway and dropped into the existing spillway.

A number of rotary core holes had been drilled through the spillway at the investigation stage, partially to investigate if voids were present. A fault runs along the spillway, which may be the cause of the voids upstream of the overflow. The fault may form springs below the spillway which could result in local loss of ground. Although no voids were found and there was no evidence of voids downstream of the overflow, this was considered to remain a risk, so the spillway was designed to span a 3m void at any point. The design involved installing a key at movement joints at 18m intervals along the spillway. The formation was exposed at each key allowing further investigation of the void risk.

The spillway discharges adjacent to the water treatment works where erosion could cause concern. A number of options to direct flow down the valley and dissipate energy were considered. To inform this an ISIS-TUFLOW model was completed. This demonstrated that the valley downstream of the spillway will be sufficiently flooded at each flow rate to provide the necessary energy dissipation. The model also confirmed that flows would not circle back around the discharge point to erode the dam toe.

To prevent the high velocity flows leaving the lower spillway where it turns sharply, a twin box culvert design was used. There were concerns that the boxes could become drowned at high flows, but from the physical model study it was found that the hydraulic jump was retained within the culvert at all flows, even with the downstream exits fully submerged.

Figure 5. Physical model of the spillway and discharge culvert.

CONCLUSIONS

A number of safety concerns at Shon Sheffrey reservoir have been resolved. The labyrinth spillway will lower the reservoir PMF water level and provide sufficient freeboard to contain wave action. Replacement of the masonry spillway in its entirety ensures that flows are contained and the risk of masonry blocks being plucked leading to erosion removed.

It would have been interesting, and possibly helpful for future designs, to complete a CFD of the final arrangement and compare this with the physical model. The accuracy of the labyrinth physical model and CFD model could be compared to consider if a CFD approach to designing labyrinth weirs is suitable. This may allow a greater variation of weir shapes to be considered. For this project, however, the physical model contractor (CRM, Bolton) were able to modify the model where required quickly and to such accuracy that the perceived benefits of CFD over physical modelling may not have been realised. With the physical model it was possible to model one of the labyrinth bays being blocked and the downstream double culvert being half blocked very quickly, albeit at the cost of a soaking.

Figure 6. Modelling blocked labyrinth V and discharge culvert.

Refurbishment of the drawoff works was completed in 2013 to facilitate lowering the reservoir level and beginning construction early in 2014. Completion is expected late 2014. The spillway works were designed by Black & Veatch for client Dŵr Cymru Welsh Water and constructed by BAM Nuttall Ltd.

Maintaining the Safety of our Dams and Reservoirs
ISBN 978-0-7277-6034-0

ICE Publishing: All rights reserved
doi: 10.1680/mdam.60340.369

The issues associated with the discontinuance of impounding reservoirs

G PICKLES, J N Bentley Ltd, Leeds, UK
D REBOLLO, Mott MacDonald Ltd, Leeds, UK

SYNOPSIS   The challenges underpinning the construction or amendment of large impoundments are broadly acknowledged. This is rarely the case when returning submerged landscapes close to their original state. This paper will describe the challenges faced when discontinuing an impounding reservoir, including the environmental, hydraulic and construction sequencing constraints and the strategies used to overcome these challenges. It will also cover the lessons learnt, the importance of early contractor involvement and the need for effective third-party liaison to achieve timely and cost-effective delivery of the works.

SETTING

The two reservoirs discussed in this paper were completed at the turn of the 20th century for public water supply. Both comprise earth-filled embankments, impounding 510,000m^3 (Beaver Dyke Reservoir) and 58,600m^3 (Oakdale Upper Reservoir) of water respectively, with overflow weirs and masonry spillways. Following Section 10 inspections of the reservoirs, the QCE recommended that the overflow and spillway arrangements be reconstructed and that works be undertaken to improve the draw-down capacity. The undertaker reviewed the feasibility of the works required. This resulted in the proposal to discontinue the reservoirs due to the fact that neither is required for water supply.

The Design-Build approach

Both schemes have constrained programmes owing to their compliance dates set against the lengthy process of applying to vary the existing statutory provisions (Acts of Parliament) of the impoundments.

At one of the sites the surrounding land is privately owned and used commercially which only permits a six month window to carry out the works each year. These factors combine to make it paramount that the programme is as lean as possible. This was achieved by two key methods:

- Carrying out early site works (enabling works, such as drainage; temporary haul road; demolition) shown in Figure 1. This ensured that there are no impediments to the excavation works.
- Whilst, traditionally, the design would be fully completed before construction began, both schemes were planned such that the design of the main works continued in parallel to construction of the enabling works. Although there was an inherent risk in this approach, in that the design may change, the risk was reduced by having submitted and approved an outline design with the local planning authority during the preliminary design stage of the project.

The above approach allowed for the contractor's site manager to be involved from the onset of the project, helping to identify site constraints (access, surplus materials, etc.) early and enabling the design to be tailored to the specific site conditions and the safest construction method.

Figure 1. Haul roads construction on the upstream face of the embankment taking place in parallel to the design of the main works

Design philosophy
At outline design it was proposed that the reservoir volumes be reduced to 5,000m^3. This will ensure that neither reservoir falls under the ambit of the Flood and Water Management Act 2010 (HMSO, 2010) amendments to the Reservoirs Act 1975 (HMSO, 1975). Both discontinuances entailed excavation of the embankment and construction of a permanent raw water channel at the lowest point, perpendicular to the dam itself, as shown in Figure 2.

From a design perspective, the principal issue associated with the proposed approach to discontinue the reservoirs is the surplus of materials resulting from the required excavation: both earth-fill and pitching stone.

Furthermore, both sites are in remote upland catchments and which make for difficult access. This led to a design philosophy that maximised re-use of salvaged materials. Additional site constraints included flood contingency: scour and supply pipes had to be maintained operational throughout the works.

Figures 2a and 2b. Before and after visualisation of one of the reservoirs viewed from the downstream side.

Hydraulic design
The proposed approach to discontinue the reservoirs allowed for the vertical channel alignment and cross-section to be adjusted to provide optimum hydraulic conditions. To capitalise on this, an initial process of optioneering of the raw water channel was undertaken aiming to reduce construction time, minimise export of materials off site and maximise re-use of material already on site. This included: concrete U-channel, concrete trapezoidal channel, stone-lined trapezoidal channel and various vertical alignments evaluated using 1-D hydraulic models developed in HEC-RAS.

At both sites, a wide-based trapezoidal channel allowed for an increased contact surface and low hydraulic radius (area of flow per unit wetted perimeter). This, in addition to slackening the vertical alignment, helped reduce channel velocities allowing the channel lining to be of salvaged stone rather than concrete, masonry or other imported materials, as indicated in Hemphill & Bramley (1989) and May *et al* (2002).

Control structures along the channels were designed to make use of salvaged coping stones. At one of the sites the gradient of the channel was such that a small weir was incorporated at the downstream end to maintain subcritical flow in the channel. This allowed the use of salvaged rip-rap stone instead of reno-mattress for almost 90% of the channel. This was expected to result in significant improvements to the visual and environmental impacts of the works as well as providing substantial cost and programme savings.

At Beaver Dyke the scour pipe and washout from the supply pipe underlie the proposed channel and discharge into an existing scour channel. A flume was designed to replace the existing headwall such that the new channel invert could be raised clear of the pipework. This allowed the site team to keep the scour and washout from supply pipes operational, maintaining a safe water level in the reservoir until the invert of the flume was constructed.

Flood Contingency
Underpinning a lean delivery programme on site is the ability to manage flood conditions and minimise stand-by time. To achieve this, a range of flood prevention measures was deployed at both reservoirs to minimise inflows. These include: turning out catchwaters; making use of the scour facilities to maintain drawdown; using existing pipework for incoming streams to bypass the reservoir; and drawing down the upstream reservoir to impose a lag on the inflow hydrograph at one of the reservoirs.

Figure 3 shows the valve shaft at Oakdale Upper. The original scour arrangement did not have adequate capacity to prevent a significant rise in reservoir levels during a flood. To increase the capacity of this arrangement the pipework was be removed from within the existing valve tower (excluding the scour draw-off), and several tower sections removed to reduce the height of the tower. The valve tower and tunnel thus acted as an auxiliary overflow in an emergency.

Figure 3. Valve shaft to be modified to act as an auxiliary overflow in the event of a flood

In both cases the guidance in Floods and Reservoir Safety (3rd Ed) Chapter 6 (ICE, 1996) was adopted, indicating the design flood events should have 1% probability of exceedance over the duration of the critical construction period. The critical construction period was taken as the excavation of the embankment, thus, a 1 in 20-year and 1 in 10-year flood formed the basis of the flood contingency plans at Beaver Dyke and Oakdale Upper respectively. The 1 in 100-year event was also run through the model to gain an understanding of the sensitivity of the reservoir level to more extreme floods than the design event.

The constraints for flood contingency at the sites were two-fold:

- Firstly, the measures for flood contingency at both reservoirs needed to reflect the changing response as the excavation progressed (smaller surface areas and faster fill time), and
- Secondly, there needed to be an allowance for variation in the level of the reservoir at the onset of a storm.

Hence, an operational risk matrix was developed (Figure 4) with four risk categories (Table 1) reflecting available freeboard and expected fill time for the design event. Elevations for each risk zone at 1m intervals of excavation were issued to site so that trigger levels could be set out at each stage of the excavation.

In the upper parts of the reservoirs, events rarer than the design flood (1% probability of exceedance) were mitigated against by maintaining a minimum freeboard greater than the design flood rise. In the lower parts of the reservoir, storage is limited and the design event was expected to overtop the works. A minimum width of excavation was specified such that overtopping flows up to the design event could be safely conveyed by the downstream grass face of the embankment as indicated in Hewlett *et al* (1987). Above the design event, scour that could occur on the downstream face would be controlled by storing salvaged pitchings near the downstream grass face of the embankment, which were to be used in the event of an emergency to mitigate scour on the downstream face of the embankment.

At one of the sites, where access was only permitted during a six month window, the full works could not be completed in one available period. In the interim period, an impermeable liner was used to protect the excavation, in addition to the flood prevention measures outlined above.

Table 1. Risk categories during excavation at Beaver Dyke

Risk Category	Description
Blue	At least 3m freeboard is available between excavation level and reservoir level. The 1 in 100-year storm is not expected to overtop the excavation level. Overtopping probability is very low, but potential impact of breach is high.
Green	At least 2m freeboard is available between excavation level and reservoir level. The 1 in 20-year storm (1% probability of exceedance) is not expected to overtop the excavation. At lower levels the overtopping probability increases whilst the potential impact of breach decreases.
Amber	At least 1m freeboard is available between excavation level and reservoir level at the onset of a storm. The 1 in 20-year storm (1% probability of exceedance) is expected to overtop the excavation level in more than 2 hours. Overtopping probability is high (increasing at lower levels), the potential impact of breach decreases at lower levels.
Red	Less than 1m freeboard is available between the excavation level and reservoir level at the onset of a storm. The 1 in 20-year storm (1% probability of exceedance) is expected to overtop the excavation level in less than 2 hours. Overtopping probability is very high (increasing at lower levels), the potential impact of breach decreases at lower levels.

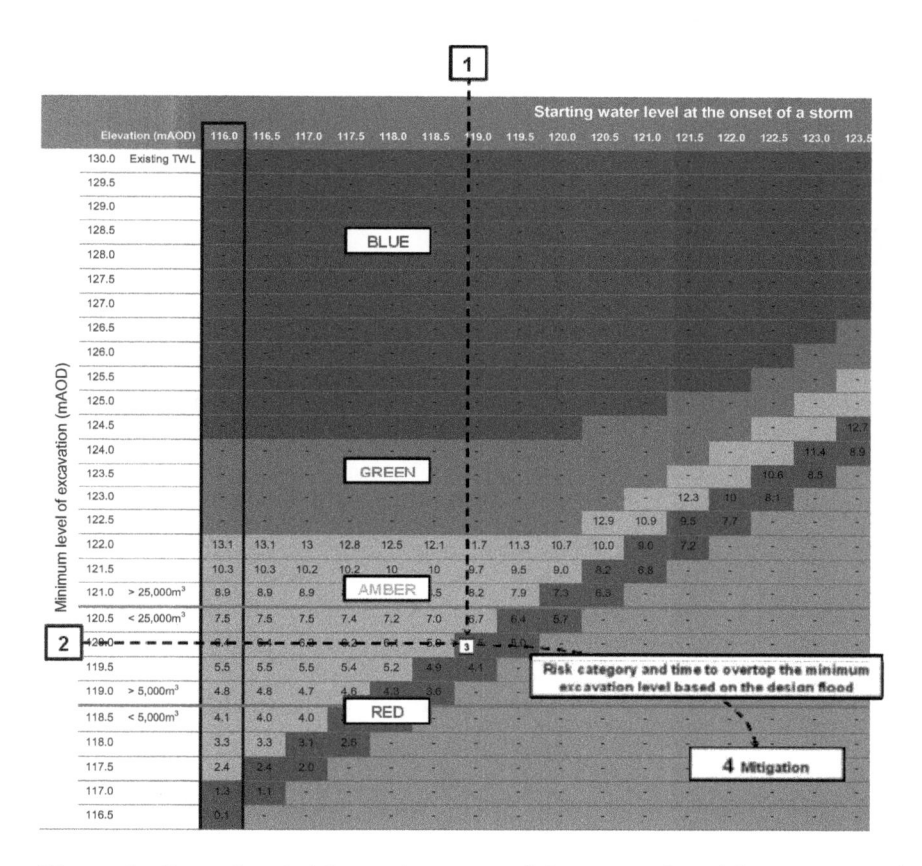

Figure 4. Operational risk matrix at one of the reservoirs: risk category and time to overtop based on level of excavation (vertical axis) and starting water level in the reservoir (horizontal axis).

Environmental consents and constraints

Licencing for the discontinuance works
The start of discontinuance works was dependent on a number of consents, including the variation of the existing statutory provisions for the impoundments (Acts of Parliament from the late 1890s and early 1900s) from the Environment Agency (EA).

The variation of the statutory provisions was required under Section 25 of the Water Resources Act 1991 (HMSO, 1991) for the construction or alteration of an impoundment structure (exemptions do apply as outlined in Environment Agency (2012)). This entailed a four month determination period, including 28 days public consultation.

Preliminary surveys
Environmental surveys were commissioned during the investigation phase at both sites. The findings of these included:

- Suitable habitats for crayfish were found at both sites. Surveys identified white-clawed crayfish (protected under Schedule 5 of the Wildlife and Countryside Act 1981) to be present downstream of one of the sites.
- Fish stocking had been undertaken at both reservoirs.
- Minimum water quality parameters were established at both sites for compliance with environmental standards.

The survey findings were reviewed in consultation with the Local Planning Authority (LPA) and the Environment Agency through various licence applications discussed above. The approach to mitigating the impact of the works was agreed as follows:

- Crayfish: Biosecurity measures were implemented to avoid movement of crayfish plague in uninfected areas. During the draw-down, no crayfish tracks or crayfish were found. On account of this and the evidence of algal blooms in the past, the team of ecologists recommended the site could not be used as an "Ark" site for White Clawed Crayfish without major catchment management changes upstream.
- Fish stocking: The reservoir was initially drawn down to allow for the fish to be rescued and translocated to neighbouring reservoirs within the same catchment.
- Water quality: a series of silt traps were installed within the existing scour channel and downstream becks. Turbidity was monitored regularly throughout the drawdown process and additional control implemented as required.

Water quality during the discontinuance works
Water quality was a major concern on both schemes as both reservoirs have retained silt from the surrounding moorland for over 100 years. Preliminary ecological investigations estimated between 1m to 2m of silt within the reservoir basins. This was despite there being a small head pond upstream of one of the sites.

In addition to the traditional straw bale silt traps, an improved silt trap was designed and installed consisting of wooden pallets wrapped in geotextile (Figure 5). Sacrificial layers were built into the silt trap such that they could be removed once the trap was heavily silted. The silt traps work in two ways: filtering fines from the drawdown process and providing a stilling effect for suspensions to settle. This silt could then be dug out and placed in settling ponds where water evaporated and percolated through the ground. The remaining solids could be used as a landscaping material within the basin.

The Environment Agency was a key consultee from the onset of both schemes and continuously throughout the discontinuances with a view to collaboratively addressing the issues with water quality.

Figure 5. Silt traps being installed within the scour channel at Beaver Dyke.

Long-term impacts
The long-term impacts and mitigation measures for the schemes were confirmed with a Water Framework Directive (WFD) Assessment. This assessed the impact of the proposals on the WFD Ecological Status of the impacted surface water bodies and groundwater.

In order to assess WFD compliance, it was necessary to ensure that the following three criteria were met:

- No deterioration in water body status;
- The works must not prevent the achievement of the water body objectives for good ecological potential; and
- The work should contribute to the delivery of the River Basin Management Plan (RBMP) where it is feasible to do so.

Construction and site programme
The critical activity in the construction programme was the excavation, as it would have the largest impact on the completion and compliance of the works. The works have been planned logistically; that is to say the programme was dictated by machine bucket volume as well as the speed and

distance of travel of plant. This determined the optimum number and size of machines as well as the activity duration which in turn informed the return period for flood contingency planning. Working with the Flood Contingency Plan, the excavation was been planned around the flood event with a 1% probability of exceedance.

Planning the works logistically ensured that the movement of the excavated material was carried out in the fastest time possible given the site constraints: tight working space; shorter days due to winter working; and access issues, among others. This approach also meant that a change in the working method, e.g. size or number of machines, could easily be fed back into the program to review the impact on the output and the overall programme.

The construction program also included a risk element to account for inclement weather and the probability and expected delays following a flood overtopping of the excavation. The excavations were be monitored not only from a site safety point of view but to learn from the approach taken so that it can be used again on future schemes.

3D models
The use of 3D models (Figure 6) greatly aided the programming process, and was integrated into the construction itself. The excavation was carried out in 1m vertical benches, cutting into the existing embankment whilst leaving a 1 in 4 slope at either side of the final V-notch. These benches were excavated with a backfall such that water drains back into the reservoir. This approach was difficult to convey through a conventional 2D platform due to the complexities of angles and slopes. With a 3D model, the various slope angles can be interrogated on site or in the design office at each stage of the works. The model can be used to explain the construction sequencing, viewing it from different angles. Furthermore, snapshots can be taken from the model to convey the excavation sequence and progression to the site team and external parties.

Excavated material
At the two sites 22,500m^3 and 8,200m^3 of excavated material respectively were expected. To reduce the environmental impact of the works, as well as minimising traffic impacts and reducing the carbon (and visual) footprint of the schemes, all the excavated material was to be retained on site for landscaping. From the 3D model, the excavated volumes were accurately determined. Existing topographical and bathymetric surveys of the basin allowed the design team to estimate fill volumes for landscaping.

The challenge, however, was that these areas have been under water and had changed dramatically from previous surveys. Additional surveys were carried out in the areas of greatest uncertainty using GPS. These data were then fed directly into the 3D model to update surfaces and fill volumes.

In addition to the changing landscape, a large proportion of the exposed reservoir basin was covered by silt. The excavated material could not be placed directly onto the silt as it would be prone to slope failure. To overcome this, the reservoir basin was be monitored during the works to ensure that the silt was dried out sufficiently in the areas where the landscaping fill was to be placed. This again was dictated by the weather conditions and could considerably reduce the unsaturated surface where fill could be placed. A range of scenarios were run with varying available surface areas to respond to this uncertainty.

Figure 6. Screenshot from the 3D model of Oakdale Upper showing completed excavation and progress on the formation of channel.

Setting out information
The setting out information was also produced from the 3D model. This information was uploaded to the GPS which was then used to set out the works on site. The reverse process will be used to complete the as-built information; data will be collected on site, stored and converted to Ordnance Survey reference within the instrument and then uploaded into the model to produce the as-constructed model. This automatically updates the drawing views, accelerating the completion process. The challenge associated with this is that the sites are in remote areas with limited mobile and internet signal. The GPS, therefore, cannot be used in the normal way and instead relies on a radio signal boost.

Conclusions
In summary, the two discontinuance schemes discussed in this paper show the programme efficiency improvements that stem from the Design-Build approach. Carrying out early enabling works allows for a condensed and efficient delivery of the main construction works.

The use of 3D models has provided a means for the design team to understand the specific site conditions as informed by the site team. It is important to make the most of the knowledge that the contractor has of the site, in addition to that of the client and operator. A collaborative approach to the schemes has allowed the design office to tailor the design to the specific site constraints. This included, among other features, adjusting the channel cross section and making the best use of salvaged materials to achieve the flow regime required to deliver capital and operational savings.

The success of the schemes hinged on a comprehensive flood contingency plan which mitigated effectively against programme delays and the health, safety and environmental risks that could arise from a severe flood during the discontinuance.

ACKNOWLEDGEMENTS:
The authors would like to thank Andrew Hobson, Dave Ellis, Chris Raistrick, Peter Down, Tony Jackson, Richard Robson, Neil Kempton and Ian Carter for their contributions.

REFERENCES
Environment Agency (2012) *Low risk impoundment: Our regulatory position and guidance on when you need to apply for an impoundment licence. Version 1, March 2012.* Environment Agency, Bristol, UK

Hemphill R W and Bramley M E (1989). *Protection of River and Canal Banks, a guide to selection and design.* CIRIA, London, UK.

Hewlett H W M, Boorman L A, and Bramley M E (1987). *Design of reinforced grass waterways CIRIA 116.* CIRIA. London, UK.

HMSO (1975). *The Reservoirs Act 1975.* HMSO, London, UK

HMSO (1991). *The Water Resources Act 1991.* HMSO, London, UK

HMSO (2010). *Flood and Water Management Act 2010.* HMSO, London, UK

ICE (1996). *Floods and Reservoir Safety (3rd Ed).* Thomas Telford, London, UK

May R W P, Ackers J C and Kirby A M (2002). *Manual on scour at bridges and other hydraulic structures. CIRIA C551.* CIRIA, London, UK.

Maintaining the Safety of our Dams and Reservoirs
ISBN 978-0-7277-6034-0

ICE Publishing: All rights reserved
doi: 10.1680/mdam.60340.381

Construction of a pre-cast concrete service reservoir using BIM

S RUSSELL, Mott MacDonald Ltd, Leeds, UK
T JACKSON, Mott MacDonald Ltd, Leeds, UK

SYNOPSIS Bramley service reservoir (SR) is a 16ML, twin-compartment, new-build that was required to replace an existing structure that had reached the end of its asset life. The reservoir is constructed of pre-cast concrete wall and roof units with an *in situ* concrete base, wall infills and roof screed. It is part of a standard product design used on more than ten reservoirs by MMB in the AMP5 period. The existing reservoir is situated within a public park and one compartment had to be demolished to allow for the construction of the new reservoir while the other compartment had to remain in service throughout the construction of the new SR to maintain water supplies.

Due to the land constraints within the park the new reservoir was located partially on the footprint the existing SRE and partially over a backfilled quarry. This paper will summarise how the challenges of the demolition and support of the existing structure were met; the varied foundations and how the piling design was incorporated into the already established precast design; and the benefits of BIM to the project.

SETTING
Bramley service reservoir is an integral part of the clean water distribution network in Leeds serving up to 35,000 properties operated by Yorkshire Water Services. The original reservoir comprised two compartments which shared a common dividing wall. The No.1 compartment was a 7.2ML brick barrel vaulted structure built in 1880 and the No.2 compartment was a 26.8ML mass concrete structure with brick lining built in 1912. Bramley SR is located within a public park and adjacent to a large telecommunications mast, a rugby pitch, a recreational area and a primary school.

Optioneering was carried out on locations for the offline build and led to the location shown in Figure 1. The available space for the new reservoir was limited. The area to the north was ruled out owing to a significant drop in ground levels and the area to the east was ruled out owing to it being a recreational sports field. This left the area to the south including partially building on the footprint of the No.1 compartment.

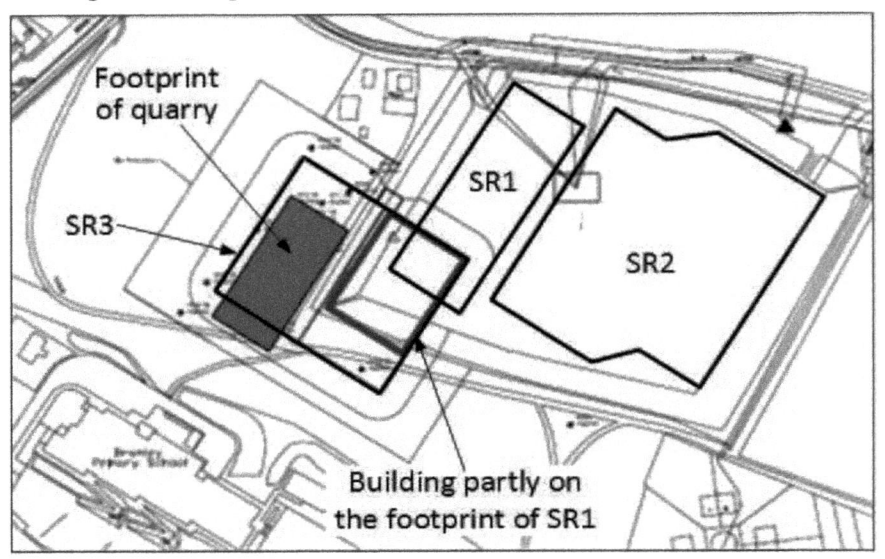

Figure 1. Plan of Bramley SR site

SUPPORT OF THE EXISTING STRUCTURE

<u>The Need to Encroach on the Existing Reservoir</u>
The initial primary constraints of the site were the land ownership boundary of the client and the existing assets, namely Service Reservoirs Nos.1 and 2. This gave some scope for siting the new reservoir, No. 3, to suit access, minimise pipe runs and optimise construction / demolition sequence.

However, the investigation of the site revealed that part of the area where the new service reservoir needed to sit was over a previous quarry that was then used as a landfill site for household waste in the 1940s. The need to accommodate this feature within the design dictated the position of the new reservoir with little tolerance.

At the time SR1 was not in service and the supply needs were being met from SR2. It was the initial intention to leave these structures alone until the new twin compartment SR3 was built and commissioned and then decommission SR2 and demolish SR1 and SR2. However, the new restraint on locating SR3 meant that the footprint needed to encroach on the embankment and structure of the original reservoir SR1.

Stability of the Existing Structures

The idea of early demolition of a redundant structure does not, on the face of it, cause any concern; however, there was a chain of structural support and supply requirements that complicated the issue.

The records showed SR1 to be an outer wall of concrete with a layer of puddle clay and an inner structure of barrel vaulted brick lining with an arched roof. Figure 2 shows a simplified version of the connecting wall drawing.

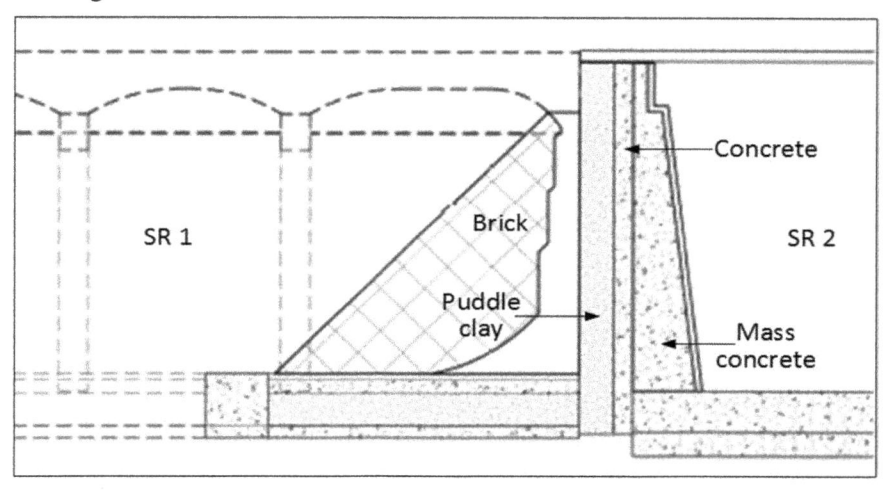

Figure 2. Interconnecting wall of SR 1 and SR 2

The later structure, SR2, was built with mass concrete and brick lining but was built in direct contact with SR1 along the common boundary. There was a suspicion that over the years these two structures had become mutually supporting and, whilst analysis of SR2 showed that the wall was stable without SR1, there was a danger that loss of the adjacent reservoir might allow rapid relaxation of a brittle structure, already considered to be 'asset expired', thus causing cracking and loss of containment.

It was uncertain how much of the rigidity of SR1 depended upon the arched roof and, in turn, how much the roof depended upon the abutment support of the external walls. It was, therefore, decided that SR1 needed to be demolished for its full width, although not necessarily its full length, in order to enable the new construction. To address the possible loss of support to the remaining SR2 the rubble from the demolition was used to form a bund against the common wall whilst SR2 remained in commission.

Other Restraints

Between the demolition of SR1 and the completion of the support bund the water level in SR2 was held down to relieve stress on the wall. In order to secure water supply this required that another new service reservoir on the other side of the city had to be completed and in service before this operation could be undertaken.

Because of the possible domino effect of removing part of the arched roof of SR1 the order of demolition and the required temporary support was worked out in some detail with the demolition contractor. Figure 3 shows the barrel vaulted construction of SR 1.

Figure 3. Demolition of SR1 of barrel vaulted construction

FOUNDATIONS OF NEW SERVICE RESERVOIR

Concept of the Precast Service Reservoir

The concept of the precast service reservoir has been explained in detail in other papers and so is only summarised here to provide an understanding of the structural issues.

The walls of the reservoir are principally precast units, in this case nominally 5.5m high and 2m wide, cast with feet to provide temporary support. These are set out on thin concrete strips and then joined together by in-situ sections 450mm wide to give continuous wall construction. Precast columns with similar support feet are also set out on thin concrete pads and then walls and columns are connected by an insitu slab to form a jointless box. The roof is then formed from precast flat slabs with an insitu screed to secure the structure and make watertight.

The structural concept is that all loads from the walls and columns are transferred to the base slab and distributed over the formation to give a more or less even distribution of load.

Discovery of the Landfill
The preliminary view of the site suggested a layout that would be relatively easy to construct without too much interference with the existing services and structures. However, the desk study identified an old quarry that had subsequently been used as a landfill site and this drove a realignment of the new structure.

Unfortunately the whole of the proposed footprint of SR3 was not available for investigation as it lay under the existing service reservoir SR1 and could not be accessed until the site works had started and the partial demolition of the old reservoir completed. The extent of the quarry was estimated from records but this could not be confirmed at design stage and some flexibility in the design was required to accommodate what might have been discovered during subsequent investigation.

Once the site was available a grid of dynamic probes was set out over the site to confirm the plan area of the quarry and the depth of the fill. The surprising result from these probes was that the quarry floor appeared to be very uneven. It was considered that the most probable explanation was that either debris in the fill was preventing the penetration of the probes or that heaps of quarried debris had been left in the quarry before infilling began. Neither situation presented a good foundation for supporting the piles therefore further information was required before finalising the design.

The options were to undertake a detailed ground investigation (GI) with further boreholes or to gather information from the pile boring rig and adapt the design of the piles as this information was gathered, in other words a dynamic design that required active communication between the site team and the design office. A GI would have delayed programme by several weeks, so after detailed discussions with the piling contractor as to what information could be gathered from the pile boring rig during operation it was decided to follow the second option. As part of this decision making the 3D model developed as part of the BIM package was invaluable and was updated on a daily basis.

Change to the Structural Concept
The discovery of the quarry / landfill site determined a change to the load path through the reservoir. It was proved through the GI that the plan area of the quarry was confined within the central area of one compartment of the twin compartment tank. For this compartment the roof loads went into the columns, which were supported on piles taken to the bottom of the

quarry. The walls were connected into the slab in the usual fashion and therefore the slab was ground supported around the edges.

An adaptation to the standard design was required in that over the quarry the slab was designed as a suspended slab from the edge of the quarry to the pile / column locations.

Differential Settlement
The change in the structural concept sounds to be relatively simple but the site now presented three different types of foundation:
1. Piled foundation above the quarry / landfill.
2. Weathered rock and engineered fill to the south and east of the site.
3. Solid rock to the north and west of the site.

The challenge has been to predict the settlement for each foundation type, to design the structure to accept the differential settlement and to ensure that the support conditions changed gradually to prevent any sudden stress change that might locally overstress the base slab. This has required detailed collusion between the pile designer, the structural designer and the geotechnical engineer, which has been a testament to collaborative working.

Anticipated Problems
As the site works began the anticipated problems were:
1. If the uneven floor was due to piles of quarry waste a large rock might show refusal to the pile boring rig but this rock could have been supported on compressible material, which over time would lead to loss of support.
2. The boundaries of the solid rock / weathered rock / engineered fill / piles had to be understood and managed to control differential settlement.
3. The edges of the quarry had to be anticipated as the pile bore could have been deflected off an inclined face with potential to damage the rig, or the pile toe could have been too close to the rock edge with the risk of the rock shearing off when the load came on to the pile.
4. The extent of the quarry / landfill may extend further than anticipated and encroach under the dividing wall.

Solutions were planned for all these scenarios but would need to be fine-tuned as details were gathered. Solutions included casting piles away from the quarry edge and constructing bridging beams onto the adjacent sandstone, or constructing additional piles if there was a question of confidence in a pile depth that had been achieved.

Alongside the development of an efficient design solution for the foundations the quality assurance during the pile construction was of paramount importance as each pile depth that was constructed was assessed by the engineer on site to ensure any problems were identified.

BENEFITS OF BUILDING INFORMATION MODELLING
The BIM model was integral to giving the design team confidence in the depths being achieved by the piles. In producing the contingency plans for the issues identified above, the creation and updating of the BIM model using daily data from site helped greatly in developing the different solutions.

During the construction of the piles information was sent to the modeller every day in order for him to model the profile of the rock and the rock sockets being achieved by the piles. The model was assessed regularly by the design team in order to identify any problem areas where quarry debris may have been hit. The model can be seen in Figure 4.

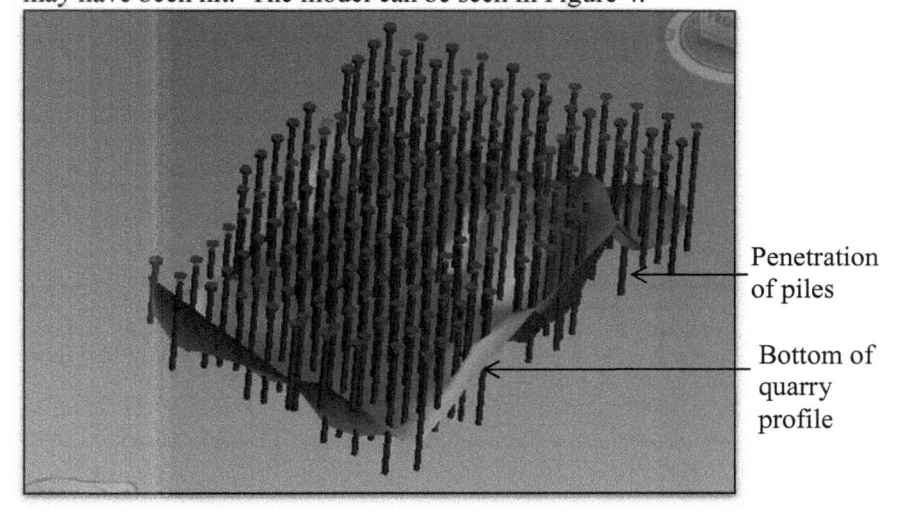

Figure 4. Quarry profile shown in BIM model

The model showed certain areas where there was clearly a ridge in the quarry, which explained a number of very short piles in the middle. This gave confidence in the lengths achieved by those piles and no additional works were required.

In addition to the as-built modelling of the piles, BIM played a large part in the initial optioneering for the scheme, including pipework layouts, orientation and clash detection. The model was used regularly on site for briefing the construction team to give them a clearer understanding on the activity they would be carrying out. In particular it was used for a sequencing drawing for the demolition of the No.1 reservoir to show clearly the order in which the reservoir needed to be demolished to avoid progressive collapse.

CONCLUSIONS

Bramley SR has been commissioned and is now in service. Despite the complex foundations, changes to standard design and site constraints the project was a success. Any other location for the reservoir would have caused significant impacts on the local area, however, with a sensible approach to the difficult foundations the reservoir was able to be built in an area that minimised impact on the public.

The scheme was a good example of collaborative working with the client and within a large design team with different expertise and influences on the project. The difficulties anticipated with the quarry were well managed through the use of BIM and careful quality management which is a good demonstration of the importance of design input during the construction works.

ACKNOWLEDGEMENTS:
The authors would like to thank Andrew Hobson, Dave Ellis, Dave Axe, Simon Roxby and Henry Hewlett for their contributions.

SECTION 6:
MONITORING AND INCIDENTS AT DAMS

Maintaining the Safety of our Dams and Reservoirs
ISBN 978-0-7277-6034-0

ICE Publishing: All rights reserved
doi: 10.1680/mdam.60340.391

Emergencies, monitoring and surveillance, and asset management – a new approach

A K HUGHES, Atkins, Epsom, UK

> SYNOPSIS   This paper will describe how monitoring and surveillance programmes are changing. A series of case histories where emergencies have occurred and electrical resistivity surveys have been used to identify leakages and then target repairs will be described. These repairs have varied from grouting to diaphragm walling and the cases will demonstrate how the technique has, in some instances, saved millions of pounds.
>
> A number of companies have investigated a number of their dams where leakages are thought or are known to exist. A number of approaches will be described.   This is a new approach to surveillance and reservoir safety management – a proactive approach rather than a reactive approach.
>
> The techniques are now being used as an Asset Management tool, with surveys being carried out every five years allowing a direct comparison in the condition of the dam at intervals, therefore enabling deterioration to be measured or confidence to be maintained.

BACKGROUND
Fact – dams are subject to incidents and accidents at regular intervals – there are four or five major incidents a year in the UK. Our legislation seeks to protect persons and property against an escape of water and yet legislation will not prevent or help to prevent a failure – it is one of the tools and it is the 'tools' which are assembled as part of a dam safety programme which serve to try to prevent failure.

In 1864 the jury reporting on the Dale Dyke failure said that every dam "should be subject to regular, frequent and sufficient inspection" – an enlightened jury, I would suggest, which set down the basis for our current surveillance and monitoring.

SURVEILLANCE
Visual surveillance of dams in the UK has been the backbone of our reservoir safety system.

Inspection by an Inspecting Engineer, a Panel Engineer, was a concept introduced under the Reservoir (Safety Provisions) Act 1930 - the '1930 Act'. The 1930 Act was conceived, enacted and commenced after the failures at Skelmorlie, Eigeau and Coedty in 1925 – again a reaction to events.

The review of the 1930 Act in 1966 by an ICE appointed committee recognised that 10 years was a long period of time between statutory inspections and that a trained pair of eyes provided by the Supervising Engineer should visit high risk or high consequence dams at least twice a year. This was a very important development in safety management systems which became law in 1987 when the Reservoirs Act, 1975 – the '1975 Act' – was commenced.

The most important element of visual surveillance at dams which seems to have formally developed alongside the role of the Inspecting Engineer was the reservoir keeper. In the 1930s there were 'keepers' who lived at reservoir sites, particularly where reservoirs were feeders to canals with locks which had lock keepers.

Reservoir keepers, particularly in the water companies of the time, were trained individuals - although there were no formal training courses at the time - who visited their reservoirs very frequently and who hopefully identified problems before the failure mode progressed to a point of no return. They literally lived and breathed their reservoirs – living on site, carrying out maintenance, cutting the grass – they had pride in what they did and were respected members of the team.

However, during privatisation of the water industry and the period following it, this level of surveillance became eroded. In the 1970s reservoir keepers lived on site perhaps looking after four or five reservoirs in a cascade. Privatisation resulted in staff reductions, houses being sold to private individuals and outsourcing.

By the end of the 1980s and into the 1990s some of the companies had reduced the frequency of their visits to two times a week, Monday and Friday, driven by accountants and cost rather than based on engineering science or judgement.

In the early part of 2002 United Utilities was examining their reservoirs every Monday, Wednesday and Friday. At the time of the Rivington incident, February 2002, I stated that if that dam had been in a different area where the dams were looked at only twice a week then the dam would probably have failed, because the failure process would have progressed too far and there would have been insufficient time to react to prevent failure.

After the Rivington incident United Utilities was so concerned about reservoir safety and had seen the benefit of visual surveillance that they put in place a surveillance regime where dams were visited at least every 48 hours.

Thames Water, because of the form of construction of some of their dams, with thin clay cores and gravel shoulders sitting on gravel above the London Clay, where failure due to piping or internal erosion would be fast, walk these dams every day.

A recent exercise for a water company reviewing the frequency and type of reservoir surveillance considered the type of dam, the consequence of failure, failure modes and therefore the speed of failure, detectability, the means of responding etc and then sought to try and allocate different forms of dams into groups. Those groups would be then subject to different surveillance and monitoring regimes. Some, the higher risk reservoirs, would be visited every 48 hours whereas others might be visited monthly – a risk-based approach in line with current thinking around the world.

Even when visits are planned every 48 hours the question must be asked - what do you want a reservoir keeper to do? Just take water levels? Access the shaft and tunnel? Operate valves? Read instruments? The exercise in one company resulted in different tasks being assigned to the three visits carried out each week. Each visit therefore took different periods of time. This information was then entered into a resource model to understand how many reservoirs could be visited in any one day; how many operatives were required on each 'patch' taking due regard of holidays, training requirements etc; and thus it enabled adequate staffing levels to be set. The process also enabled a review of vehicles to be undertaken – whether vans could carry out the task or whether four wheel drive vehicles were necessary. The company also embarked on a comprehensive programme of 'training' - right from the awareness of the Board on issues of reservoir safety down to a City and Guilds accredited course for reservoir keepers. This led to increased levels of investment on reservoir safety management.

Formal City and Guild accredited courses are now available for impounding reservoirs, non-impounding reservoirs and service reservoirs. Training at this level has been found to be very beneficial. The individuals have been motivated; some even to strive to become Supervising Engineers - they now understand why they are carrying out the activities that they are carrying out; why it is important to take good quality readings from instruments; that failures do occur and that they can happen relatively quickly at some sites.

INSTRUMENTATION
Instrumentation is an essential tool within the owner's toolbox. Instrumentation types are many and varied and include peizometers, V-notch weirs, levelling pins, inclinometers, extensometers, total pressure cells, etc.

Instrumentation arrays are usually designed for particular reasons; for example at the time of construction and the early life of a dam to check the actual behaviour of the dam against the assumptions made at the time of design, or perhaps to monitor a wet patch and/or check whether movement is taking place within a particular area on a dam.

Instrumentation is important in identifying trends or changes in behaviour which might be an indicator of a problem - which could be part of a failure process. Thus it is vitally important to collect good quality readings from the instruments and this applies to a level pin and a peizometer as much as it does to an inclinometer or extensometer. It is also vitally important to analyse the date received from the instrument array in a timely manner.

However, attempts are often made to read instruments which are well beyond their working life; conversely readings continue to be taken from instruments which have already provided all the information for which they were installed.

Some would say that it is a 'brave' Panel Engineer who ceases the readings of instruments, but in reality there is no need to read instruments where steady state conditions have become established or where the information sought has been obtained.

The instruments need to be read for a sufficient period of time to cover the behaviour of the structure over a number of seasons to be able to understand whether there are any 'cyclical features' in the performance of the structure and also be able to remove the effects of, say, rainfall or temperature on the structure. Again it is important to obtain good quality readings to be able to identify any significant changes or trends.

In my opinion, leakage measurement is often the most important indicator of dam behaviour. Leakage measurement can be measured using observational techniques, volumetrically using a stopwatch and container, or by a V-notch weir for example. Any form of leakage measurement should ideally include measurement of flow and turbidity.

Even when apparently clear a useful check on a leakage flow is to put a sample in a clear glass container and leave it to stand for a day or two to see if any sediment drops out of the apparently clear liquid. Often deposits settle out which are not visible to the naked eye.

Real time monitoring is used in a number of countries where access to sites is difficult during periods of bad weather; also for dams where there is a very high consequence of failure. However, I would strongly suggest we should never lose the benefit of trained staff visiting our dam sites to carry out visual surveillance. One must never forget that the zone of influence of an instrument is often about 150mm – anything outside that zone would not be detected. For example, there are numerous instances where level pins have not moved and yet depressions in the crest have occurred between the pins e.g. Greenbooth Reservoir in the 1980s.

LEAKAGE DETECTION METHODS

Not all leakages can be seen and visually detected. In some forms of construction the leakage disappears into the downstream shoulder or even the foundation without being seen. This was certainly the case with Greenbooth Reservoir, where the dip of the rock was downstream (Fleming, JH and Rossington, DT (1985)).

The dam was heavily instrumented. The suite of instruments included V-notch weirs, none of which registered any increase in flow despite the fact that a significant amount of material had been removed from within the embankment by a significant flow of water causing a form of internal erosion.

For dams with thin cores and gravel shoulders sitting on sand and gravel foundations it is very unlikely that leakage will be seen if there is defect in the thin clay cores. Once through the core the seepage/leakage will disappear into the shoulder or foundation and may never be seen.

In such instances different methods of leakage detection are required. In the past, even when one knew where leakage was leaving the basin and where the leakage/seepage outflow was visible on the downstream toe, it was not possible to decide the route of the leak. This resulted in remedial works often being over-estimated and therefore the works cost more than they needed to. Often the whole of the dam and the foundation was treated.

Nowadays there are a number of techniques which allow leakage routes to be pinpointed. These allow remedial works to be targeted and costs of repairs reduced. Sometimes the cost of the survey is considered high, but when compared with the certainty of knowing what is being repaired, and the associated peace of mind, the surveys are extremely good value when measured as part of the overall cost of repair (the cost of survey and the targeted repair).

Leakage detection techniques commonly used include temperature probing, where the temperature change in water in the dam by the leaking water is detected by direct temperature measurement (Kappelmeyer), and magnetic resistivity, where a magnetic field is applied and the field is changed by the presence of water (Willowstick).

EXAMPLES

Spade Mill, Lancashire

At Spade Mill in Lancashire the temperature probing technique promoted by Kappelmeyer was used to good effect.

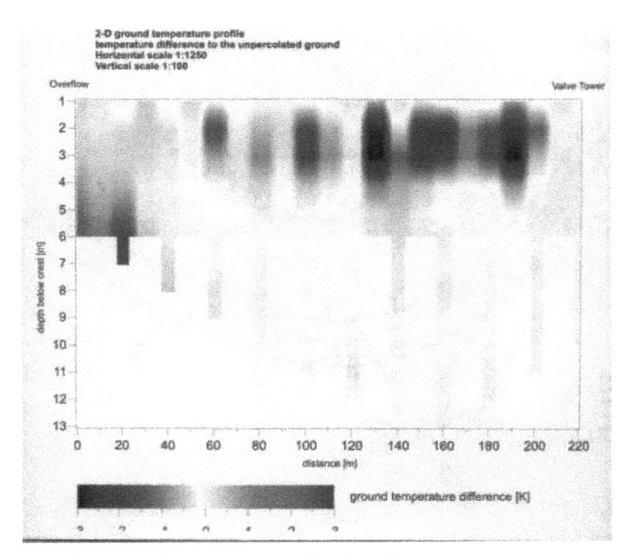

Figure 1. Kappelmeyer Temperature Probe Output

The output from the study showed leakage beneath the weir, which was subsequently cut off via the installation of a diaphragm wall.

Samanalawewa, Sri Lanka
At Samanalawewa in Sri Lanka some £53M had been spent between first filling in the 1990s and 2005 in trying to stem a significant leak.

I was engaged to answer a number of simple questions; is the dam safe? Is the situation deteriorating? How can the dam be repaired?

Figure 2. Samanalawewa Dam

Rather than read hundreds of reports on the investigations carried out over 15 years, I engaged the services of Willowstick. The survey, carried out over a period of just two weeks, identified two areas of leakage, which could easily be repaired at an estimated cost of £2M.

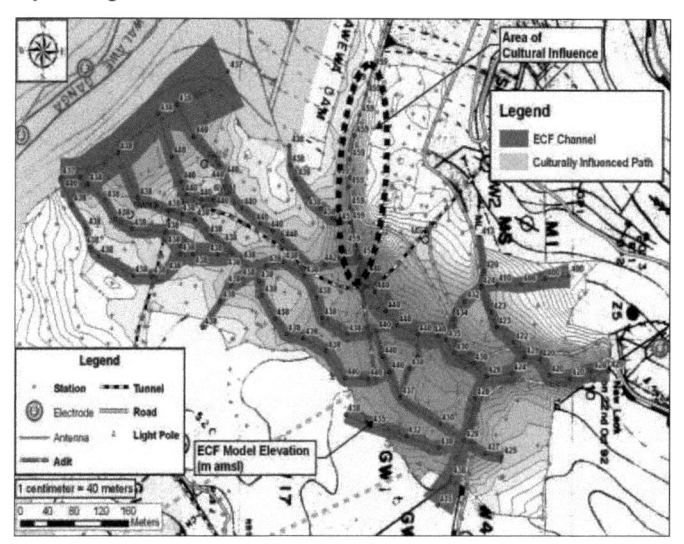

Figure 3. Leakage Plots at Samanalawewa

King George V

At King George V Reservoir, owned by Thames Water in the UK, a leak suddenly appeared on the downstream face, at about mid height, apparently for no particular reason. The water level was subsequently reduced and the leak appeared to stop when the water level was one metre below top water level. Because of previous experiences, the owner suspected desiccation of the core in the upper part of the core. However, being a prudent owner, Thames Water decided to carry out a leakage survey of the whole of the embankment to establish if there were any other leaks, before carrying out a more detailed survey in the area of the known leak.

Thus, Willowstick was engaged, firstly to check the rest of the embankment for leaks – more than 5km - and then to study the leakage area in some detail. The survey showed no other leaks through the embankment but did show that the leakage at the point identified was not high up in the core, but that the leakage path was at the base of the core. As the driving head was reduced by lowering the water level it appears that the leakage/seepage was not driven to the surface but disappeared into the foundation. Research after the Willowstick survey found the defect to be on the route of the old watercourse.

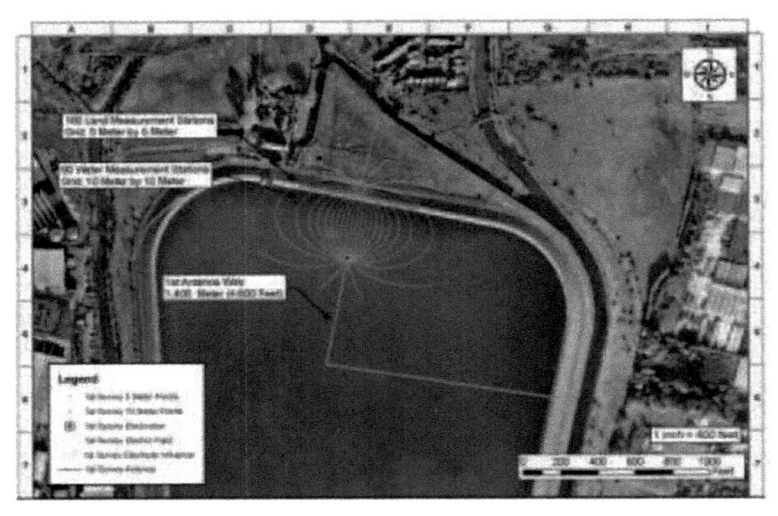

Figure 4. Survey layout

Figure 5 - Reservoir in drawdown state showing route of old water course

The subsequent remedial works, which entailed core replacement techniques using diaphragm walling, involved repairing only 100m of the core at a very low cost of £100k. The Willowstick survey saved the owner many hundreds of thousands of pounds via targeted remedial works.

Upper Rivington
Another example was the case of Upper Rivington, which in 2002 caused an emergency to be declared and rapid actions to draw the reservoir down to avert a failure. Willowstick was again used to identify seepage paths and

help with the direction of a grouting programme designed to stem the leakage through the core and repair the dam.

Figure 6 – Rivington Lower and Upper (in the distance)

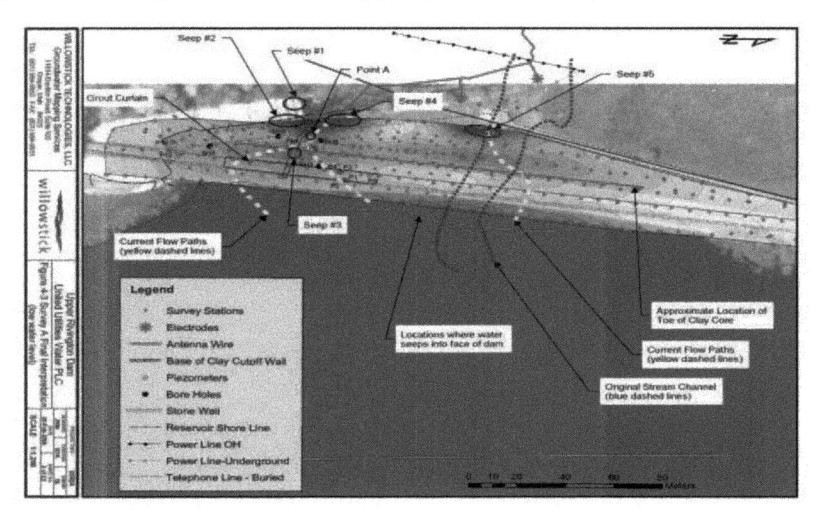

Figure 7 – Rivington Leakage Paths

In this case an emergency was narrowly averted because a vigilant reservoir keeper had noticed a discoloured flow in the stream downstream of the embankment.

ASSET MANAGEMENT

The use of the leakage tracing techniques, particularly where they do not involve intrusive works at the dam, can be used as an asset management tool. Checking for leakage at, say, five-yearly intervals can identify areas where deterioration is occurring or give the owner confidence that deterioration is not occurring, which is a useful asset management tool. Tools like the USBR Seepage toolbox can be used to determine probabilities of failure due to seepage and information from these types of surveys can inform the process.

EMERGENCIES

We have of the order of four or five incidents or accidents a year where intervention by a Panel Engineer is required to try to avert failure. We have been very fortunate that no failure involving loss of life has occurred in recent times. However, very few, if any, Panel Engineers have received any training in dealing with an emergency. Perhaps it is obvious that the first priority is to get the water out of the reservoir, but often the situation is much more complex. It can be a stressful time for all involved, and in particular for the Panel Engineer! How people cope with stress obviously varies from person to person but it is essential that people stay calm and the correct direction is given. Panel Engineers will have to be prepared not only to direct actions on the ground to avert failure but to answer difficult questions from the owner, the police and the media, including:-

- Is the dam going to fail?
- When is the dam going to fail?
- Who should we evacuate?
- When should we evacuate?

Of course with the benefit of dambreak analysis and inundation mapping some of these questions are more easily answered than they used to be. However, these are still difficult questions to answer with any certainty, and if the Panel Engineer gets the answers wrong this can have serious consequences for:

- those in the flood plain if failure occurs earlier than predicted or when not foreseen;
- the company if the emergency evacuation is called too late or not at all;

- the company if the emergency evacuation is called and then failure does not occur.

In these cases confidence and reputations can be damaged.

In these stressful conditions the Panel Engineer can only use his best endeavours and experience to give the advice he thinks is correct. Often in an emergency a Panel Engineer is acting without a formal contract in place and in that respect could put himself or his company at risk. Perhaps Panel Engineers should prearrange contracts with the major owners to cover situations where they may be called upon to give advice in an emergency.

CONCLUSION

Incidents and accidents occur four or five times a year, but are often stopped from progressing to failure of the dam by effective surveillance and monitoring by trained reservoir keepers; by instrumentation arrays; by analysis of results and by asset management tools.

An owner who suffers a failure will be compared with other owners in terms of the staffing he has in place; the reservoir safety systems he has in place; and his track record in looking after his dams.

It is therefore very important for reservoir owners to employ their best endeavours to ensure reservoir safety; to employ sufficient resources; to employ the latest and comprehensive surveillance techniques and to employ trained and competent staff.

Emergencies can still occur. Improvement in inundation mapping and emergency planning will help with management of these events but it must never be forgotten that humans, with all their failings, are involved!

REFERENCES

Fleming J H and Rossington D T (1985). Repair of Greenbooth Dam following Localised Settlement in the Crest, *Transactions of International Congress on Large Dams*, New Delhi, Vol 2 pp 233-245.

The monitoring and performance of rock anchors at Seathwaite Tarn Reservoir, Cumbria

D E JONES, United Utilities, Warrington, UK
C D PARKS, United Utilities, Warrington, UK

SYNOPSIS   Seathwaite Tarn is a large reservoir situated within the Furness Fells west of Coniston in Cumbria. It is impounded by two dams; the main dam is of composite construction with a mass concrete section and an earth and rock-fill section with a concrete core and cut-off; the smaller, a mass concrete subsidiary dam, is located to the north of the main dam. Both dams were completed in 1907. The subsidiary dam was modified to become the Auxiliary Spillway (AS) dam in 1960. In 1975, 35No rock anchors (c3.6m c/c) were installed, each with a Design Working Load (DWL) of 540kN. Eight anchors were installed in the AS dam in 1991 each with a DWL of 640kN. Load checks and testing of the anchors has been performed since installation. There have been several modifications to the main concrete section, including closure of the original spillway, removal of an architectural masonry facing and the application of an upstream membrane, each of which may have influenced the performance of the dam and anchors.

This paper describes aspects of the anchor performance monitoring and how the anchor head detailing and conditions have influenced the check lifting process and testing. Finally there is discussion regarding the on-going collated data and how these may influence continued reliance on the anchors and future strategy.

LOCATION AND GEOLOGY OF THE SITE
Seathwaite Tarn reservoir is situated within the Furness Fells, approximately 5km west of Coniston, Cumbria (NGR SD250986). The reservoir is located on the Fern Beck, a tributary of the River Duddon, which flows NE to SW through the valley between the Troutal Fell to the northwest and Seathwaite Fells to the south-east and approximately 3.2km northeast of Seathwaite village. The catchment has an area of 4.9km², mainly upland moorland, and is mountainous in character.

The solid geology of the area is the Birker Fell formation consisting of the Borrowdale Volcanic Group of the Ordivician Age comprising andesitic lavas with interbeds of volcaniclastic sandstone, breccia and lapilli tuff. There are numerous geological faults in the area.

HISTORY OF SEATHWAITE TARN RESERVOIR

The reservoir is impounded by two dams; the main dam is of composite construction, comprising a mass concrete section and an earth and rock-fill embankment section with a concrete core and cut-off. A smaller, mass concrete, AS dam, is located to the north of the main dam. Figure 1 indicates the main features of the dams. The reservoir has a capacity of 2.95Mm³ and a surface area of 254,900m² when at top water level of 374.93mAOD.

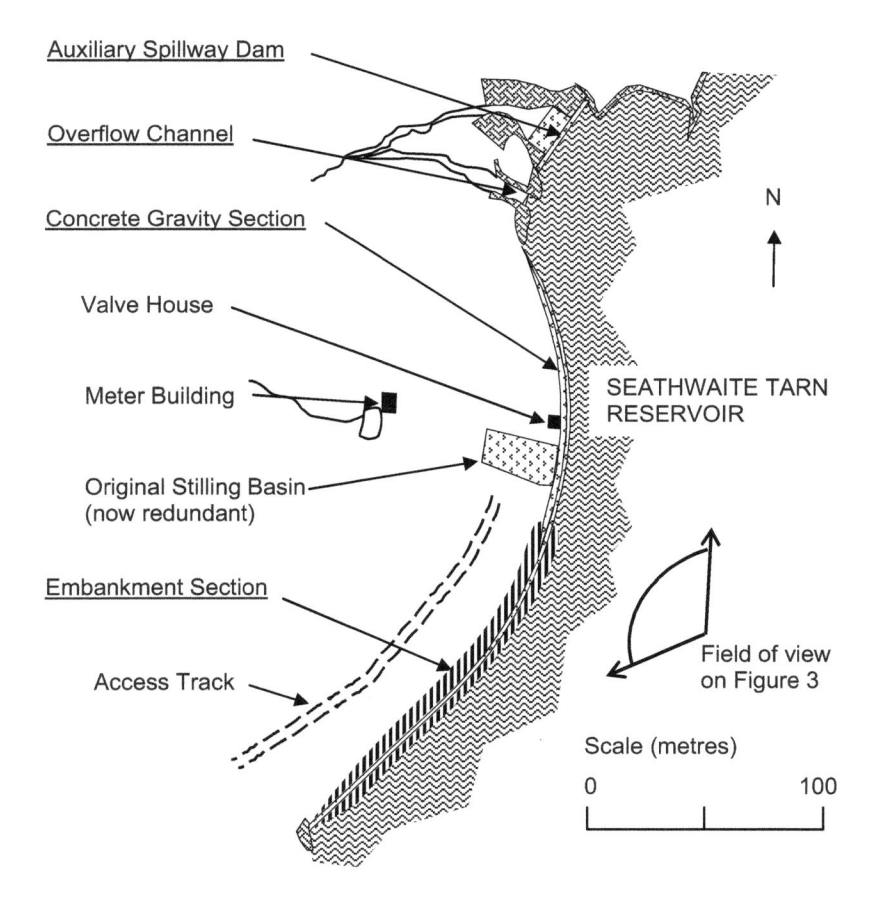

Figure 1. Layout of Seathwaite Dams

The reservoir was constructed in 1906/07 for the Barrow in Furness Corporation with concreting being halted during the winter of 1906. The main dam is over 300m in crest length; the concrete gravity section has a maximum height of 8.3m and is approximately 136m long, and the embankment section is 177m in length. The AS dam was constructed as a concrete gravity dam at the same time as the main dam; it was subsequently converted to a spillway. The AS dam lies to the north of the main dam and is 42m in length with a maximum height of approximately 5m. An additional overflow channel was formed in 1934 through the rock knoll that separates the two concrete dams. Although not greatly reported in technical papers, it featured as a "Dam of the 20th Century" in the 9th Geoffrey Binnie Lecture presented by T A Johnston (Johnston, 2000). He explained these were not the century's most important dams; they are nine 20th Century dams selected from his personal knowledge that help illustrate particular facets of development chronologically through the century. Notes in that paper of shortfalls in the Seathwaite dam's design and construction, regarding leakage, stability and particularly the anchoring, are revisited in this paper. He regards Seathwaite as one of the dams that demonstrated expensive masonry facings were not essential and that shortcomings in concreting would be resolved on future dams.

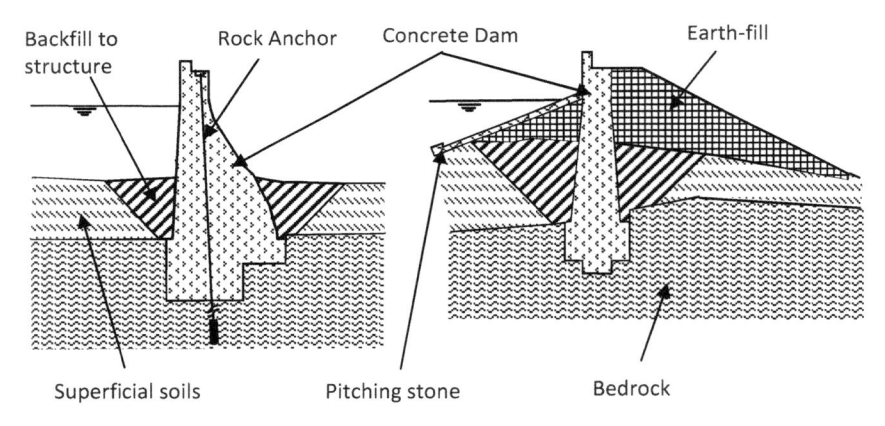

Figure 2. Typical sections through the Main Dam: Gravity section (left) and Embankment Section (right)

When constructed and upon impounding the dam was, indeed, prone to leakage through fissures within the mass concrete. The majority of these fissures were thought to be the result of thermal expansion and contraction along construction joint weaknesses. Some of these weaknesses may have been exacerbated by the break in construction during the winter of 1906. In 1955 a dry-stone masonry facing was constructed on the front face of the main dam to protect the concrete from the thermal effects of direct sunlight. Subsequently grouting was used in an attempt to control leakage, although this was unsuccessful. In 1960 a new overflow was constructed along the AS dam and the original spillway, in the face of the main concrete dam, was filled with masonry. A new valve house was constructed in 1970/71. Twelve pneumatic piezometers were installed in 1987 and a further four pneumatic and two standpipe piezometers were installed in 1998. These were drilled into the concrete dam, the foundation and the rock immediately downstream of the toe. The piezometers were to provide information on hydrostatic pressures affecting the main concrete dam and foundation.

In 1974, 35No rock anchors were constructed along the length of the main gravity section of the dam at approximately 3.6m centres in order to improve mass stability of the gravity section. In 1991 8No rock anchors were installed in the AS dam to improve stability, and a new stilling basin was constructed immediately downstream of that dam. The following year the masonry facing to the main dam was taken down and the rock was used to fill the original stilling basin and bring the ground level up between the valve house and the embankment section of the dam, leaving the maximum exposed height of the main concrete dam at approximately 7.3m. The masonry infill to the original spillway was removed and brought to crest level with mass concrete.

The reservoir was drawn down to scour level in 2008 and the structure and integrity of the concrete was assessed. A watertight geo-membrane was supplied and installed by Carpi Tech of Italy to the upstream face of the main and auxiliary concrete dams up to approximately 650mm from the top of the wave wall (the wave wall is approximately 950mm above the crest) (Figure 3).

Figure 3. Seathwaite Tarn Reservoir drawn down for the application of the watertight membrane in 2008

Cracks and irregularities on the upstream face were pressure jetted, raked out and in-filled with flexcrete, prior to placing the membrane. A large amount of stone armouring was placed on the shoreline adjacent to the left hand side of the south flank to provide erosion protection.

ROCK ANCHOR DETAILS

The 35 rock anchors were installed in the main dam in 1974 by breaking out a pocket at each location approximately 600mm square and 200mm deep, later to be covered with an airtight inspection cover. In the centre of each pocket holes were drilled to depths of between 15m and 20m below crest level. The tendon for each anchor comprised a 40mm diameter Macalloy bar. The fixed length of the anchors varied from 4.7m to 5.4m and the bar was grouted over this length with no record of protective encapsulation. In the anchor-free length, through the concrete dam section and the first few metres of the underlying rock, the Macalloy bar was de-bonded within a grease-filled plastic sleeve. A 2002 testing report confirmed the apparent free tendon length of the anchors equal to the depth of the dam concrete section plus about 3m in each case.

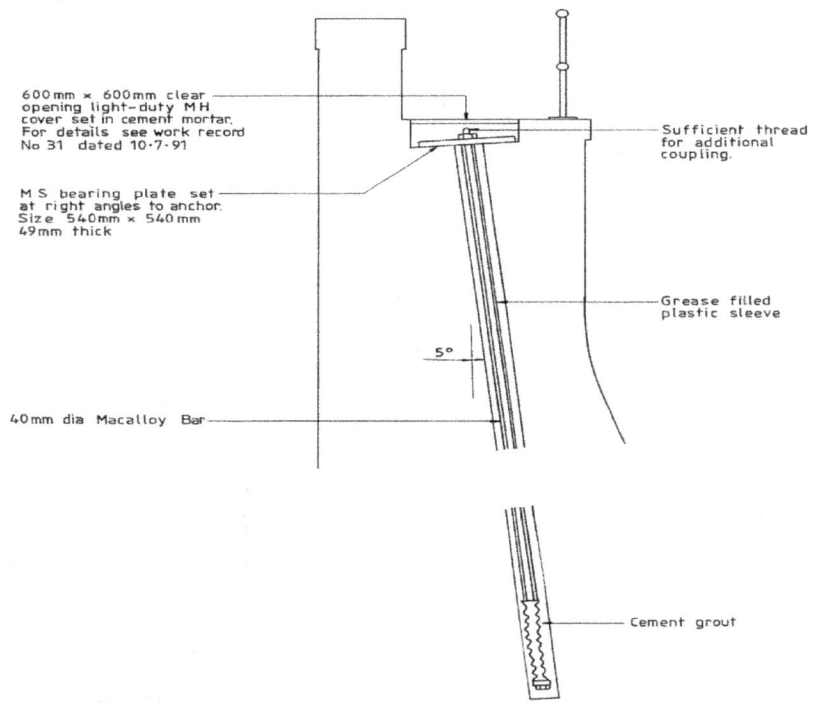

Figure 4. Main Gravity Dam Rock Anchor detail

The sketch of the anchor construction on a North West Water drawing of 1991 (Figure 4) indicates that a threaded nut was in place at the termination of the anchor bar; such load transfer compression fittings were not uncommon components of the anchor fixed length at that time. The head detail comprised a 540mm square, mild steel bearing plate, set perpendicular to the tendon long axis, with load transfer provided by a washer and nut. The plastic sleeve is shown as terminating at the underside of the bearing plate, although it has not been possible to verify the construction detail; there was no dedicated under-head protection. The protruding threaded bar remaining above the nut after installation is very short and this has proved a constraint in terms of subsequent testing and de-stressing options.

It is understood that the anchors had a Design Working Load (DWL) of 540kN; it is likely that they were locked off at 590kN (WL x 1.1; a typical lock off load) in 1974, however, in that year the utilisation of bar tension capacity was increased for dam structural reasons to allow future pre-stressing to 700kN. The resulting grout to ground bond factor of safety was likely in the order of 2.3 and probably significantly higher. In 1991 it is reported that the anchors were re-stressed to loads of between 655kN and 725kN. The design and construction of the 35 anchorages in the main concrete dam preceded the introduction of BS Draft for Development Recommendations for Ground Anchorages (BSI, 1982) and, understandably, BS8081: Code of Practice for Ground Anchorages (BSI, 1989).

In contrast the eight anchors in the AS dam, generally of a similar design, were installed in 1991 after the introduction of BS8081. The detailing of the anchorages reflects that development; the plastic sleeve is detailed as corrugated and the head detail includes a 530mm square mild steel bearing plate, 50mm thick, on a mortar bed and trumpet to allow below head application of corrosion protection measures after stressing. There is also a separator plate and a grease-filled mild steel cap over the lock nut to provide above head protection. The anchor sketch detail is indicated in Figure 5. The bar diameter is the same and the working load and adjustments since installation are similar to the main concrete dam.

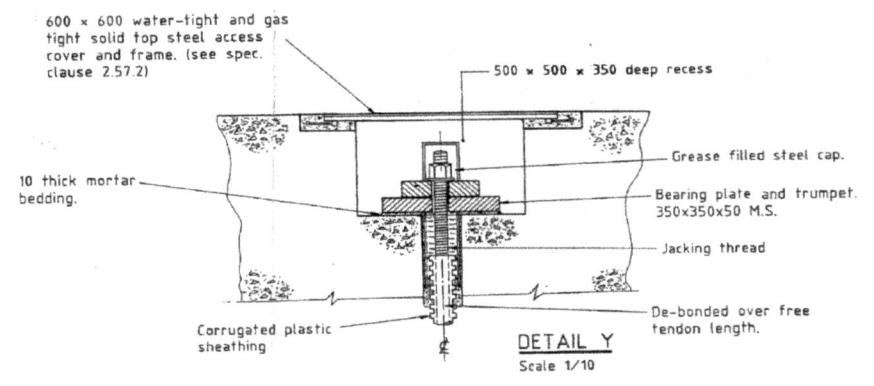

Figure 5. Auxiliary Spillway Gravity Dam Rock Anchor head detail

SUMMARY OF DAM CONDITION AND STABILITY ASSESSMENT
A stability analysis of the concrete gravity section of the dam was undertaken by Babtie Shaw and Morton (BSM) in 1988 (Babtie, 1988). This report indicated that the earlier assessment undertaken in 1974 by G H Hill and Sons, the designers of the anchors, was confirmed as satisfactory for above ground sections of the dam, although no calculations were noted for the assessment of below ground sections. The BSM report assesses the dam for flood, earthquake, top water level, ice loading and rapid drawdown, on a number of cross sections and levels within the dam. It concluded that the dam was stable in all cases and at all sections and, significantly, that the anchors were necessary in order to maintain stability at an acceptable Factor of Safety. It was recommended that the anchor loads be increased to 700kN to reduce the theoretical tension loads in the upstream face under extreme loading conditions. It is noted that this required an anchor load above that in BS8081 for permanent anchorage tendons.

A stability analysis of the embankment section was carried out in May 2004 and the lowest factor of safety for shallow slips was 1.53 and for pseudo-static analysis the minimum for safety was 1.21. Deep seated slips for the static condition had a Factor of Safety of 1.69 and for the pseudo static case 1.26. The stability of the embankment is therefore considered to be adequate under all load conditions.

Further investigation of the concrete dam was carried out in 2003. This found some evidence of alkali silica reaction in cracks and voids in the concrete. A study and stress analysis was carried out for the base joint and lift joint on each section. The upstream stresses were found to be always positive and the factors of safety for sliding for all load cases (normal, flood and seismic) were found to be adequate but confirmed the continued need for the anchors.

Seismic risk is assessed as part of each statutory inspection and the dam is in Category II based on the criteria within "An Engineering Guide to Seismic Risk to Dams in the United Kingdom" (BRE 1991). During the 2006 inspection the risk was not considered to be sufficient to pose a significant risk to the dam, particularly as the dam has been anchored (Hughes, 2007).

With the reservoir drawn down during 2008 it was possible to inspect the concrete to the upstream face and the stone pitching on the embankment sections of the dam. The concrete to the upstream face was considered to be generally in good condition; however, there was extensive cracking and crazing with associated efflorescence over the exposed concrete surfaces, particularly towards the top 2m of the face, with honeycombing of the concrete in places on the horizontal lift joints. Three main vertical cracks were observed, one 5m from the left hand end of the wave wall and the others 3m and 40m from the right hand side of the wave wall. Exposed aggregate was visible in the concrete in the central section of the dam. The wave wall had, in many areas, moved towards the reservoir, particularly the north end of the embankment section just south of the concrete dam. The stone pitching was in good condition.

ROCK ANCHOR TESTING
There are no records to show that the rock anchors were tested between installation in 1974 and 1991. The rock anchors were tested in 1991; the testing regime comprised "lift off" tests of all 35 anchors on the main dam to determine prevailing anchor load compared to the original "lock off" load. The anchors were then locked off at loads of between 655kN and 725kN on the main dam. In two cases it was not possible to turn the lock nut and in these cases and for one further anchor, shims were inserted to achieve the lock-off loads. The AS dam anchors were installed in 1991 and locked off to a specified load of 705kN.

The rock anchors were tested in 2002. The specification for the testing was based on BS8081. The testing regime comprised "lift-off" tests of all 35 anchors on the main dam and all eight on the AS dam to determine the prevailing anchor load. It was intended to de-stress the anchors to zero load to perform an on-site acceptance test to 150% DWL, then perform a further lift-off test before locking off. Undertaking on-site acceptance tests proved impossible on many of the anchors: 18 on the main dam and all eight on the AS dam. In some cases it was impossible to turn the lock-nut and in the remainder of cases there was insufficient exposed thread on the anchor bar for the drawbar coupler to be safely engaged. For these anchors it was agreed to perform extension of the anchors from lift-off load to 150% DWL. Residual load-time behaviour tests were performed on six of the anchors on the main dam and one on the AS dam. The anchors were locked off at loads of between 623kN and 735kN on the main dam and to between 679kN and 723kN on the AS dam.

The rock anchors were tested again in 2012 to a similar specification to that used in 2002. Lift-off tests of all 35 anchors on the main dam and all eight on the AS dam were performed. On-site acceptance tests were performed on seven anchors on the main dam. It was not possible to perform on-site acceptance tests on nineteen of the anchors on the main dam and all eight on the AS dam. The anchors were locked off at loads of between 600kN and 787kN on the main dam and to between 719kN and 758kN on the AS dam.

Anchor Test Results
Figure 6 illustrates the comparison of the lock-off and lift-off loads in 1991 to 2002 and 2002 to 2012. The results show that for the main dam there has been effectively no load loss; the mean lift-off values in 2002 were on

average 8kN higher than the lock-off values in 1991. The anchors were then locked off about 4kN higher than the lift-off load. In contrast to earlier check showing little variation in load it is interesting to note that the data from the lift-off loads in 2012 were, on average, 49kN higher than the 2002 lock-off loads.

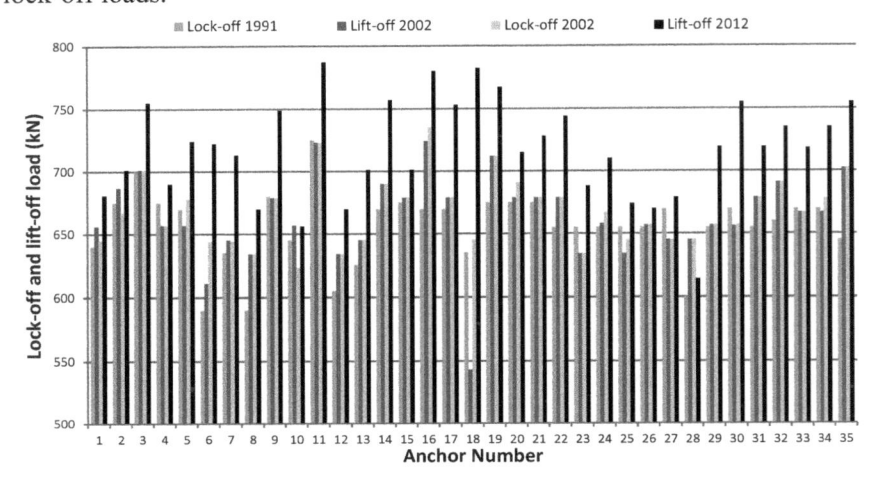

Figure 6. Lock-off and lift-off loads between 1991 and 2012

On the AS dam a similar picture emerges. There was a mean drop of 16kN in lift-off load when tested in 2002 and a mean increase in the lift-off load measured in 2012 of 43kN when compared with the lock-off load of 2002

Table 1 summarises the changes from lock-off load to lift-off checks approximately at 10 year intervals. The load losses in 1991 from the original and again in 2002 are in line with the 110% lock off load from BS8081. The increase in load to 2012 of, on average, about 7%, warrants future consideration.

Table 1. Comparison of main dam lock-off and lift-off loads between 1974 and 2012

Date	Load	Average Load (kN)	Change in Load (kN)	% Change in Load
1974	Lock-off	590 (assumed)		
1991	Lift-off	Not recorded	-	-
1991	Lock-off	656		
2002	Lift-off	664	+8	+1.2
2002	Lock-off	668		
2012	Lift-off	717	+49	+7.3

The individual anchor load changes are shown in Figure 7 which highlights the peak increase in load between 2002 and 2012 of 137kN on Anchor 18, which also exhibited the greatest loss of load between 1991 and 2002 of 92kN. The average change of +49kN is reasonably representative along the dam with a couple of higher outliers and a few anchors exhibiting a lower increase, for example numbers 25 to 28. A load increment of 50kN for these anchors represents a theoretical extension of about 6mm with associated increased tensile stress within the concrete and possible opening of fissures or upward movement of the dam face.

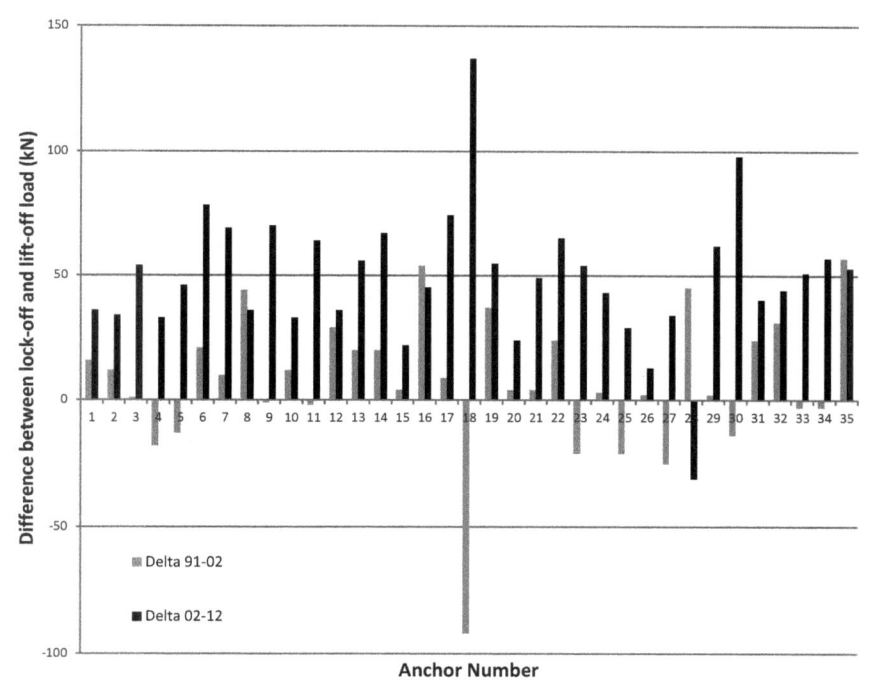

Figure 7. Comparison of the change in lock-off and lift-off loads between 1991 and 2012

The reasons for an increase in anchor load may include:
- Anchor testing / recording anomalies
- Reservoir level at the time of lock-off and lift-off testing
- Changes to dam structure

The same testing company carried out the testing in 2002 and 2012. The calibration certificates were recent and valid, the same equipment was used on both occasions and one of the operators in 2012 had been part of the 2002 team. This suggests that a case of calibration error was very unlikely. The reservoir water level at the time of testing in 1991 and 2002 could not be verified, although in 2012 the water level was at TWL. A water level difference may well have influenced the recorded lift-off loads. One major difference to the dam during the intervening period was the addition, in 2008, of the upstream membrane.

Consideration of the Anchor Increased Load
It seems that, from the above reasons, the most likely factor influencing increase in anchor load is associated with the membrane installation, with the possibility that reservoir levels between lock-off and subsequent lift-off checking are affecting the results. Prior to installation of the membrane the dam was known to be leaking along thermal expansion cracks and fissures within the concrete. Grouting in the 1950s had proved ineffective in stemming the leaks. Visual evidence of leakage on the downstream face and the installation and data from 'v-notch' weirs at the toe measured the flow rate; however, installation of the membrane substantially reduced the leakage of water through the main dam.

It is possible that changes in flow paths have affected hydrostatic pressures along the fissures within the dam with potential changes in uplift pressures, now that water leakage has been minimised. It is possible that this has, in turn, led to changes in the prevailing forces leading to increased loading of the anchors.

In reviewing of the existing piezometer data and the hydrostatic profile it has not been possible to establish a pattern of behaviour linking water pressures within the dam foundation and rock beneath with reservoir level change. This perhaps suggests that while there has, prior to membrane installation, been obvious leakage through the dam above the downstream toe level the frequency of open fissures and fractures through the lower dam and foundation has not been sufficient to allow easy interception by piezometer boreholes. It is possible that consideration of further installations will be required to facilitate understanding of the behaviour of the dam post membrane installation and for consideration of the future of the ground anchors.

While the 7% increase in load will be considered further it is noted that over the decades since installation a number of lift-off checks and on-site suitability tests have demonstrated reasonable anchor behaviour. The anchors are now 40 years of age since installation in 1974 and although testing is proving on-going functionality, it is significant that they were installed with limited corrosion protection measures, and prior to when UK codes provided increased levels of standardisation. The tendons are also loaded at a significantly higher stress level than the original design intended and for a time were above code requirements; the introduction and guidance of BSEN1537 (BSI, 2013) has somewhat redressed that situation.

CONCLUSIONS

Associated with many structures over 100 years old there can be an interesting history that includes assessment of performance, modifications to address design or construction shortfalls, or changes in operational requirements. In the case of Seathwaite it is clear that few of those changes are made in isolation from other aspects of the dam performance. Considerations concerning the existing anchors are linked to the concrete structure under a number of design load conditions; this is linked to the 'sub-dam' and internal hydrostatic pressure and potential changes caused by the membrane installation. Further piezometer installation and monitoring may be justified.

Data from the piezometers and a further review of the anchor load behaviour will form part of consideration of the ongoing reliance on the existing anchors. Further monitoring of the anchors is clearly prudent although the age of the anchors associated with a pre-UK code of practice approach to corrosion protection are factors that will feed in to decisions regarding the direction of future investment.

REFERENCES

Babtie (1988). *Seathwaite Tarn Reservoir Report on the Stability Analysis of the Gravity Section of the Dam.* Internal UU Report.

BRE (1991). *An Engineering Guide to Seismic Risk to Dams in the United Kingdom: BR 210.* Building Research Establishment, Watford, UK

BSI (1982). *DD81: Recommendations for ground anchorages.* British Standards Institution, London, UK.

BSI (1989). *BS8081: Code of Practice for Ground Anchorages.* British Standards Institution, London, UK.

BSI (2013). *BSEN1537: Execution of special geotechnical works. Ground anchors.* British Standards Institution, London, UK.

Hughes, A K (2007) *Seathwaite Tarn Reservoir Report on an Inspection under Reservoirs Act 1975, Section 10(2) of the Act.* Atkins Water.

Johnston, T A (2000) Taken for Granted. The Geoffrey Binnie Lecture 2000. *Dams and Reservoirs Dec 2000, Vol 10 No 3.* British Dam Society, London, UK.

Maintaining the Safety of our Dams and Reservoirs
ISBN 978-0-7277-6034-0

ICE Publishing: All rights reserved
doi: 10.1680/mdam.60340.414

Liquefaction failure in a Derbyshire fluorspar tailings dam

M CAMBRIDGE, Cantab Consulting Ltd, Ashford, Kent

SYNOPSIS   In the early morning of 22 January 2007 the survey team on the site of a fluorspar mining operation in northern Derbyshire was disturbed by the noise of plastic barrels being propelled across the mine site and on to the public road. Within half-an-hour the residents of the nearby village of Stoney Middleton found their doorsteps awash with a mixture of tailings and rainwater. The flood emanated from a classified tailings facility in which a liquefaction failure had occurred. The liquefied tailings had overtopped the confining wall, leading to the discharge of water and tailings off-site. The tailings were rapidly contained and further off-site discharges prevented, although the limited outflow had been sufficient to cause traffic chaos and to lead to some environmental degradation of houses and gardens in Stoney Middleton and along the River Derwent. Though the initial impact was perceived to be devastating, the rapid clean-up operation, which commenced within hours, enabled the village to return to normal in a short space of time, the only long-term remediation required being that along the banks of the River Derwent.

This paper presents the technical assessment of the failure and the postulated mechanism for both the liquefaction of the tailings and the overtopping which followed. The paper summarises the work subsequently undertaken to prevent any further such incidents on the site and concludes with the lessons to be learnt with regard to emergency planning.

BACKGROUND

Glebe Mines Ltd, a subsidiary of INEOS in 2007, owned and operated the mining and processing facilities at Cavendish Mill in northern Derbyshire. Associated with these operations are some four mine waste disposal facilities, originally constructed between 1945 and 1999 to receive wastes from the processing of fluorspar ore from both mining and reprocessing operations on the site. The disposal facilities are all registered under the Mines and Quarries (Tips) Act 1969 and were designated as active classified tips. In 2007 only Dams 3 and 4 were used for the disposal of tailings and process water, Dams 1 and 2 being excavated for reprocessing as part of the restoration works on the site.

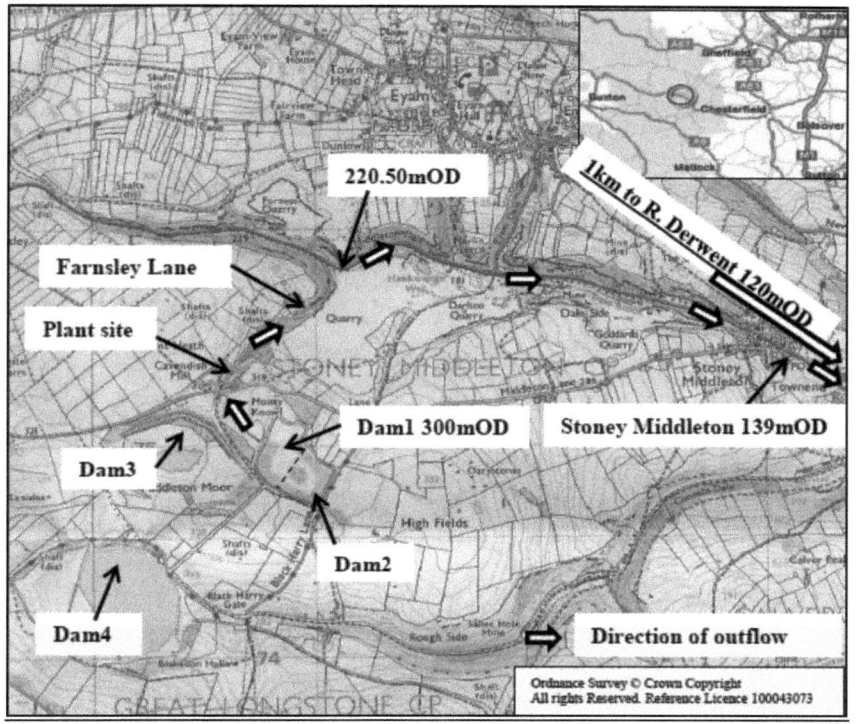

Figure 1. Location plan

Dam 1
Dam 1, the facility of interest, is located to the south-east of the Cavendish Mill Process Plant and comprises a rectangular paddock dam originally confined by 12-15m high embankments (Figure 1). This facility was constructed in the 1960s to receive wastes from the process plant and was subsequently extended to the south-east by Dam 2. The confining embankments used mine waste and local fill to form a semi-homogeneous dam section with no distinct zoning. During the disposal operations in Dam 1 hydrocyclones were used to create piers of the coarse tailings (underflow) with intervening zones of finer overflow material. This deposition system leads, subject to the location of the decant, to the development of a laminated deposit with alternating fine and coarse layers. These depositories were used continuously up until 1972, when they were closed and disposal transferred into a new mine waste facility (Dam 3) on the same site.

On the northern perimeter of Dam 1 a small lagoon, the Head Wrightson Dam (HWD), had been constructed to receive runoff from parts of the aggregate stockpile area and from the downstream slopes of Dam 1. This lagoon was confined by an embankment of some 4m in height and had an estimated capacity of about 8800m³. The HWD confining embankment had been buttressed during 2004/2005 to ensure the stability of the wall and to increase its resistance to erosion.

Dams 1 and 2 were reclassified as closed classified tips under the Mines and Quarries (Tips) Act 1969 at closure in 1972, and the surfaces of the depositories restored to coarse sheep pasture. In 1994 these facilities were reclassified as active classified tips in order to permit the phased extraction and reprocessing of the deposited waste material. It was intended that all the tailings confined within the combined depositories would be removed and reprocessed prior to final rehabilitation. The reclassification of these facilities required compliance with the Quarries Regulations 1999 and necessitated a geotechnical assessment of the depository. The geotechnical characteristics from the 2001 investigation are summarised in Table 1 and typical CPTU logs shown in Figure 2. The deposited material was shown to be highly variable, consistent with a laminated deposit, exhibiting a wide range of relative densities and shear strengths but with predominantly low values in the upper 4m.

Table 1. Geotechnical characteristics of tailings samples from Dam 1

Tailings sample	% Sand	% Silt	% Clay	Specific Gravity	Liquid Limit %	Plastic Limit %	Plasticity Index
Clay/silt	5	63	32	3.34	41	20	21
Sand	66	29	5	2.70	Na	Na	Na

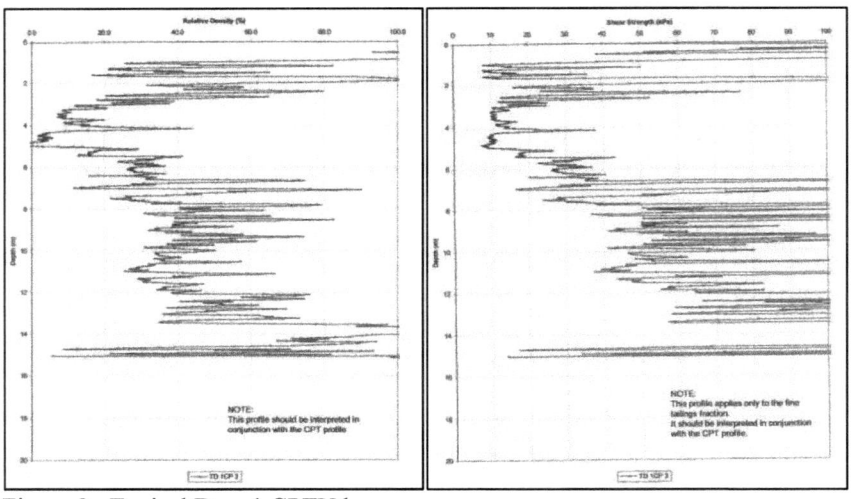

Figure 2. Typical Dam 1 CPTU logs

During the Geotechnical Assessment the implications of a minor failure in 2000, which had resulted in localised liquefaction following heavy rainfall, were also reviewed. This liquefaction event resulted in tailings flowing across the deposit, but was for the most part contained within the confines of

Dam 1. It was concluded that some of the materials investigated "could experience flow type failure during the excavation process", and that:

- modifications to the excavation rules were required to ensure that this operation could be undertaken in safety;
- a rock bund should be constructed along the north-western periphery to ensure that, should any face collapse and the material liquefy, the tailings would be confined within the dam periphery; and
- surface water drainage management within the depository should be modified to prevent ponding, as far as practicable, within the vicinity of the exposed faces.

THE 2007 LIQUEFACTION INCIDENT

Pre-incident conditions

By the end of 2006 tailings extraction for reprocessing had created an excavation face which extended in an arcuate shape over the full width of the facility (Figure 3). The exposed tailings faces were for the most part concave in profile, between 7m in height in the west and 4m on the eastern side, and the preliminary excavation void had been retained as a sump for surface water control. The surface of the depository was thickly vegetated, principally with coarse marshy grasses in the centre with some willow stands on the periphery, and had been grazed by sheep until 2004.

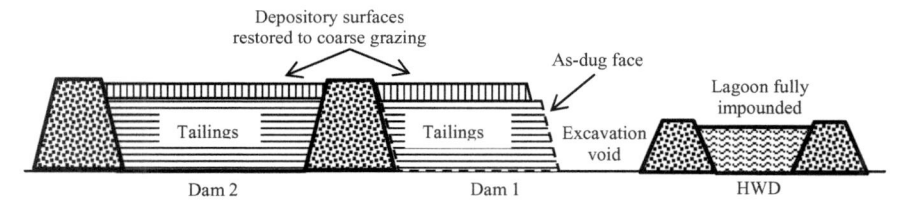

Figure 3. Pre-incident section through Dams 1 and 2

At the beginning of 2007 the monitoring records indicated that site drainage was operating effectively, and on 5 January no standing water was reported on the surface of the depository or against the excavated slopes.

The first three weeks of January were characterised by very heavy rainfall with some 196mm occurring during the period. At the end of the third week of January there was a sudden cold period with a very sharp frost on the night of 21/22 January, with subsequent snowfall giving a cover of approximately 75mm across the surface of Dam 1. The surface temperature during the hours immediately preceding the incident was below freezing and it is unlikely that any significant thawing would have occurred by the time of the event. The presence of the thick snow cover on the surface of the depository may, however, have prevented some of the ponded water from freezing, as was evident during the site visit on the morning of 23 January.

The liquefaction incident - 22 January
The sequence of events leading to the discharge of tailings from Dam 1 is summarised below.

The site records and diaries indicate that the initial failure occurred sometime before 8.15am and took the form of a shallow rotational slip failure in the old excavation face on the south-western side of the facility. At 8.15am the Quarry Manager was contacted by an operator who reported that an initial failure had led to liquefaction and to the development of tailings flow through the excavation face in the location of the original slope movement. The tailings immediately upstream were failing progressively, causing more material to liquefy and adding to the tailings mass then flowing through this narrow gap. It was noted that the tailings mass had not flowed in the direction of the sump but had veered towards the north-west.

The volume of the mobile tailings overwhelmed the north-western protective rock bund, resulting in a discharge into the fully-impounded HWD. At 8.30am the HWD embankment was overtopped at two locations, namely in the south-west corner and along the southern wall. The displaced water and tailings, together with debris and vegetation from the surface of the HWD, flowed across the area used for storing empty reagent drums and breached a 2m-high earthfill bund in the process, carrying these plastic barrels and other debris down the main site access road. A significant volume of tailings was discharged into the plant yard, inundating it to a depth of 600mm. The remainder of the discharge flowed into the process plant site and on to Farnsley Lane, the noise of the debris flow being heard in the offices at around 8.45am. At about the same time the emergency services and the Environment Agency were informed of the incident, and key personnel left site to drive to the junction of Farnsley Lane and the A623, the main Buxton to Sheffield road, in order to prevent traffic from entering Stoney Middleton.

The discharge flowed directly down Farnsley Lane, reaching the junction with the A623 within ten minutes and arriving in the village of Stoney Middleton by 9.25am. Due to the steep gradient down Farnsley Lane the initial flow reached damaging velocities, causing severe surface erosion to the tarmac road (Figure 4). The dislodged tarmac dammed the lane, causing the subsequent flow to be diverted into a semi-restored quarry and thence into a pond at the junction of the Lane and the A623. This pond forms the headpond for the Stoke Brook roadside diversion channel and, in addition, receives all discharges from one of the historic mine drainage adits, the Watergrove Sough. Tailings were deposited in this pond, blocking the inlet to the diversion channel and causing the combined flow from Dam 1, Stoke

Brook and Watergrove Sough to be diverted on to the A623 through Stoney Middleton and on towards the River Derwent (Figure 5). The combined flow along the A623 led not only to its closure but also to the inundation of a number of houses and gardens in Stoney Middleton. By mid-morning the police, fire brigade, Environment Agency, Food Standards Agency, District Council and County Council emergency services were all involved in assisting in the aftermath of the incident.

Figure 4. Farnsley Lane   Figure 5. Stoney Middleton

From around 8.30am onwards the Company transported limestone from the Longstone Edge Quarry to the site in order to seal the breach in the Dam 1 rockfill bund at its narrowest point. By about 10.45am the principal flow had been stemmed by these works. Progressive calving-type failures and liquefaction flow continued from the residual tailings surface inside Dam 1 for approximately a further twelve hours. From late morning onwards any water and silt continuing to flow down Farnsley Lane comprised only runoff, bleed water from deposited tailings and water draining from the roadside. By mid-morning the principal outburst was fully contained and the initial clean-up operation had commenced (Figure 6).

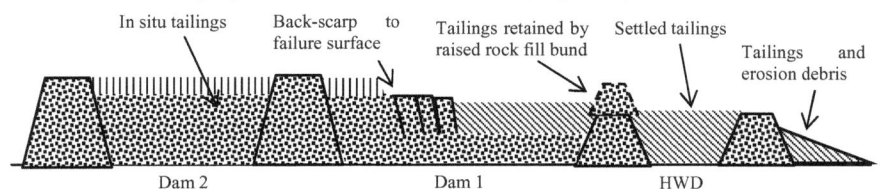

Figure 6. Post-incident section through Dams 1 and 2

Post incident investigation – 23 January 2007
The Author was requested to visit site and investigate the incident in order to make immediate safety recommendations and to prepare a report for the HSE. The site inspection was undertaken on 23 January and was followed by a meeting with Derbyshire Emergency Planning Unit and the Environment Agency in order to brief all parties on the proposed investigation and preliminary safety measures.

Though access on to Dam 1 and the HWD was limited for safety reasons it was evident that the failure had occurred in the southern excavation face. The surface of the initiating failure formed a narrow v-shaped channel of 3-4m in width and of similar depth through which the main tailings flow took place (Figure 7). Despite the volume of material involved in the flow, and its velocity, there was apparently little, if any, subsequent lateral development of the original failure declivity. The post-failure configuration of the tailings surface formed an irregular arc of some 100m in diameter radiating from the gap in the excavation face (Figure 8). The back-scarp was irregular in profile and some 3-4m high, with a series of sub-parallel ridges relating to each failure and liquefaction phase. The failure surface was similarly irregular, with upstands of intact, more-competent material where more-resistant strata had been encountered by the flowing tailings. The configuration of the surface indicated that failure had been initiated at the excavation face, and had progressed southwards as more of the tailings failed, liquefied and began to flow. The limited erosion resulted from the characteristics of the excavation face, which remained relatively stable throughout the event and which prevented any lateral development of the failure surface. This evidence parallels that of the operators, who confirmed the face to have been generally stable throughout its previous two-to-three years' exposure.

Figure 7. Initiating failure surface   Figure 8. Back-scarp

The materials subsequently mobilised were, as far as could be determined, restricted to the upper 4m of tailings, consistent with the CPTU data. The underlying material appeared to have suffered little disturbance, indicative of a more-competent deposit. Tension cracks, which developed naturally in the tailings surface during excavation, were noted throughout the periphery at a distance of up to 5m from the back-scarp. The inspection also showed that some of the drainage ditches were flooded and that there was ponding on the surface of Dam 1.

The mobile tailings involved in the flow and subsequently retained by the rockfill bund had formed a generally level surface, with some ponding evident as they settled. The material placed by the Company to seal the breach had covered the original rockfill bund, which was no longer visible. The tailings mass had overtopped the HWD, the discharge continuing for at least two hours and causing some erosion along the southern face. However, the embankment remained intact and was not breached, the works previously undertaken by the Company to improve the stability of this embankment being instrumental in preventing both its failure and a more significant discharge of material off the site (Figures 9 and 10).

Downstream of the HWD the flow path across the site was evident from the tidemarks in the stockpile area, the erosion of the dam access road and the extensive deposits of settled tailings in the plant yard at the toe of Dam 3. Across the stockyard area the depth of flow was probably of the order 300-500mm, but appears to have reached a maximum of about 600mm in the plant yard.

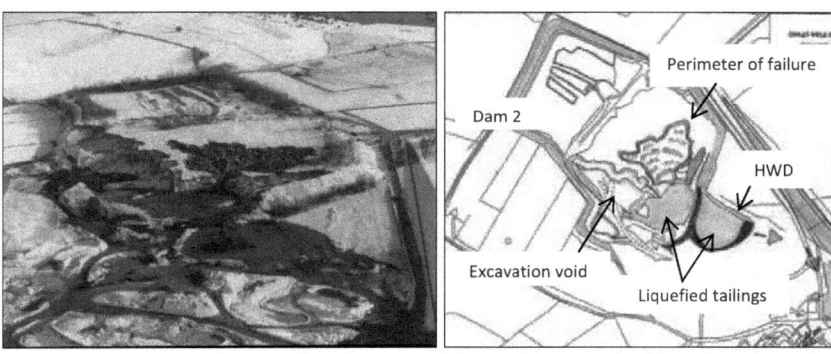

Figure 9. Dams 1 and 2 post-incident   Figure 10. Post-incident location plan

Farnsley Lane was not accessible in safety during the visit and by the time the A623 was inspected the flow was minimal, leaving the road surface covered in thin residual deposits of tailings and other debris. By the end of 23 January clean-up of the affected houses in Stoney Middleton was well underway.

CONSEQUENCE OF THE FAILURE

Analysis of the pre- and post-incident surveys, together with the Company's data records, enabled the volumes involved in this event to be estimated. It appears that the rate of discharge from Dam 1 was constrained by the width of the initial slope movement, by the competence of the underlying surface and by the existing rockfill bund. Both the rockfill bund and the HWD provided buffer storage for the failed material and, together with the prompt action taken by the Company on the morning of 22 January, were instrumental in preventing a significantly larger volume of material from leaving the site.

Further analyses indicated that the majority of the tailings were retained on site within Dam 1 and the HWD. The flow down Farnsley Lane comprised water displaced by the tailings deposited in the HWD and local runoff, together with the residual tonnage which had not been deposited in the Cavendish Mill plant yard. Only a relatively small proportion of the tailings ultimately reached Stoney Middleton, it being estimated that some 5% of the material displaced from the dam was deposited in the village and along the A632. However, though this small volume of material was spread widely, a number of houses, gardens and fields downstream were inundated and localised damage caused by the deposition of silt. There were, however, no reported incidents of injuries to local inhabitants. The Company initiated immediate clean-up measures along the section of the A623 and in the affected houses in Stoney Middleton, effecting full remediation. The confluence of Stoke Brook and the River Derwent, the ultimate receiving water, took some months to rehabilitate and return to pre-existing conditions.

POST-INCIDENT SAFETY MEASURES
The preliminary results of the investigation were presented to the HSE and the Environment Agency on 16 February 2007 and included a review of the incident, a preliminary assessment of the tailings discharge and recommended measures required to ensure that no tailings would be released from the site should there be further heavy precipitation. The following safety measures were initiated to ensure that the risk of any future discharge from Dam 1 was minimised:

- The height of the perimeter rockfill bund and all points on the Dam 1 periphery were to be raised to a level commensurate with the need to confine any potentially liquefiable tailings within the depository.

- The HWD embankment was to be extended to ensure full containment of the deposited tailings.

- The existing site drainage system within Dam 1, the HWD and the downstream area was to be modified to improve control of surface runoff and prevent any untoward flow off-site.

- All operating rules, inspection and monitoring routines and record sheets were to be reviewed and updated to reflect the changed nature of the depository prior to any further excavation of tailings.

- The Operating and Maintenance Manual was to be updated to reflect the changes implemented, and all operating and inspection routines for other operational areas were to be reviewed and modified to reflect the lessons from the incident.

- Permanent survey beacons were to be installed at key locations throughout the crest of Dam 1 and the HWD to ensure that any movement could be detected at an early stage and remedial measures instigated.
- A review of the measures recommended in the report was to be undertaken by the Competent Person (the Author) within three months from the date of the incident.

CONCLUSIONS

The observational records and photographs taken on 23 January, together with additional confirmatory sampling and basic laboratory testing, were used in the analysis of the slope failure and its aftermath. It was concluded that the configuration and outcome of the incident owed much to circumstances, its extent and impact being affected by a number of individual and disparate parameters. Though further precautionary steps could have been taken, it was concluded that reasonable mitigating measures had been put in place by the Company. The principal conclusions arising are summarised as follows:

The initiating movement in the southern excavation face was of a small-scale nature, being v-shaped, shallow and probably involving considerably less than 50m^3 of coarse tailings from the pier separating the upper deposits of finer material from the excavation face. It is evident that this face retained a layer of finer low-density tailings which failed as a result of the removal of lateral support following the initiating slope movement. This finer material was evidently susceptible to uncontrolled flow and liquefaction, and the movement initiated progressive failure and the development of flow across the more competent basal deposits.

The cause of the initial minor slope movement in a previously stable face is most likely to have been induced by a change in local drainage conditions caused by the relatively intense precipitation over the preceding days. This rainfall is indicated to have overwhelmed the drainage system and is likely, either through surface ponds or the overflowing drainage ditches, to have resulted in the ingress of water into cracks adjacent to the area of interest. The localised nature of this failure, which was not replicated anywhere else in the extraction area, suggests that the excavation face and the drainage conditions were specific to this location. The shallow nature of the failure is also consistent with changes in drainage induced by tension cracks, which are known to have been of limited depth in these tailings. Localised ponding above the excavation, possibly feeding tension cracks above the face, may have given rise to increased pore pressures or to the development of concentrated seepage paths. Local stability will also have been affected by the rapid change in temperature and the freezing conditions over the previous twenty-four hours. This would have been instrumental in initiating the small-scale slope movement in the previously intact material. A minor failure of this nature would normally have been considered of limited consequence in the course of the extraction and excavation process, as evidenced by earlier slope movements. However, the juxtaposition of the initiating slope failure and material susceptible to flow resulted in the more serious event of 22 January.

The CPTU and subsequent sampling indicated that the depository was characterised by a relatively competent thin vegetated cap of exposed fine tailings. This cap was underlain by a mixture of fine and coarse deposits, the finer material exhibiting very low density and high moisture content. The material which liquefied and flowed was derived from the upper tailings layers on which surface water was ponding. Progressive failure of the surficial tailings occurred in a generally southerly direction, feeding more liquefied material into the flow. Once liquefaction had been initiated, flow would have continued until a resistant stratum or confining structure were encountered or drainage from the base of the flow occurred, increasing shear strength and preventing further movement. During this event the tailings mass moved progressively across the depository, carrying floating islands consisting of small blocks of intact cap material. This flow would have been expected to have followed the maximum gradient towards the excavation void, and thus be contained. However, in the event the slide material was diverted to the north-west, away from the void and towards the rockfill confining bund. It is unclear whether this resulted from the configuration of the deposit or from the cap material blocking the drainage path.

The tailings level rose behind the Dam 1 rockfill bund, eventually overtopping this structure and overflowing into the HWD where it displaced the stored water, further adding to the moisture content of the tailings mass. The Company put immediate measures in place to stem the flow and to infill the breach in Dam 1. This prompt action resulted in the breach being sealed within two hours and thus prevented further discharges of tailings off-site, although the flow continued to develop within Dam 1 for the next 12 hours. These and other emergency works to retain the tailings mass were well-executed and prevented further discharges of solids and water from the site in the post-event period. The remediation of Stoke Brook and the River Derwent, as agreed with the Environment Agency and other stakeholders, was satisfactorily completed with no discernible impact, as confirmed by the independent ecological survey undertaken in May 2008. This event evidenced that downstream impacts from failures of mine waste facilities may, in many instances, be of short duration only – particularly where a catchment has been affected by local geology/mineralogy over a long (>1000-year) period.

Finally, the failure in Dam 1 did not involve the total volume of material contained, with only the finer material and that susceptible to liquefaction being discharged over the top of the HWD and on to the mine site. It is conjectured that, without the presence of the free water impounded by the HWD, the consequence of the discharge would have been significantly less. The topography between the HWD and the plant site, and the nature of the strata, would have led to increased shear strength in the liquefied tailings, resulting in the majority being deposited upslope of the mine access road. The only material leaving the site via Farnsley Lane would have been that eroded by surface water runoff and would probably have been wholly contained within the Stoke Brook channel rather than inundating Stoney Middleton. As discussed elsewhere (Cambridge, 2014), such incidents are important in the development of emergency plans for mine waste facilities. The failure at Cavendish Mill emphasised the difference in emergency planning requirements between a mine waste facility and a large raised reservoir, particularly in terms of failure mode and volumes of material involved post-incident. Further, it is evident that the presence of water-bodies downslope increases the extent and impact of any downstream inundation. It is therefore important that the lessons learnt from the Cavendish Mill and from other historic failures be considered in the development of emergency plans for all mine waste facilities.

ACKNOWLEDGEMENTS
The author would like to thank INEOS plc for its permission to publish this paper.

REFERENCES
Cambridge M, Hill T J and Harvey P (2014), Emergency planning for mining waste facilities in England, *Proceedings of the 18th BDS Biennial Conference, Belfast, September 2014.* ICE Publishing, London, UK

Cambridge M, The Cavendish Mill (2013). TD1 incident, - the use of historic tailings dam incidents in the development of emergency plans. *Proceedings of ICOLD Conference, Stockholm, November 2013.*

ICOLD (2002). *Tailings Dams, Risks of Dangerous Occurrences, ICOLD Bulletin.* ICOLD, Paris, France

Health & Safety Commission (1999). *Health and safety at quarries, Quarries Regulations 1999.*

Penman A D and Charles J A (1990). The safety of tailings dams and lagoons in Britain; *Proceedings of the 6th Conference of the British Dam Society, Nottingham, September 1990.* British Dam Society, London, UK.

HMSO (1971). *The Mines and Quarries (Tips) Regulations, 1971.*

HMSO (1969). *The Mines and Quarries (Tips) Act 1969.*

Maintaining the Safety of our Dams and Reservoirs
ISBN 978-0-7277-6034-0

ICE Publishing: All rights reserved
doi: 10.1680/mdam.60340.426

Improving the overtopping resistance of existing flood detention reservoirs

J D GOSDEN, Jacobs UK Ltd
T AMBLER, Jacobs UK Ltd
A P COURTNADGE, Jacobs UK Ltd

SYNOPSIS   Recent Reservoirs Act inspections identified a number of concerns with the ability of several flood detention reservoir embankments to accommodate overtopping flows during extreme flood events without significant risk of failure. The issues identified arose from a number of sources, including changes during construction to accommodate adjacent landowners' concerns; construction outwith required tolerances and improved understanding of risk. Lessons learned from these will be summarised to provide guidance for investigations to be considered during future inspections.

While the works required in the interests of safety identified were in principal straight forward – raising embankments to prevent overtopping, lowering embankments to create spillways, installing crest beams, and using different types of reinforced grass e.g. Ankalok blocks, Salix Vmax^3 and Enkamat – the application of these presented several challenges to the designs, including:

- Causing an increase in flood risk downstream of the reservoir over a range of intermediate frequency flood events, through lowering an auxiliary spillway to prevent the main embankment overtopping in an extreme event.

- Balancing environmental concerns and cost with different erosion protection systems, embankment slope and resulting landtake.

- Controlling a short term potential increase in risk of failure while a new protection system, to achieve greater long term security, is established.

During the construction stage several difficulties were encountered in achieving a satisfactory interface with the existing erosion protection, contractor self-supervision and maintenance once the contractor had completed the initial construction. The paper concludes with lessons learned for the benefit of future projects.

INTRODUCTION
Since the publication of *CIRIA Report 116 – Design of reinforced grass waterways* in 1987 the use of overtoppable spillways on reservoir embankments has been increasing. This is particularly the case at flood detention reservoirs of limited height, between 5m and 10m. In many cases where re-evaluation of the consequences of a breach or re-assessment of hydrology has led to an increase in spillway design flow, the tolerance of reservoir embankments to overtopping is assessed.

Such spillways normally have one of four forms of construction: plain grass, reinforced grass, interlocking concrete blocks or Grasscrete. In order to achieve the design intentions, two particular aspects are sometimes overlooked: care and attention to detail in construction, and the subsequent short term maintenance. Particularly with the general reduction in site supervision across much of the industry, or contractor self-certification, the requirements for satisfactory performance are not always well understood.

This paper discusses a number of issues which have been found during Section 10 Inspections and have required subsequent works. The performance of several spillways during flood events is also described. A range of issues which have arisen during design and implementation of remedial works are discussed. The paper concludes with recommendations for future consideration.

ISSUES IDENTIFIED DURING SECTION 10 INSPECTIONS
Two examples of Section 10 Inspections are described, where the spillway appeared to be in good condition with well-maintained grass coverage but detailed assessment of the composition and spillway dimensions showed that the spillway capacity would not perform satisfactorily during the design flood event. The other main reason for works to spillways, i.e. to increase spillway capacity where a re-assessment of consequences of failure or hydrology led to a re-assessment of the spillway capacity, will not be discussed in this section.

Cox's Meadow FSR
Cox's Meadow FSR is a 3.5m high flood storage reservoir protecting the town of Cheltenham. The first inspection following issue of the Final Certificate included a level survey of the spillway and embankment crest. The results of this survey are compared with the intended design in Figure 1.

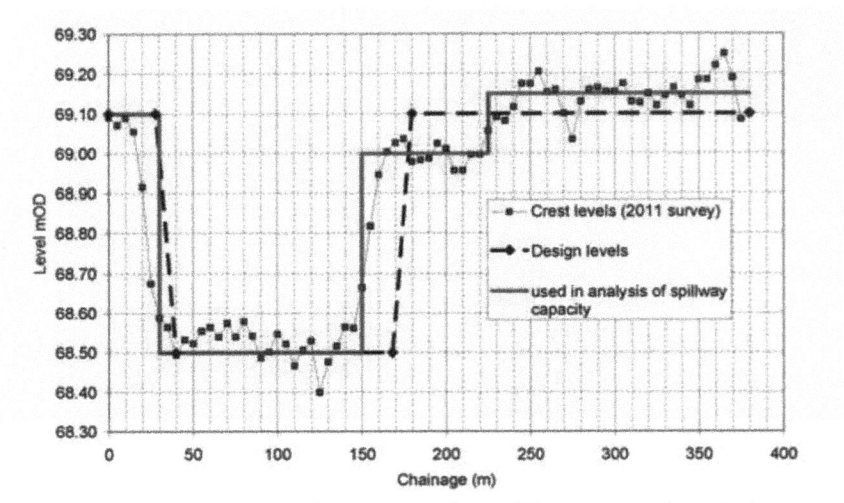

This showed two issues of concern; the spillway crest length is around 110m, much less than the design value of 128m, and the lowest point on the spillway crest is 150mm below the mean. Both of these will result in increased velocities on the spillway slope and expected velocities in excess of the allowable limit for reinforced grass. The reduction in crest length apparently resulted from an objection from an adjacent property owner who wanted a new hedgeline on the embankment crest extended to provide more privacy. The variation in spillway crest level is presumed to be a result of inadequate construction control.

Halstead FSR
Halstead is a 4.5m high on-line flood storage reservoir on the outskirts of Halstead in Essex. The overflow spillway of Halstead FSR is comprises open cell concrete blocks and discharges the 1 in 2,000 year flood event before the rest of the embankment crest overtops. Away from the spillway the embankment crest and downstream face is covered with reinforced grass to provide the necessary erosion protection. During the first Section 10 Inspection a shallow hole was dug to check for the presence of the geotextile reinforcement. This discovered the geotextile to be below 200mm thickness of topsoil and none of the grass roots had penetrated to this depth. Subsequent systematic survey showed that the depth to the geotextile on only 3 of 22 holes was less than 150mm and in five holes it was in excess of 250mm. Clearly, with no connection between the grass roots and the geotextile, the erosion protection system was only plain grass. The allowable velocity for plain grass would be exceeded during the design flood. This is believed to be again a result of inadequate construction control.

OBSERVED PERFORMANCE OF GRASS SPILLWAYS

During the last few years the Author has experienced overtopping of three grass spillways and this section summarises the observed performance of each.

Gipping FSR

Gipping FSR comprises a low earth embankment either side of a bottom hinged tilting flood gate on the River Gipping, just upstream of Stowmarket, Suffolk. It is a Category A dam, capable of storing up to 85,000m^3 to alleviate flooding.

The embankments are up to approximately 2m high with upstream and downstream slopes of approximately 1v:2h. The embankments adjacent to the tilting gate are designed to overtop over a length of approximately 130m in floods greater than 1 in 25 annual exceedance probability (AEP). As such, the crest and downstream slope are reinforced with Enkamat. A plastic geogrid has also been laid on the surface, presumably as a deterrent to burrowing animals.

Following a relatively dry winter, April 2012 was particularly wet with two and a half times the average rainfall for that time of year in Southern England. Following particularly heavy rain on Thursday 3 May 2012 the tilting gate automatically closed to detain flood water. There are doubts over the accuracy of the telemetry data at the time but images from the webcam indicate the dam was overtopped by around 400mm. The duration of overtopping is also uncertain but is believed to have been several hours. On this basis, the velocity on the downstream face is estimated to have been 5.7m/s, which is the limiting velocity for Enkamat for two hours duration. It is likely that this duration of overtopping flow was exceeded.

The Supervising Engineer (Andy Courtnadge) visited the reservoir the following day once water levels had receded and found that the downstream face had eroded in several locations. The worst erosion was approximately 15m left of the control structure, where a small scour hole had developed (Figures 1a and 1b). Signs of small animal burrows approximately 50mm diameter (possibly vole or shrew) were observed along the back of the hole which may have contributed to the failure (Figure 2).

There also appeared to be a lack of grass reinforcement over this area which is likely to be a significant factor to the localised damage in this area. Subsequent investigations have shown that Enkamat is relatively continuous over the downstream face but is up to a depth of 75mm below the surface, which is likely to make it ineffective at binding the grass roots together.

Figures 1a & 1b. Gipping FSR: Scour hole to left-hand side of outlet structure

Figure 2. Gipping FSR: Animal burrow within scour hole (also note Enkamat)

Figure 3. Gipping FSR: Poor performance of weeds against overtopping

There was also clear evidence to support the fact that good grass cover is far more tolerant to overtopping flows than areas with a predominance of broad leaved weeds (Figure 3).

In summary, the evidence at Gipping FSR suggests the published guidance is correct, i.e.:

- Enkamat needs to be within the top 20mm to 30mm of the surface to be effective at binding grass roots together.
- Even small diameter burrows can weaken the resistance of an embankment slope to overtopping flows.
- Broad leaved weeds offer significantly less resistance to erosion from flowing water than uniform grass cover.

Works are currently in progress to upgrade the overtoppable embankment with an interlocking concrete block reinforced grass system.

Rattlesden FSR

Rattlesden Flood Storage Reservoir is similar to the Gipping FSR, also on the outskirts of Stowmarket in Suffolk, but located on a different water course, namely the River Rattlesden. As such it was also subjected to the flooding on 3 May 2012. The dam category, height and length are all very similar and the downstream slope angle is only marginally shallower at between 1v:2.75h and 1v:3.5h. However, the key difference is that the overtoppable grass spillway on the Rattlesden FSR dam is protected with precast concrete blocks (non-interlocking type).

Telemetry at this site was more reliable and showed that the spillway overtopped by up to 400mm depth for approximately 17 hours. With 400mm depth of overtopping at the crest, the flow velocity on the downstream slope would have been around 5.2m/s, which is within the limiting velocity for non-interlocking concrete systems (6m/s).

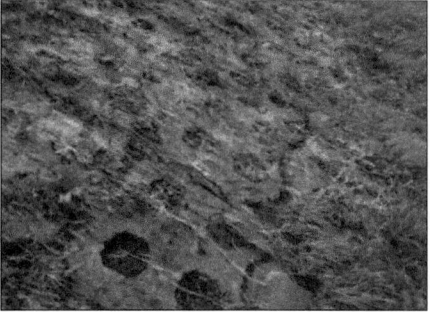

Figure 4. Rattlesden FSR: View from right abutment

Figure 5 – Rattlesden FSR: Concrete blocks on left-hand side of downstream face

Inspection the following day showed that the concrete blocks performed well against the overtopping flow and there was no damage to the embankment (Figure 4). In one shaded area beneath a large tree, there was very limited grass cover growing within the precast voids before the event - but even without the grass, the precast blocks performed satisfactorily with no significant damage to the embankment (Figure 5).

Curry Moor

Curry Moor is a large offline flood storage area on the Somerset Levels which is filled when the water level in the River Tone rises above a controlled inlet spillway along its right-hand bank. The storage area is then emptied by pumping. It is a Category A dam. There are five overtoppable sections on the perimeter embankments but the main spillway is designed to overtop by 870mm in a 1 in 500 AEP flood. The spillway is an unreinforced grass spillway approximately 20m wide with a slope angle of approximately 1v:20h and a total drop of approximately 2m.

During the recent high profile floods, from December 2013 to February 2014, Curry Moor spillway operated continuously for over nine weeks with a typical overtopping depth at the crest of 500mm, and with a maximum overtopping depth of over 800mm at the peak of the floods.

Figure 6. Curry Moor Spillway: View from the crest over the downstream slope during overtopping

The typical depth of flow on the downstream slope was measured as 180mm on the left hand side but appeared to be greater on the right-hand side (approximately 300mm depth) where there was also an area of turbulence which slowly extended up the spillway through the flood duration. The typical flow velocity on the slope was estimated, using 'Pooh sticks', to be on average 2.8m/s in the centre of the flow but again faster on the far right-hand side. At the peak of the floods, with 800mm depth of flow overtopping the crest, the peak velocity on the downstream face is estimated to have been 3.4m/s.

Grass cover on the slope was typically around 150mm high, tufted arable grass.

CIRIA Report 116 suggests that the limiting velocity for good grass cover after 50 hours duration of flow is less than 2m/s, so it was expected that the prolonged discharge at higher velocities than this would have caused erosion to the underlying embankment. However an inspection following the floods revealed surprisingly little damage, apart from in the area of turbulence (Figure 7). In the lower part of the slope, where the depth increased due to the tailwater, the grass stems had died during the prolonged submergence but with the root system presumably intact is likely to recover quickly.

Over most of the slope the grass was still healthy and provided good coverage, with only small areas of bare earth but no significant erosion.

Figure 7. Curry Moor Spillway: View of surface after flood on the left hand side

Figure 8. Curry Moor Spillway: View of surface after flood on the right hand side

In the area of turbulence (Figure 8) the grass and topsoil had been washed away and some of the underlying stony subsoil eroded. This had been extending up the spillway but, because of the shallow slope and the length of the spillway, there was still a long way to go before it could result in a reservoir breach.

DESIGN OF GRASS SPILLWAY IMPROVEMENT SCHEMES

For all slope protection methods except interlocking blocks, achieving sufficient grass coverage is an integral part of providing protection against overtopping flows. The slope is much more vulnerable to erosion before grass has established, spread roots and formed a strong composite with the protective matting and subsoil below. Therefore the key issue with such remedial works is to minimise the risk during the period of increased vulnerability between the removal of the existing erosion protection and the full development of the improved erosion protection. This has been handled in different ways for different situations.

At Cox's Meadow FSR the general solution was relatively straightforward – to raise and lower sections of embankment as required to match the original design. However the works to the existing spillway crest were difficult due to the presence of fully established Enkamat protection, which when removed would leave the spillway vulnerable to erosion. It was not possible to exclude water from the reservoir or use pre-turfed Enkamat, for which as well as being more costly, there was insufficient time remaining to establish grass off-site and then establish it on the spillway. As a result it was decided to keep as much of the existing grassed Enkamat as possible and replace the minimum area in zones of lower velocity i.e. just on the crest and not extending onto the downstream slope. This required the existing Enkamat to be cut and folded back during construction then pinned over the new. The method of folding and pinning was chosen to avoid a weak longitudinal joint between the two Enkamat layers, which would be susceptible to pulling up by overtopping flows.

Kidderminster FSR

At Kidderminster FSR no works were required to the spillway but the existing grassed embankment outer slopes needed replacing with reinforced grass. As the embankment is not overtopped until a 1 in 1,000 year flood event, the risk is limited. To minimise the risk of increased vulnerability an open mesh geotextile interwoven with a biodegradable fabric was used, to provide a better erosion resistance than bare earth or plain open mesh geotextile.

Saintbridge 2 FSR

Saintbridge 2 FSR is contained by two embankments which both have variable grass cover which provides insufficient erosion protection during extreme flood events. Overtopping at the current level of the embankment starts at around 1 in 200 annual probability. The solution preferred by the client aimed to limit works to the western embankment alone, and avoid works to the southern embankment altogether to reduce disruption to residents living just beyond the toe of the southern embankment. To reduce the risk of overtopping of the southern embankment to an acceptable level required lowering the crest of the western embankment and protecting the downstream slope from erosion. Protection would either be interlocking concrete blocks on the current embankment slope or reinforced grass on a shallower slope.

Hydraulic modelling of the lowered western embankment to provide an auxiliary spillway, showed that operation would start at 1 in 25 AEP flood event. This would result in an increase in flows and levels downstream of the embankment for this relatively frequent flood, which was not considered acceptable.

The analysis showed that both embankments would be able to pass the design flood (in this case the PMF) with good grass coverage alone, and only a limited amount of slope slackening. This also required removal of a number of trees, including their roots and backfilling the resulting void. In order to provide some degree of erosion resistance to bare areas while the grass is establishing, the solution specified was to use coir matting pinned to the topsoiled and seeded ground surface. This provided the added benefit of warming the soil, so that germination would be initially faster once the geotextile was removed.

CONSTRUCTION OF GRASS SPILLWAY IMPROVEMENT SCHEMES

Issues during construction

To perform effectively, reinforcement matting must be laid flat on a smooth surface free from voids and raised patches, in order reach their full protective potential. While this is easy to achieve on large open areas, it is difficult on improvements to existing schemes, where the areas are more restricted and an overlap with existing material is required.

At Cox's Meadow FSR rolling back (Figure 8) and replacing the existing Enkamat reinforcement on the crest, in order to pin new material underneath, proved to be particularly difficult. It did not prove possible to reliably remove the wrinkles when replacing the reinforcement so that the depth of topsoil required to full cover the reinforcement varied. The contractor's intention was to re-use site won topsoil, which proved insufficient and some topsoil was imported. While there may have been no reinforcement visible on the surface when the topsoil was placed, topsoil consolidation uncovered Enkamat in a number of locations. (Figure 9). In future the following might be considered; cut the Enkamat, close-cut the existing grass followed by laying and pinning the new reinforcement over the top; import all topsoil to place above the Enkamat with site-won topsoil only used below.

Figure 9. Rolling back of existing reinforcement at Cox's Meadow FSR

Figure 10. Patches of reinforcement visible through topsoil at Cox's Meadow FSR

At Saintbridge 2 FSR, once the vegetation had been cut on the western embankment it became clear that a greater length of the downstream slope would need re-profiling and that the concentration of weeds was such that once they had been treated there would be few vegetated areas. The decision was taken to remove all vegetation cover and topsoil and seed the entire slope following re-profiling. This was completed and the coir matting pegged on top. However, over the weekend before the site visit, vandals had removed large areas of the matting (Figure 11). It was decided to cover the replaced matting with topsoil to reduce the temptation to repeat the vandalism.

Figure 11. Saintbridge 2 FSR western embankment

Grass establishment
Both initial grass growth and subsequent grass maintenance proved to be an issue on all sites. Although establishment and maintenance of the grass was specified in the contract, civil engineering contractors, often with bases remote from the site, are not used to carrying out regular operations once the initial construction works have been completed. Our visits to Cox's Meadow site found extensive weed growth in the works areas, and a lack of grass establishment (Figure 12). It was clear that no maintenance was being undertaken. At this site conditions improved markedly once the Local Authority took back control of maintenance and established a regular cutting regime (Figure 13).

Figure 12. Cox's Meadow FSR poor grass maintenance

Figure 13. Cox's Meadow FSR improved grass maintenance

CONCLUSIONS

The issues encountered above provide lessons to implement on similar schemes, where grass establishment is an essential part of the overall embankment protection. These lessons include:

- The implementation of the details of the design in the construction stage are fundamental to the reliability and safety of the scheme. An adequate level of site supervision is required and the site supervision team need to be briefed on the important aspects.

- While grassed spillways are tolerant and resistant to flowing water, they are particularly vulnerable to damage if any defects are present. This includes broad-leafed weeds and small diameter burrows.

- When repairs are made to grassed surfaces attention needs to be paid to the period of increased vulnerability while the improved surface develops. A range of means are available to provide protection.

- Connecting new reinforced grass surface to existing ones needs to be done with great care.

- Most civil engineering contractors are not comfortable with intervention to establish comprehensive grass cover and returning regularly to a site to maintain the grass. Consideration should be given to an alternative approach whereby the grass establishment and maintenance is carried out by a landscaping contractor who is more used to tending vegetation.

Maintaining the Safety of our Dams and Reservoirs
ISBN 978-0-7277-6034-0

ICE Publishing: All rights reserved
doi: 10.1680/mdam.60340.438

The Rhymney Bridge Incident

A K HUGHES, Atkins, Epsom, UK
T WILLIAMSON, DCWW, Newport, Monmouthshire, UK

SYNOPSIS   In the early part of 2013 the spillway at Rhymney Bridge 2 reservoir in South Wales failed due to uplift pressures after a storm event.

Unfortunately the incident occurred in the winter during a period when there were a number of storms following one after the other and more than one metre of snow on the ground.

The paper will describe how the incident was managed during a period of extreme weather that caused the reservoir, which had been emptied, to fill and overflow.

It will describe the emergency repairs, how judgement had to be practised on the ground, how significant investment had to be provided by the owner and how the Panel Engineer planned measures for the possible situation where the dam might reach a condition close to failure.

The paper will conclude with a description of the permanent repairs at the site.

DESCRIPTION OF RESERVOIR
Rhymney Bridge No. 2 Reservoir is situated about 3km north of Rhymney in the County Borough of Caerphilly. The reservoir is located high in the foothills in South Wales just off the Heads of the Valley Road at an elevation of about 356mOD.

Rhymney Bridge N° 2 reservoir was completed in 1901 to supply the local water undertaker and has continued as a water supply reservoir since then although ownership has changed a number of times. Raw water is pumped from the reservoir to Shon Sheffrey reservoir about 3km to the east-northeast, from which it is passed for treatment to the adjacent Nantybwch WTW.

The dam is about 220m long and has a maximum height above the downstream toe of about 19m, with the grassed crest only 2.5m wide, following raising of the dam crest in 1989.

Figure 1. Location Plan *(© 2014 Google)*

It has a 60° change in direction about 90m from the right abutment (concave on the upstream side). The longest axis of the reservoir, which is about 250m long, is oriented approximately north-northwest to south-southeast, with the dam at the southern end. The surface area is about 2.9ha, the capacity 191Ml and the maximum depth about 16.5m.

The upstream face has a slope of about 1 in 3 and is protected by riprap. The slope of the grassed downstream face is about 1 in 2.5. The dam has a puddle-clay core, below which there may be a puddle-filled or concrete filled cutoff trench.

Improvements undertaken in 1989 include an L-shaped reinforced concrete retaining wall that stands 2.4m above its foundation level, with a 2.2m wide base slab located directly beneath the dam crest. The top of the wall is nominally 550mm above the dam crest level. The design top level of the wall is 358.60mOD, which is 2.80m above the spillway crest level of 355.80mOD.

The catchment area is 656ha comprising steep slopes draining moorland and farmland into reservoir. Extensive areas of karstic limestone in catchment.

GEOLOGY

The geological survey, the dam is located mainly on orthoquartzites and mudstones of Millstone Grit, which are below the mudstones and siltstones of the Lower Coal Measures that outcrop on the hill slopes above and downstream of the reservoir. The right shore of the reservoir and the right abutment of the embankment lie on Glacial Till.

The solid geology of the area comprises a cyclic sequence of mudstones, siltstones and sandstones with clay ironstones, coals and seatearths. The coals are thin and generally impersistent and there are a number of marine bands. The beds are shown with a shallow (7°) dip to the south.

The strike of the regional steeply inclined normal faulting in the area is generally NNW/SSE. There is one fault shown immediately upstream of the reservoir that does not extend downstream of the embankment. The downthrow is towards the southwest. A second fault is marked downstream of the embankment on the right abutment.

The drift material that lies under the right shore of the reservoir and the right abutment and on the hill slopes above the reservoir is shown on the map as Glacial Till.

OVERFLOW WORKS
The overflow works, which are located at the left (east) abutment, comprise:

- a curved approach channel, approximately 30m wide, from the reservoir basin;
- a 30m wide overflow sill, in line with the dam crest centreline, leading to a channel (with a concrete floor and brick and masonry walls) that slopes gently, gradually reduces in width down to about 15m and turns about 60° to the right;
- a brick-lined spillway channel with a dished invert (which is presumed to have a concrete backing beneath and behind the brickwork), that runs steeply down the left mitre of the dam, then flattens, gradually reducing in width down to about 10m; and
- a tight curve of about 60° to the left into the downstream river channel.
- There is no stilling basin, but there are masonry and brick sidewalls to the river channel extending for a short distance downstream of the curve at the downstream end of the spillway.

The outlet works are located at or near to the original streamcourse at the highest part of the dam. They comprise:

- a deep narrow forebay that extends over most of the height of the upstream shoulder of the dam;
- a circular wet-well brick-lined shaft located just upstream of the dam crest;

- a brick-lined horseshoe-shaped tunnel, from the shaft to the downstream toe, founded on the west side of the valley;
- a 380mm diameter cast iron standpipe within the shaft, having three 380mm diameter penstocks on horizontal branches at different levels; and
- an outlet main that passes through a concrete/brick plug at the point where the culvert meets the shaft and thence along the bottom of the tunnel.

Outside the tunnel portal, the outlet main divides into 350mm and 300mm supply mains. There is a 300mm scour branch off each of the supply mains, both of which discharge to the river a short distance downstream of the dam.

About 600m downstream of the reservoir the river passes under the A465, which is carried across the valley on a high embankment, beyond which is the village of Llechryd. The river continues through Rhymni, New Tredegar, Bargoed, Ystrad Mynach and several other towns before discharging to the Severn estuary near Cardiff, about 35km south-southwest of the reservoir. The dam is deemed to be a Category A Reservoir as defined by Floods & Reservoir Safety: An Engineering Guide and with many hundreds of people at risk is a high risk situation.

In the 1999 Section 10 Report the Inspecting Engineer stated:

'Because the capacity of the reservoir is small in relation to the run-off from the catchment, the reservoir overflows for long periods. Consequently every year it is necessary to replace significant areas of brickwork which have been removed from the steeply sloping channel invert by fast flowing water. The undertakers propose in the near future either to replace all the remaining brickwork with mesh-reinforced concrete, or to place a layer of this concrete over the top of the remaining brickwork. I prefer the second option, despite the fact that there will be a marginal reduction in freeboard provided by the side walls of the channel. ... The brick/masonry side walls downstream of the steeply sloping section are in danger of being undercut and de-stabilised by floodwater and they need to be inspected annually by the Supervising Engineer, and repaired as necessary.'

The Inspecting Engineer in 2009 recommended a programme to implement the long term plan to improve the overflow works and spillway by the date of the next inspection in 2019 with the investigating works carried out within three years of the inspection.

The summer PMF has been calculated as having a peak inflow of $110 m^3/s$ and a peak outflow of $107 m^3/s$ with a flood surcharge of 1.8m. Wave surcharge was calculated at being 0.29m. The drawoff facilities provide a useful drawdown capability - the calculated drawdown rate being of the order of 3 metres a day assuming no inflow.

The summer PMF has been calculated as having a flow of $50m^3/s$ and an outflow of $146m^3/s$ with a flood surcharge of 2.2m. Wave surcharge was calculated at being 0.29m. The drawoff facilities provide a useful drawdown capability - the calculated drawdown rate being of the order of 3m a day assuming no inflow.

HISTORY
The dam was designed by Charles Hawksley and was completed in 1901. The Reservoir was built by the Rhymney Valleys Gas and Water Company in 1901 and, in 1921, the two Rhymney Bridge reservoirs were transferred to the Rhymney Valley Water Board. By 1942 it was noted that settlement of the embankment had resulted in there being insufficient flood freeboard, that the main outlet pipe from the reservoir had been undermined and exposed by floodwaters, and that 'further successive floods may cause damage to the foot of the embankment.

By 1945 the top of the puddle and the crest of the dam were raised to 2ft and 4.34ft respectively above the overflow cill, the top of the bank being protected by a new concrete slab retaining wall.

In 1954 the Inspecting Engineer found the spillway channel to be in good order 'the lower end having been rebuilt recently after damage by flood'. However the Inspecting Engineer considered the overflow weir to have insufficient flood capacity and recommended the following measures:

- raising the puddle core to 3ft above overflow level;
- raising the embankment crest to 6ft above overflow level; and
- lengthening the overflow to 100ft, widening the upstream approach to suit.

According to records these measures were implemented between November 1959 and June 1960 and a 'Certificate as to Execution of Works' covering them was issued in September 1960.

In 1964 the Inspecting Engineer recommended maintenance work on the pipework and further repairs to the brickwork lining of the spillway. There is no record of whether this was done at the time. In 1970 the next Inspecting Engineer found that some of the bricks were missing or had lifted in the spillway, with repairs required.

In his 1973 report, the Inspecting Engineer again found damage to the brickwork lining of the spillway, including 'in particular there is a heaved and cracked section of brickwork at the bottom of the spillway on the left hand side'. He again specified a three-year interval to the next inspection and made the following recommendations 'in the interests of safety':

- repairs to the brickwork lining of the spillway and clearance of vegetation;

In November 1976, the Inspecting Engineer found that brickwork in the lower part of the spillway on the left side had been removed in recent floods and that temporary repairs had been effected using mass concrete. He suspected that the breakup of the brickwork was initiated by uplift and recommended that drainage layers be provided beneath future repairs. I am not sure whether this was done but the remedial measures recommended in the 1985 inspection report were implemented in early 1989 and comprised the following works which suggest the works were not done in 1976/1977.

- the construction of a new reinforced concrete L-shaped retaining wall along the whole length of the dam crest, having a top level 2.80m above overflow level and backfilled over the slab by a raised dam crest that is nominally 2.25m above overflow level;
- a porous concrete drainage pipe in gravel surround running along the RC slab and discharging via a chamber near the right abutment; and
- remedial works to the floor of the brick-lined spillway channel, including the installation of a reinforced concrete slab, pressure-relief holes and under-drainage at the downstream end, and pressure-relief holes and a 750mm deep concrete 'cutoff beam' near the upstream end.

In 1999 the Inspecting Engineer made no recommendations in the interests of safety, but recommended that the remaining brickwork in the steeply sloping section of the spillway be replaced or covered by a layer of mesh-reinforced concrete. This was not done.

It is clear this spillway has had a number of problems and perhaps no one thought realistically about the system i.e. that uplift was a problem perhaps caused by water getting under the floor in the upper part of the spillway.

THE INCIDENT OF 2013
The winter of 2012/2013 was a particularly wet winter with record levels of rainfall and significant snow fall.

On 26 January 2013 the reservoir filled and the spillway came into action. The flood peaked with a head of more than 350mm passing over the weir at the top of the spillway. The spillway coped well with the flood but as the flow over the spillway ceased the concrete panels at the base of the spillway lifted and 'floated' down the channel leaving the dam vulnerable as the spillway had broken up and the integrity of the spillway could not be guaranteed.

The owner declared an emergency. Failure of the floor and possibly the walls of the spillway could easily divert flows onto the mitre and toe of the embankment possibly leading to failure.

A plan to meet the short term needs to protect the dam and the longer term plans to stabilise the spillway were quickly developed. With heavy rain predicted at frequent intervals and the uncertainty associated with snow melt the owner was able to use the valves to empty the reservoir. Instructions were given to a contractor to remove the slabs which had been lifted, to excavate beneath them and to lay a significant base layer of no fines concrete to act as a drainage layer and then replace the concrete slabs in the short term.

Figure 2. Damaged concrete slabs to spillway

Figure 3. Damaged slabs and brickwork

Figure 4. Ongoing repairs

This was done over a period of some five days although conditions were very difficult. Drainage holes were left through the slabs and unfortunately within days the reservoir filled overnight and started to overflow again. The slabs survived but individual bricks were plucked out both from upstream and downstream of the 'repaired' panels, again putting the spillway and its repairs at risk.

Figure 5. Spillway in operation with emergency provision should flows be diverted on to the mitre

Large pumps brought specifically for the emergency were installed to try to control water levels as the valves themselves seemed to be incapable of doing so. This was due partly to the extreme rainfall coupled with the snow melt but also because it was discovered that at the last inspection large stones were left upstream of the outlet pipe to act as a filter for smaller stones! Attempts to empty the reservoir and clear the stones were hampered by the reservoir repeatedly refilling due to the storms. The pumps were positioned on the crest and because of the topography had to discharge across a 'Site of Special Scientific Interest' (SSSI).

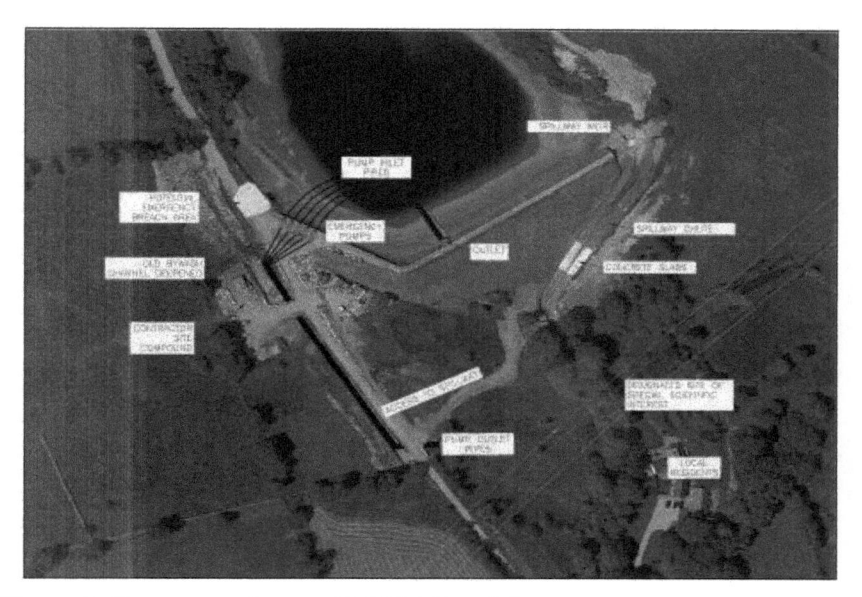

Figure 6. Emergency layout *(© 2014 Google)*

Figure 6 above shows the layout achieved for the emergency drawdown pumps. The concerns that the spillway might break up were so serious that a site was chosen to breach the upper part of the embankment should a serious situation develop as detailed below. This was at the right hand end of the dam onto the bedrock. Water would follow the route of the of the delivery pipes of the pumps to a point well clear of the site.

The weather conditions were so severe contingency plans were also put in place for a situation where the spillway failed taking with it the spillway walls and the toe of the embankment. The elements included:

- Discussions with the emergency services with respect to evacuation requirements
- The construction of a bulkhead to lower from the 'Heads of the Valley Road' down to the culvert beneath it to use the road embankment to 'hold' the reservoir water
- The deepening of the old bywash channel left but abandoned on the right hand side of the dam so a controlled breaching of the embankment could take place to take the top 2 or 3 metres off the 'top' of the reservoir.
- Large cranes (200t+) had to gain access to the site and 24 hour working was undertaken with three shifts of operatives.
- A quarry was also opened to supply stone to the site to form site roads and working platforms.

- A mini bus was kept on standby to evacuate the inhabitants of the one house between the dam and the Heads of the Valley Road.

Figure 7. Pumping installation on crest

Figure 8. Discharge pipes to downstream area   Figure 9. Discharge of pumps to stream

THE FUTURE
The reservoir level has been controlled and all temporary and emergency facilities removed. The temporary repair has survived a number of floods since its repair. A model test of the spillway arrangement for the Probable Maximum Flood has been carried out and a new reinforced concrete spillway chute designed to replace the whole of the brickwork and repaired sections. The channel is actually smaller than the existing and incorporates a 'bus shelter' at the base to turn the flow around the bend at the base of the spillway. Construction work is likely to be undertaken this year.

CONCLUSIONS

A significant effort by Welsh Water's own staff involving 24 hour working by more than 50 men, and involving a significant expenditure on the part of the company averted a dam failure which would have put hundreds of lives at risk.

The incident provided a useful reminder for owners and engineers to plan for emergencies, and to try and foresee the unexpected!

Inspecting Engineers should look for 'clues' indicating the previous behaviour of the dam and its appurtenant structures.

Safe access to reservoirs in winter

A W D ROSS, Scottish Canals, Inverness, UK

SYNOPSIS   Scottish Canals has 19 operational reservoirs. Most of these reservoirs are visited weekly by operational staff to control discharges and to carry out surveillance inspections of dams.

In the past, staff have been unable to reach some high level and remote reservoirs due to snow and ice. Scottish Canals has prepared a plan to provide staff with a safe means of accessing dams in such conditions.

This paper highlights the risks associated with accessing reservoirs in severe winter weather and describes the options for safe access, including the use of tracked all-terrain vehicles (ATVs).

Regulations and guidelines covering the safe use of ATVs are listed, together with advice on the purchase and hire of such vehicles.

INTRODUCTION
Scottish Canals currently operates 19 reservoirs across Scotland (Figure 1). These reservoirs supply water to the canal network as follows;

- **Crinan Canal** – Loch a Bharain, Cam Loch, Daill Loch, Gleann Loch, Loch an Add, Loch Clachaig, Loch na Bric, Loch na Faoilinn, Loch nam Breac Buidhe, Lochan Duin
- **Caledonian Canal** – Loch Dochfour (including Loch Ness), Loch Oich, Loch Lochy
- **Forth & Clyde Canal** – Townhead, Birkenburn
- **Monklands Canal** – Black Loch, Lilly Loch, Hillend
- **Union Canal** - Cobbinshaw

Access to these reservoirs is quite varied. There are some which are close to habitation and easily accessible by foot, whereas others are in remote Highland areas at a considerable elevation.

Experience has shown that there are some occasions when severe winter weather precludes the use of two- or even four-wheel drive (4WD) vehicles to access some of the more exposed reservoirs. Difficulties can arise due to localised flooding, snow, ice on steep ground, or storm damaged trees blocking access tracks.

However, it is recognised that localised flooding is usually of short duration, and that wind-blown trees can usually be cleared from access tracks soon after storms have abated. On the other hand, prolonged periods of very cold weather can lead to icy conditions and deep snow blocking access routes for many weeks. In these circumstances 4WD vehicles are not a safe means of transport to reservoirs and alternative solutions are required.

This paper is intended for use by Reservoir Managers who have responsibility for setting people to work at reservoirs in winter, typically staff that have a duty to carry out regular surveillance inspections.

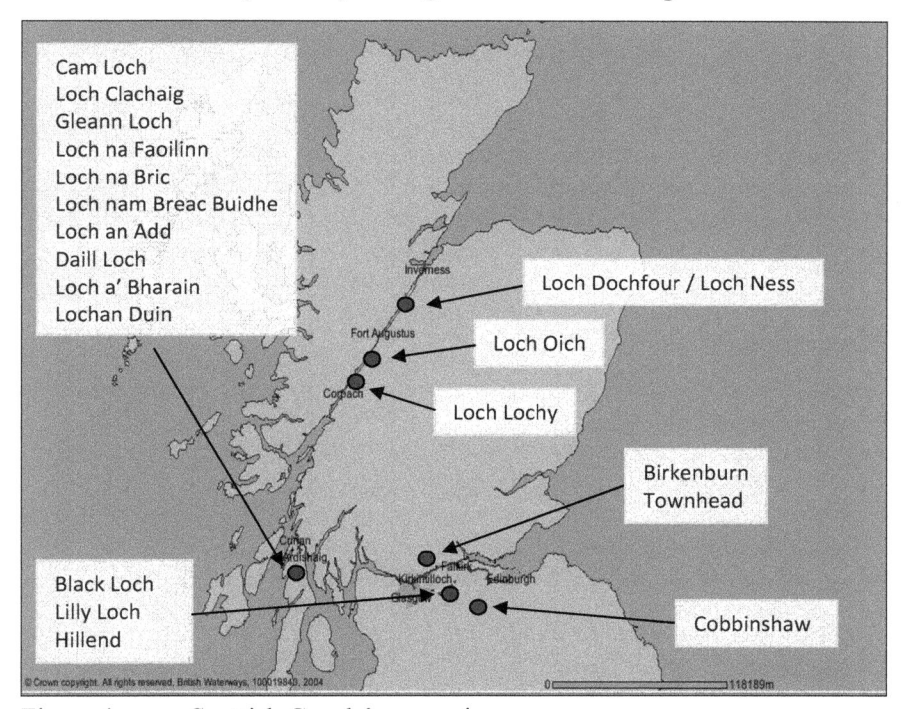

Figure 1   Scottish Canals' reservoirs

RESERVOIR RISKS IN WINTER

There are a number of reasons why it is important to maintain a level of reservoir surveillance, even in severe weather conditions.

Clearly the risks associated with flood events, high water levels, wave overtopping of dams, and leakage can exist at any time of the year.

However, additional risks that are likely to occur in winter are:

- Wind-blown trees blocking overflow weirs and spillways.

- Ice sheets, broken up by wind and waves, blocking overflow weirs and spillways.
- Waves eroding dam upstream face protection.

It is therefore imperative that reservoir surveillance inspections continue throughout the winter, including the Christmas and New Year holiday period, and that every reasonable effort is made to check the condition of dams and spillways at the frequency agreed for a particular reservoir.

FREQUENCY OF SURVEILLANCE INSPECTIONS

The frequency of reservoir surveillance inspections is often governed by the standards set by an organisation that owns or manages reservoirs. Scottish Canals has Asset Inspection Procedures that define the frequency of surveillance inspections for all its reservoirs. Most of these reservoirs are visited once a week.

Furthermore, the statutory inspection report for large raised reservoirs carried out under Section 10 of the Reservoirs Act 1975, will normally state the requirements for monitoring under Section 11 of this Act. The intervals at which, for example, Reservoir Managers are required to ensure that reservoir water levels are recorded (often weekly), may influence how often a reservoir is visited. However, it is important to consider the probability and consequence of dam failure when deciding how often to carry out surveillance inspections.

It is recognised that there will be occasions when it will not be possible to safely access reservoirs in a 4WD vehicle due to snow and/or ice. Whilst the likelihood of this happening somewhere across the UK each winter is high, the probability of these events lasting more than a few weeks at most reservoirs is relatively low. However, there are some reservoirs, particularly in the Highlands of Scotland, where snow is experienced most years. For these reservoirs, the probability of having access difficulties is much higher.

PLANNING FOR WINTER

There are a number of measures that can be taken to reduce the risk at reservoirs and the risk for surveillance staff going to these reservoirs in winter.

Reservoir partial draw-down

Where there is the potential for downstream flooding or blockage of weirs and spillways, a partial draw-down of the reservoir may be appropriate. Scottish Canals currently draws down two reservoirs by 1m below weir crest level from the end of October to mid-March.

Use of four wheel drive vehicles
Additional training for staff in the use of 4WD vehicles in winter conditions will be appropriate for general access in poor conditions. This training will also ensure that staff can recognise when not to attempt to drive to a reservoir in severe snow and ice conditions. Consideration should also be given to the use of snow tyres and/or snow chains.

Winter Personal Protective Equipment (PPE)
It is important that staff have the appropriate PPE for winter conditions, and that their Risk Assessments for working at exposed locations, such as high level reservoirs, specify suitable clothing.

Satellite phone
Where mobile phone reception is poor, unreliable, or a signal cannot be obtained on route to a reservoir or at the reservoir, the purchase of a satellite phone should be considered. Scottish Canals has two such phones for use by the Reservoir Attendant whilst carrying out weekly surveillance inspections of reservoirs in the Knapdale Forest above the Crinan Canal in Argyll.

Snow poles
At high altitude reservoirs, where access is, for example, over remote and exposed moorland that can be covered with deep snow in winter, it is useful to erect snow poles along the access route.

REMOTE MONITORING
Reservoir Managers who have remote reservoirs with known access difficulties in winter may wish to consider remote monitoring as part of their dam safety measures.

Scottish Canals has one such system at a high altitude reservoir with a very steep access track. Winter access is a known problem. A toe drainage system has therefore been designed to pick up any leakage from the dam and from a redundant outlet pipe. The drains have been installed below the level of this pipe, and discharge is directed through a measuring flume. The water level in the flume is monitored by a SCADA (Supervisory Control and Data Acquisition) unit. The discharge is monitored continuously and transmitted, in real time, from the reservoir to a national receiving centre. Flow parameters can be set such that high level alarms are sent to a duty officer or duty engineer.

However, it is still essential that regular surveillance inspections are carried out by trained and competent staff.

OPTIONS FOR SAFE ACCESS

Walking to remote reservoirs through deep snow in unforeseen weather conditions is not considered to be a safe means of access. Equally, the use of a helicopter is unlikely to be a practical means of access, particularly when consideration is given to landing and taking off in difficult terrain and potentially adverse weather conditions.

The use of a tracked All-Terrain Vehicle (ATV) is therefore likely to be the most appropriate option. Whilst there are many multi-wheeled ATVs available, these vehicles are generally best suited to access over soft ground or shallow snow. Tracked ATVs are more suitable for access up and down steep icy tracks and through deep snow. Furthermore, these types of vehicles can usually accommodate at least two personnel within an enclosed cab with a roll cage. Due to the infrequent nature of circumstances requiring the use of such a vehicle, hiring a tracked ATV with a fully trained and certified operator is likely to be the most appropriate solution for most reservoir operators. However, if appropriate, say for operators of large numbers of remote reservoirs, ATVs can be purchased and garaged at strategic locations across the network.

REGULATIONS

The following regulations apply to the use of ATV's;

- Provision and Use of Work Equipment Regulations (PUWER)
- Road Vehicles Construction and Use Regulations
- All Terrain Motor Vehicles Regulations

Reservoir Managers should take cognisance of these regulations when planning the use of ATVs.

ATV SAFETY

The following sources of information are recommended for planning a safe means of access using an ATV:

- HSE Agriculture Information Sheet No.33 – *Safe use of all-terrain vehicles (ATVs) in agriculture and forestry.*
- ATV Safety Institute – www.atvsafety.org

TRACKED ATVS FOR WINTER ACCESS

Reservoir Managers planning safe access to reservoirs in winter may wish to consider the following makes and models of tracked ATV's for purchase or hire:

- Hillcat 1700 tracked personnel carrier
- Scot Trac 2000R tracked personnel carrier
- Hagglund BV 206 tracked personnel carrier

Figure 2. Tracked all-terrain vehicle in severe winter conditions

PREPARING AN ACTION PLAN

Scottish Canals has prepared an action plan for safe access to reservoirs in winter. The plan lists actions to be taken whenever a reservoir surveillance inspection has not been possible in winter by 4WD vehicle due to deep snow and/or ice. Reservoir Managers may wish to include the following items in such an action plan;

(i) Look at weather reports and forecasts daily.

(ii) Using a 'Point of Work Risk Assessment' procedure, make daily assessments as to whether staff can safely access the reservoir using a 4WD vehicle.

(iii) For reservoirs close to public roads and housing, use a 'Point of Work Risk Assessment' procedure to make daily assessments as to whether staff can safely access the reservoir on foot. In this instance, a Safe System of Work must be adopted. For example, two persons with proven communication equipment (mobile phones and/or satellite phone) to walk to the reservoir, with one other person remaining in a 4WD vehicle for back-up assistance.

(iv) If a routine reservoir surveillance inspection is not possible, for whatever reason, inform the Reservoir Supervising Engineer immediately.

(v) Define a method by which the Reservoir Manager and the Reservoir Supervising Engineer can easily agree when to instigate an alternative means of access to a reservoir, say after a specified number of missed routine surveillance inspections, dependant on the dam category. The following table is used by Scottish Canals to inform this decision.

Table 1. Trigger points for instigation of action plan

Dam category	Consequence of dam failure	No. of routine surveillance inspections missed due to severe winter conditions
A	High	2
B	High	2
C	Medium	3
D	Low	4

(vi) List contact details for local companies that hire tracked ATVs. Agree terms and conditions of hire in advance, including mobilisation time, transport costs to and from site, day rates for tracked ATV and operator, etc.

(vii) List details of reservoirs for emergency extraction of personnel. Include a map of the access route, Ordnance Survey grid coordinates of the reservoir, post code of the nearest property on the access route for use of vehicle satellite navigation systems, emergency contact details (internal and external).

SUMMARY

Reservoir Managers should consider what actions will need to be taken in the event that surveillance inspection staff cannot reach a reservoir in prolonged severe winter weather conditions. Advance preparation of an Action Plan is recommended, with trigger points for instigation of an alternative means of access such as a tracked all-terrain vehicle.

ACKNOWLEDGEMENT

The author would like to thank Mr Darran Mellish of West Cost Tool and Plant Hire, Ardrishaig, for use of his photograph.

Maintaining the Safety of our Dams and Reservoirs
ISBN 978-0-7277-6034-0

ICE Publishing: All rights reserved
doi: 10.1680/mdam.60340.457

The monitoring systems of the dams of the Baixo Sabor Hydroelectric Development

G MOURA, Dams Department, EDP (Electricity of Portugal)
D S MATOS, Dams Department, EDP (Electricity of Portugal)

SYNOPSIS   The Hydroelectric Development of Baixo Sabor is located in north-east Portugal on the Sabor River, a tributary of Douro River. It comprises two hydroelectric schemes and is part of the Programme of Dams under construction in Portugal. The first filling of both reservoirs will take place during 2014.

Baixo Sabor development contains the first large reservoir in a tributary of the Douro River, allowing the impounded water to feed four hydroelectric power stations at dams downstream. It is also possible, when there is surplus wind energy, to pump water into the upstream reservoir thus optimising the wind energy production. Due to its size the upstream reservoir also provides for flood control.

In addition to the usual monitoring systems, both dams will have a specific detailed seismic monitoring system to take into account a nearby large geological fault. Both dams have also an integrated automatic monitoring system for collecting observed data. For the geodetic control of the horizontal displacement measurements a Global Navigation Satellite System (GNSS) will be installed. The continuous dynamic characteristics of the upstream dam will also be monitored.

This paper presents the main features of the monitoring systems of both dams, referring to the legal regulations they must comply with and including the specific aspects mentioned above.

DESCRIPTION OF THE BAIXO SABOR DEVELOPMENT
The hydroelectric development of Baixo Sabor comprises two schemes. The upstream scheme is 12.6km and the downstream scheme 3.3 km from the confluence of the Sabor River with the Douro River.

The downstream scheme was designed to for pumping into the upstream reservoir. There is a channel from the Valeira reservoir on the Douro River to the downstream scheme, so water from the Valeira reservoir is pumped into the downstream reservoir and from this into the upstream reservoir.

The upstream dam is a concrete double curvature arch dam with a maximum height of 123m, and having a total concrete volume of 670,000m³. The crest is 505m long and has a theoretical width of 6m. The structure is divided into 32 blocks whose width varies from 15.393m on the left bank to 17m in the valley bottom and 15.731m on the right bank.

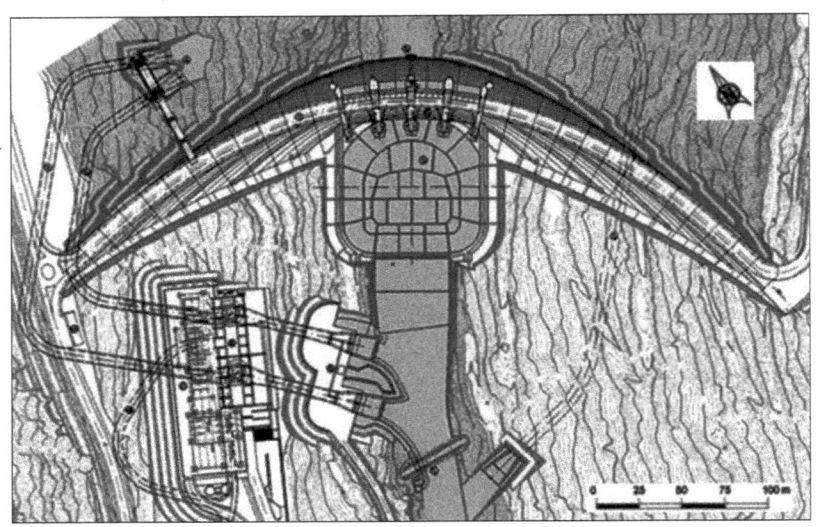

Figure 1. General plan of the upstream scheme of Baixo Sabor

Figure 2. Overall view of the upstream scheme of Baixo Sabor

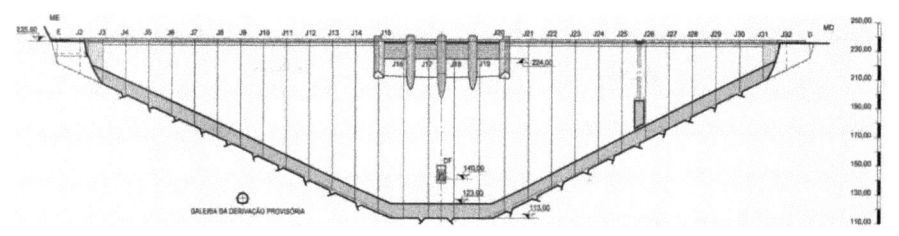

Figure 3. Upstream view

There are six inspection galleries and a drainage gallery that is divided in two in the bottom of the valley, i.e. upstream and downstream. The inspection galleries have geodetic observation systems installed and allow the monitoring devices to be accessed. Drainage and piezometric systems were installed along the drainage gallery.

The spillway, which is controlled by four radial gates, is located in the central part of the dam and comprises four 16m long bays with spillway gates separated by 5.83m wide piers. The spillway has a maximum discharge capacity of 5,000m³/s at maximum flood level. A downstream stilling basin dissipates the energy produced by the discharge.

The bottom outlet is located in the mid-section of the dam below the central pier of the spillway and it is equipped with an upstream fixed wheel gate and a downstream radial gate.

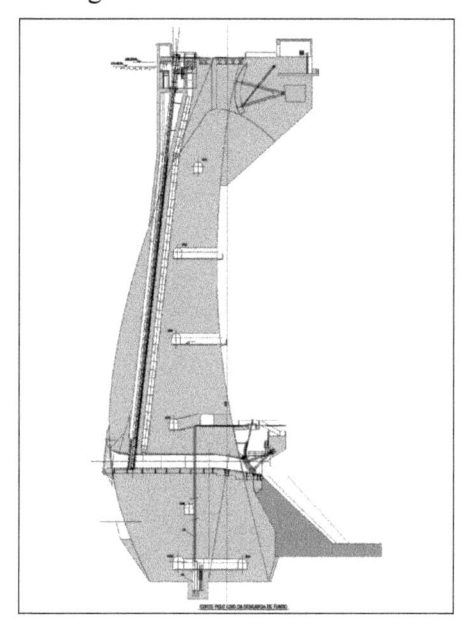

Figure 4. Cross section through the bottom outlet

In the powerhouse, an underground structure located on the right bank, there are two units each with a capacity of 81MW. The low pressure headrace tunnels, located on the right bank, have a diameter of 6.7m and are 238m and 338m long respectively, and the tailrace tunnel serving both units is 34m long.

The reservoir will have a capacity of 1,095Mm³ and a surface area of 2,819ha at normal water level (NWL). At maximum flood water level the reservoir capacity rises to 1,275Mm³ and has a surface area of 3,100ha.

The downstream scheme is a pumped storage scheme comprising a 130,000m³ concrete gravity dam; a controlled overflow spillway with four 16m wide radial gates; a roller bucket stilling basin and a powerhouse. The dam is 45m high and has a crest length of 315m which is divided in 22 blocks – eight on the left bank, seven on the right bank and seven in the valley bottom. The blocks vary in width from 21m on the left bank to 9m on the right bank. The dam has three inspection galleries.

Figure 5. Overall view of the downstream scheme of Baixo Sabor

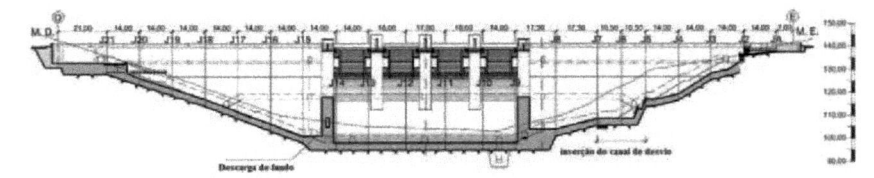

Figure 6 – Downstream view

The underground powerhouse is located on the right bank about 15m downstream of the dam. It has two reversible units with a maximum capacity of 18MW - 20MW.

The low pressure headrace tunnels, located on the right bank, have a diameter of 5.5m. One is 170m long with a 60m tailrace tunnel, while the other has lengths of 160m and 55m respectively (EDP, 2012).

GENERAL REQUIREMENTS OF THE PORTUGUESE DAM REGULATIONS

The Portuguese dam regulations, namely the Regulation of Dams (RSB, 2007) and the Guidelines for the Observation of Dams (NOIB, 2007), focus on the constitution of the Observation Plan. It is a mandatory document that establishes the general rules for the observation of the dam, underground works, foundation rock mass and reservoir.

The Observation Plan has to take into account the interpretation of the behaviour of the different structures during the different phases of their lifespans, namely the construction, the first filling of the reservoir and the operational periods. Based on observed behaviour, it will be possible to predict the future behaviour after the validation of mathematical models and project hypotheses.

The Observation Plan (NOIB, 2007) mainly concerns the definition of the variables that will have to be monitored and their frequency of observation, which depends on the height of the dam, the size of the reservoir, the safety factors built into the project itself, the characteristics of the dam foundation, the size of the population downstream and the strategic importance of any downstream infrastructure.

The Observation Plan includes the definition of the different visual inspections required and their frequencies; the definition of the monitoring system; the placement of the monitoring devices; the observation frequency of the devices of the monitoring system; the gathering and processing of the monitoring data; the reporting and communication scheme in the event of exceptional occurrences or the detection of abnormal behaviours; the reports of the installation and operation of the monitoring system; the qualifications of the staff responsible for the installation and operation of the monitoring system; the analysis of the behaviour; and the assessment of the structure's safety.

The Observation Plan must also include some general guidelines for the preparation of the Plan for the Control of the First Filling of the Reservoir. This document has already been produced for both schemes.

THE DOWNSTREAM DAM MONITORING SYSTEM

For the monitoring system of the downstream scheme the main parameters to be monitored were, according to regulations, displacements; joint and crack movements; drainage flows; uplift pressures; air temperature and air moisture. Temperatures in the concrete are also monitored. Concurrently the different loads acting on the structure, such as the hydrostatic pressures on the upstream and downstream faces, the air and the water temperatures and the seismic loads - both at the dam site and in chosen points along the perimeter of the reservoir - should be monitored. Additionally the downstream and upstream levels will be monitored with a water level indicator (EDP, 2007).

Parameters will be measured by three plumb lines; foundation extensometers; geodetic survey of the crest of the dam; joint meters and deformation meters; drains and piezometers. The local air parameters, such as temperatures and moisture, will be monitored automatically in a meteorological station installed close to the dam site.

During the construction, and taking into account the topographical characteristics of the valley, it was decided to use a new system to monitor the absolute displacements of the dam. So a GNSS will be installed in the dam in order to obtain the absolute displacement of three points on the dam crest.

The most important thermal loads in the dam will be the ones due to the hydration of the mortar and the variations of the air and the water temperatures in the reservoir. The thermal loads in the dam body will be monitored with thermometers placed to measure the temperatures along the height of the dam and through thickness and specifically near the faces, mainly during the release of the heat of hydration. Those temperatures will also be monitored with all the electrical devices installed in the dam body, namely, joint meters, strain meters and pressure meters.

The most relevant chemical actions will be controlled. Their measurement will be achieved either through the chemical analysis of the reservoir water and seepage water from the concrete and foundations, or the solid products carried in the seepage water.

The dam is located near an important geological fault (the Vilariça fault). In order to monitor seismic activity, two seismographs will be installed, one near the centre of the dam and the other in the drainage gallery.

The values of the thermal, mechanical and hydraulic properties of the concrete will be determined by laboratory tests on concrete samples. Those tests will include the determination of the coefficient of linear thermal expansion, the heat of hydration and permeability both on the dam site and in laboratory. These tests will consist of water injections in the dam concrete and on samples.

The geological and geotechnical characteristics of the rock mass were determined by the studies made during the project stage, with several tests carried out before concrete was placed in the dam. There was an onsite survey to allow for the complete mapping of the foundation footprint. This survey assessed all the discontinuities - faults and foundation joints - in order to determine their main characteristics (dip, strike and filling)

To complete the geological survey of the foundation, there will be permeability tests and geophysical tests. If needed, these tests will be performed throughout the dam's life to assess changes in the foundation conditions.

The structural behaviour of the dam will be assessed by monitoring the displacements of the structure; the relative displacements of the contraction joints; the strains; the seepage in the foundation and the physical and chemical characteristics of the reservoir and seepage water.

As mentioned above, the horizontal movements will be monitored with three plumb lines and the vertical movements with rockmeters and by levelling the crest and both drainage galleries in the valley bottom. At the same time there will be a GNSS that will measure the horizontal components of the displacement of three points of the dam crest. The following table describes the main devices of the monitoring system.

Table 1. Monitoring system of the downstream dam

	Values	Monitoring devices
1	Air parameters	Meteorological station
2	Hydrostatic pressure	Water level indicators in the upstream and downstream faces
3	Uplift pressure	Piezometers
4	Displacements	Plumb lines, geodetic survey of the crest with GNSS, levelling of the crest and drainage gallery, rockmeters, joint meters and extensometers
5	Drainage	Flowmeters in drains
6	Concrete temperatures	Thermometers and all the electric devices
7	Strains	Extensometers
8	Stresses	Stress meters

Additionally the dam will have an integrated automatic monitoring system that allows for the online assessment of some of the main values that characterise the dam's behaviour.

THE MONITORING SYSTEM OF THE UPSTREAM DAM
Due to its dimensions and the consequential importance of the dam, the upstream structure has a far more complex monitoring system (EDP, 2005).

Similarly to the above description regarding monitoring of the downstream scheme, the main values defined to be monitored were, according to regulations, displacement, joint and crack movements; drainage flows; uplift pressures; air temperature and air moisture. Concrete temperatures, strains and stresses are also to be monitored in the dam. At the same time the different loads acting on the structure, such as the hydrostatic pressure on the upstream and downstream faces, the air and the water temperatures and the seismic loads both at the dam site and in chosen points along the perimeter of the reservoir are to be monitored. Additionally the downstream and upstream levels will be monitored with a water level indicator. The following table lists the main devices of the monitoring system.

Table 2. Monitoring system of the upstream dam

	Values	**Monitoring devices**
1	Air parameters	Meteorological station
2	Hydrostatic pressure	Water level indicator
3	Concrete pressure	Concrete pressure meters
4	Uplift pressure	Piezometers
5	Displacements	Plumb lines, geodetic survey of the crest with GNSS, levelling of the crest and inspection galleries, polygonals in inspection galleries, rockmeters, joint meters and extensometers
6	Drainage	Flowmeters in drains
7	Concrete temperatures	Thermometers
8	Strains	Extensometers
9	Stresses	Stress meters

Using the devices listed in Table 2 the aim is to monitor the hydrostatic, thermal and dynamic loads. The hydrostatic load will be monitored with water level sensors upstream and by the water pressure in the concrete measured in six concrete pressure meters. The thermal action will be assessed by measuring the air temperatures at the dam site, the temperatures in the reservoir and on the dam faces and the concrete temperatures.

Dynamic loads will be assessed by monitoring the micro-seisms produced either by filling the reservoir or by the movement of soils around it, and the macro-seisms that may be experienced due to the Vilariça geologic fault that is located not far away from both dam sites. To assess that information there will be six seismographs in the dam – two in the upper gallery, three in the gallery immediately below that and one in the lowest section of the drainage gallery. In the area of influence of the development six remote stations will also be installed in judiciously chosen points along the perimeter of both reservoirs.

Any movement of the slopes upstream and downstream of the dam will be monitored using geodetic survey.

In assessing the structural response and complying with Portuguese Dam Regulations, the Observation Plan envisaged the monitoring of uplift pressures, horizontal and vertical displacements, joint movements, stresses and strains.

Six concrete pressure meters will be installed in the central block to monitor the water pressure in the concrete. The uplift pressure in the foundation will be accessed with a piezometers installed in each block.

The horizontal displacements will be monitored using GNSS at three points on the dam crest near the upper monitoring points of the three central plumb lines, the five plumb lines and the polygonals installed in three of the inspection galleries. The vertical displacements will be monitored by levelling the crest in three inspection galleries and the horizontal movement of both drainage galleries in the bottom of the valley. The rockmeters will provide information on the vertical displacements at the foundation/dam interface.

Using electrical joint meters, manual 2D joint meters, manual 3D joint meters and electric 3D joint meters it will be possible to determine joint movements in several different locations.

The monitoring of stresses and strains will be possible with stress meters and extensometers.

There will be five drains per block to control seepage along the contact between the dam and the foundation.

The state of the dam concrete will be determined using concrete properties such as compression and tensile strengths, the coefficient of linear thermal expansion, the modulus of elasticity and parameters of the creep law. Concrete tests, made on screened and normal concrete, will be performed.

The creep law will be determined through the results obtained from the six creep cells installed in the dam.

Similarly to that implemented in the downstream scheme, the dam will also have an integrated automatic monitoring system that provides information online on some of the main values that characterise the dam behaviour.

CONCLUSIONS

The upstream Baixo Sabor dam will be the second highest dam in Portugal and its reservoir will also be the second in volume. This paper describes the complete monitoring plans implemented in both dams of the Hydroelectric Development of Baixo Sabor in order to comply with Portuguese dam Regulations and Guidelines.

In addition and in compliance with Portuguese Regulations an Emergency Plan and a Warning and Alert System, that will enable the population immediately downstream of the dam to be warned in case of an incident or accident, are being implemented.

REFERENCES

EDPP (2005). *Aproveitamento Hidroeléctrico do Baixo Sabor. Projecto. Memória geral e Apêndices. Volume II – Apêndices do Escalão de Montante. Tomo 6 - Plano de Observação.* Dezembro de 2005 (*in Portuguese*)

EDPP (2007). *Aproveitamento Hidroeléctrico do Baixo Sabor. Empreitada geral de construção. Processo de Concurso. Volume III - Elementos de Projecto. C - Escalão de Jusante. Tomo C6 - Plano de Observação.* Setembro de 2007 (*in Portuguese*)

NOIB (2007). *Normas de Observação e Inspecção de Barragens, 2007. Portaria 847/93 de 10 de Setembro de 1993* (*in Portuguese*)

RSB (2007) *Regulamento de Segurança de Barragens 2007. Decreto - Lei n.º 344/2007 de 15 de Outubro de 2007* (*in Portuguese*)

EDP (2012) *The Baixo Sabor Upstream and Downstream Dams. Relevant design and construction features.* 54º Congresso Brasileiro do Concreto. CBC 2012. Outubro de 2012

SECTION 7:
REPAIRS TO DAMS

Rehabilitation of an 800 year old masonry dam

A J BROWN, Stillwater Associates
A J ELDER, Stillwater Associates

SYNOPSIS   Frensham Little Pond main dam is of masonry construction on a very loose sand foundation with a sand embankment providing lateral support to the flank dams. The reservoir is environmentally important, forming part of internationally designated heathland. The paper describes the rehabilitation works carried out in 2013, following recommendations in the interests of safety, which comprised chemical grouting of the masonry spillway structure, construction of a new concrete spillway channel to lengthen the seepage path, and surface repairs to the masonry with lime mortar.

INTRODUCTION

Frensham Little Pond was built in 1246 following failure of the first dam six years earlier. The dam retains a reservoir of 0.75Mm^3, with a maximum water depth of 2m. It is formed of masonry from local hard sandstone bands, but with much brick patching.

Major renovation works were carried out in the early 1950s to allow refilling of the reservoir after the Second World War (WWII) but since then there has been significant deterioration. This paper describes the rehabilitation works carried out in 2013, following recommendations in the interests of safety.

The reservoir is owned by the National Trust. The last section 10 Inspection and Feasibility study were by Jacobs, with detailed design and construction stage inputs carried out by Stillwater Associates. Overall project management was by the client, with individual specialists commissioned in respect of trees, reptiles and archaeology. Construction was by Webb UK Ltd in the period August to November 2013 with Grout Injection Specialists Ltd carrying out the chemical grouting as the first stage of works.

HISTORY

Frensham Little Pond was built about 1230 by "bondsmen", bound by their labour to the Lord of the manor. In 1245 the dam failed and had to be rebuilt. It is believed that this led to a labour dispute, as the bondsmen demanded payment as dam construction was not part of their tithe, resolved in favour of the labourers.

The Frensham estate was owned by the Bishop of Winchester as a fish pond to provide protein in winter, and for the church's fast days. He also promoted a nearby bridge over the River Wey at Elstead which included similar masonry construction. The fish included carp, pike and eels which after capture were landed at a quay and treated in a salting house next to the current warden's cottage.

During WWII the reservoir was drained to remove the large area of water which, with the nearby Great Pond, provided a navigation aid for German bombers. The army carried out the drainage, and accelerated the process by blowing up a section of the dam, the resulting flood wave destroying a fish farm just downstream. The reservoir was then used as a tank training ground, with Figure 1 showing a photograph taken when a special exhibition was staged for the Prime Minster, Winston Churchill and Generals de Gaulle and Sikorski, leaders of the free French and Polish forces.

Figure 1. Wartime use of the reservoir basin *(photograph from National Trust information board)*

After the war in 1949 the lake was sold to Mr & Mrs Atherton, who removed the vegetation which had grown up in the reservoir floor and built two sections of concrete wall to infill the breach formed when it was drained. Difficulties in refilling also led to construction of a clay filled trench along the right flank, and a concrete apron with upstream sheetpile cut-off around the main spillway. When Mr Atherton died in 1973 he left the lake to the National Trust, who have owned it ever since. Site inspection shows that that there has been much patching of the masonry over the years, with different phases of brick infill and facings.

The first inspection under the Reservoirs Act 1975 was in 1989, following which a 50m length of wall was added on the right flank to provide 1.3m freeboard.

SITE CHARACTERISATION

Environmental setting

The reservoir is located on the north side of Frensham Common, which is an internationally important heathland, being designated as a Special Protection Area (SPA), Special Area of Conservation (SAC), Site of Special Scientific Interest (SSSI) and Area of Outstanding Natural Beauty (AONB), with the dam being within both of the latter areas.

Geology and Hydrogeology

The dam and reservoir lie on the Folkestone beds, part of the Lower Greensand which is one of the major aquifers in the south-east of the UK. These overlay the Sandgate beds, which in turn overly the Hythe beds and are slightly artesian. Tilford Pumping station is a significant groundwater source with pumping from both the Folkestone beds and Hythe beds, the combined deployable output reported as 19Ml/day.

Window samples and dynamic probing at the main buttress spillway showed alluvium to between 2.4m and 3.2m depth, with DPT blow counts of between 0 and 1 (average of 0.75 above 2.5m depth). In the underlying solid deposits DPT increased steadily to 9 at 2m below the base of the alluvium. The sand was predominately medium sand with up to 20% gravel sized particles (cemented sand).

Dam

The majority of the dam is a sand embankment with an upstream masonry wall providing wave protection. Local to the stream the masonry wall joins two massive buttress sections with a lower arch over the old gated outlet defining the spillway and an upper arch providing strut support and originally being the platform for gate operation (Figure 2). Although the original construction was of hard local sandstone blacks, the appearance is more of a patchwork with much brickwork infilling.

THE NEED FOR WORKS

The condition of the dam steadily deteriorated with both the 2000 and 2010 Section 10 Inspections recommending works. Symptoms of poor performance included jets of water through the masonry (Figure 3) and surface leakage emerging on both sides of the stream when the reservoir was full, possibly carrying fines.

Figure 2. View from downstream prior to works

Figure 3. Jets of water through masonry into void below arch

The dam is categorised as Flood Category C, due to the presence of a road downstream. An important driver was also the need to preserve a National Trust asset, which retains a reservoir contributing to the internationally designated heathland site.

The 2010 Section 10 recommended the following "in the interests of safety":

i) a filter blanket is established in the area downstream of the spillway to prevent any further loss of fines from the dam foundations and extended to at least 2m beyond the area where seepage is currently evident

ii) the walls of the masonry spillway structure should be treated to reduce leakage and prevent any further degradation of the structure, and the lower arch of the masonry spillway is stabilised.

PROJECT PLANNING AND PREPARATION

A feasibility study and Phase 1 habitat survey were carried out in 2011, providing options for the various works and the associated constraints and costs. In addition to the mandatory items consideration was given to increasing spillway capacity and provision of a disabled access path along the north side of the reservoir. Environmental constraints meant that the lake could not be lowered below its normal seasonal level and the works would have to be carried out with a partially full reservoir (the summer level being around 0.5m below overflow). A client workshop in January 2012 fixed the scope of the works and allowed the internal funding application to proceed.

Topographic survey was carried out in 2011 with a ground investigation in autumn 2012. The Trust rangers held informal discussion with Natural England regarding works in the SSSI and with the local angling society. The county council public rights of way (PROW) officer was consulted, and provided a design for a standard footbridge, which obviated the need for consents of a bespoke design. Samples were taken of the lime mortar in the existing wall and analysed, to allow specification of similar mortar for repairs. A trial repair to the masonry wall was carried out in September 2012.

A planning application was submitted in May 2013, and a flood defence consent to the Environment Agency in June. Tender documents were issued in June 2013 with a five week tender period.

DESIGN

The design concept to control internal erosion was to infill under the existing lower arch with mass concrete and construct a watertight concrete channel downstream backed by a filter to lengthen the seepage gradient across the dam; details are listed in Table 1.

Chemical grouting was specified as a hydrophilic polyurethane grout, in recognition that this would have to seal concentrated flow paths under a differential head from the reservoir. It was to be carried out by a specialist sub-contractor with the tender drawings showing a systematic pattern of grouting to 0.3m above TWL, and extending 5m each side of the main spillway structure. Holes were all accessed from either the upstream or downstream face of the masonry. Holes were shown at around 0.6m centres, in recognition of the typical masonry block size of 0.2 to 0.3m.

The surface repairs to both faces of the 60m long masonry wall were specified as a series of work items, based on rectified photography with the works split into 20 panels on the flank walls and the main buttress (38 photos). The mortar was specified as St. Astier NHL 5 hydraulic lime with a specified 0-4mm sand source and mortar mix of 1 part lime to 2 parts sand. Works included making good root damage, wave erosion, cracks and repointing.

The design included a small on-site borrow area; this had the important benefit of avoiding import of soil to the SSSI.

Table 1. Key aspects of design of channel and filters

Aspect	Adopted	Basis/comment
Length	8m	Similar to adjacent sand dam
Plan	Taper to widen	Visual, spread out flows
Bed level	75mm below existing bed, with end sill	Allows long term naturalisation
Connection to existing	Dowelled, with movement joint	Allow differential settlement on loose foundation
Height	1.2m	Freeboard at design flood, above existing seepages
Filter	Along back of side walls	Specified as processed engineering material
Finishes	Brick cladding and coping	Match brick in existing masonry dam
Footbridge	Timber to SCC standard design	
Groundwater control	Wellpointing – Indicative design shown. Require precise settlement monitoring (± 1.0mm) using digital level and invar bar code staff	Below the water table in very loose sand, adjunct to structure vulnerable to undermining
Approach paths	Disabled access (10H:1V with resting places)	
Reinforcement design	Typically 12 mm at 150 centres	Crack control and differential settlement

ADVANCE WORKS

Advance works comprised tree removal over the winter of 2012/13, out of the bird nesting season, and detailed environmental surveys for protected species which the Phase 1 habitat survey identified as could be present; namely bats in the masonry dam and reptiles in the woodland. Additionally it was decided that the lake was not dropping as fast as it did in some years, so two 100mm diameter siphons were installed between 13 May and 2 July 2013 to accelerate the natural lowering. The public footpath over the dam was diverted, articles put in the local press and the Waverley Common user group briefed by the designer.

CONSTRUCTION

Chemical grouting

This was carried out first, to minimise leakages into the formation for the structural works. The grouting contractor proposed a slightly modified drilling pattern, based on using hand held drills (maximum 1.2m length, 19mm-25mm diameter) with access from the sides of the structure as well as faces, and a hydrophilic polyurethane grout TamPur150.

Overall grout take is summarised in Table 2, and was around 7% of the masonry volume in the buttress and 1% to 3% in the flanks. The holes were pumped with each hole being a single stage with pressure applied in several stages up to 300 psi.

Table 2. Chemical grout takes

Location	Number of primary/ total	Gout take: Total (litre)/ average (litre/hole)	Additional visit in November
Left flank	6/8	80/10	10 holes, 120kg
Left buttress	14/26	1100/44	
Weir block	2/4	100/25	7 holes, 200 litres
Right buttress	13/30	1237/41	
Right flank	6/11	228/21	
Overall	41/79	2745/35	17 holes, 320kg

On occasions during the grouting, air bubbles were observed, for example upstream of the apron in the reeds, and also downstream - through and just downstream of the mass concrete slab. Evidence for significant and interconnected leakage path lengths included:

- while injecting was in progress grout was observed to escape at points up to several metres away from the injection point

- when grouting simultaneously from two points in the same buttress, the grouters reported that the pressure on one line was influenced by the pressure on the other pump
- As grouting proceeded, the points of egress of water changed, presumably reflecting blockage of the larger paths.

On the other hand, when drilling secondaries, significant grout takes were sometimes observed in holes only 0.3m from the preliminary holes, suggesting limited spread from the primary holes. Grouting was deemed to be completed when the downstream face was dry, and there was refusal with no or minimal takes in the last stage of grouting.

Channel works
The well pointing was very effective, taking a day to install and providing a dry formation. Wellpoints were 32mm tube jetted into place within a 200mm pipe. Wellpointing was maintained for around five weeks, until the base slab had been completed. The precise levelling was carried out by the site staff using hired equipment and with training in its use. This gave results repeatable to ±0.2mm and showed the buttress settled by 1.0mm on installation of the wellpoints, with full recovery at the downstream side, but the upstream side subject to heave of about 1.3mm.

Surface repairs to Masonry
This was carried out by a local subcontractor, with close supervision by National Trust staff experienced in lime mortar repairs. The method of working was modified to include hand jet wash, and tarpaulin cover over the work area in recognition of the longer set time of lime mortar compared to cement.

Other works
Other works included construction of an access ramp to avoid damaging the roots of a mature (1m diameter) oak tree, the on-site borrow area and a flume to measure summer leakages when the reservoir was not overflowing. A conservation blacksmith was employed to refurbish two metal railings which inhibit public access onto the top of the arch, and thought to have been installed in Edwardian times. Environmental control measures are summarised in Table 3.

Table 3. Environmental aspects of construction

Aspect	Constraints	Mitigation
Ecology	SPA, SAC, SSSI	Trees - Root protection & fencing
	protected species	Reptile survey/removal/exclusion
		Borrow pit – finished ground left low to encourage self seeding as wetland

Aspect	Constraints	Mitigation
Pollution	Silt management	Reservoir flows over dam in siphon. Silt traps. Pump onto ground
Landscape	AONB	Maintain existing landscape
Archaeology	High potential	Desk study and monitoring during construction
PROW	Public footpath	Temporary diversion
Public	Health and safety, Engagement	Facebook page, website with progress, clear signage on site
Services	200mm dia 100 bar fuel pipeline	Close to site – prior notification to owner of works
Bats	Present in area	Bat boxes installed on trees around edge of area cleared for borrow area

Surprises

As with any works on old structures embedded in the ground there are often surprises at construction. At Frensham the first surprise was the sudden self ignition of an old WWII phosphorus grenade 0.5m from the public footpath, where it must have lain undisturbed for 60 years. The bomb squad arrived in force and destroyed it on site. A specialist UXO company then carried out a full geophysical search of the site and found nothing. However, a second, dummy anti-tank mine was then found whilst excavating to formation for the channel, and was removed by the army for their museum.

Inevitably when excavating to formation level old foundations were found. The first was in excavating along the flank wall to make good the upstream face at the water line, an old concrete footing was found, so the planned footing was deleted and the existing concrete used. When excavating within the existing buttress for the new channel an old mass concrete apron was found, leading to a change in design to raise the formation level for the new RC channel.

A further feature of the old structure was that when stripping off the existing weir prior to reconstruction some 150mm square vertical holes were found within the weir block extending down to the formation, thought to be old post holes for a timber sluice gate reported to have been present historically. This required 120kg of cement to grout up, prior to reconstructing the weir block.

The second surprise was that after completing the grouting and channel construction, a new surface leak suddenly emerged under the left flank wall, at the outer limit of the chemical grouting. This was dealt with by both additional grouting and lowering the reservoir sufficiently to build a mass concrete trench and brick facing upstream of the flank.

Control of Reservoir level
The lake naturally stops overflowing and drops by around 0.5m most summers. The advance works had dropped the lake to this level at the start of construction, but as autumn progressed the lake started refilling naturally, so two siphons were installed between 19 September and 2 October to lower the lake by 0.1m back to the starting level. Following the new leak on 11 November, seven 100mm siphons were installed to lower the lake by 0.4m, sufficient to install the upstream cut-off, which was achieved in 10 days. This was towards the upper limit of acceptable flow over the ford on the public highway 100m downstream of the dam. The lake refilled naturally and started spilling on Christmas Eve.

Outcome
The works were completed in November on programme and with an outturn construction cost of £210k. They are shown in Figures 4 and 5 and were well received by local residents who use the public footpath along the dam.

Figure 4. Completed channel and arch stabilisation

Figure 5. Completed flank wall with additional upstream brick facing

CONCLUSIONS

Although not a high dam Frensham Little Pond retains a large body of water which is important environmentally and in terms of landscape. The dam is a vulnerable structure comprising 750 year old masonry of overall height 4.5m built on a loose sand foundation. This meant that although renovation works 60 years after the last works were not large scale in size they were technically complex. Wellpointing and monitoring of movements of the buttress were used to facilitate construction; observed vertical movement comprising rotation towards the wellpointing and residual movements of 1.3mm. Overall average grout take was 30 litre/m^3 in the flanks and 70 litre/m^3 in the main buttress. Refurbishment of the 750 year old structure was completed successfully in late 2013, and should provide many years of safe operation before needing any further refurbishment.

ACKNOWLEDGEMENTS

The authors would like to thank the National Trust for permission to publish the paper, and Martin Archer and Tim Mockridge who assisted with delivery. Additional thanks are extended to Matt Webb and David Vaughan of Webb UK, Guy and Chris Vallings of GIS Ltd and Andrew Hodkginson of Hewson Consulting who carried out the structural design.

An investigation into the impact of a 50yr old discrepancy on the safety of Tittesworth Reservoir

O J CHESTERTON, Mott MacDonald Ltd
I M HOPE, Severn Trent Water Ltd
A M KIRBY, Mott MacDonald Ltd
J R CLAYDON, All Reservoirs Panel Engineer, Independent

SYNOPSIS   In 1963 the Tittesworth Reservoir, now operated by Severn Trent Water (STW) was commissioned to respond to the increasing demand for water supply. The reservoir comprises an earth embankment with a concrete core wall. It incorporates an older dam in its upstream shoulder and has a bellmouth spillway and tunnel in the left abutment.

In 2012 a pre-inspection identified differences between the as-built spillway and 1963 physical model geometry and raised concerns that the as-built geometry might not have sufficient capacity compared with the physical model.

After consideration of the benefits and limitations of both physical and numerical modelling, the spillway rating and performance were evaluated numerically using Flow3D, a computational fluid dynamics (CFD) package. Using the 1963 physical model as a reference, the modelling showed that the spillway gorged earlier than predicted as a result of the changes and therefore the capability to safely pass Probable Maximum Flood (PMF) outflows was reduced. It was also found that, once gorged, flows were controlled by the hydraulics within the shaft and tunnel. The stilling basin was also modelled for PMF flows.

PMF flows increased the expected flood rise and options were put forward to increase freeboard at the dam.

The use of CFD showed it was possible to model complex flow conditions with control moving between inlet, shaft and tunnel and demonstrated the complex range of flow conditions possible for bellmouth spillways.

The project has illustrated the benefits of STW's pro-active strategic approach to reservoir safety which has provided time to consider the options ahead of statutory drivers.

INTRODUCTION

STW manages over 700 dams and reservoirs, 58 of which are currently statutory reservoirs under the Reservoirs Act 1975. Key to water supply in North Staffordshire is Tittesworth Reservoir. It supplies water to the North Staffs water resource zone comprising 230,000 households or 520,000 people. It operates in conjunctive use with groundwater sources. Sited at the top of the Churnett catchment it is STW's most drought-sensitive reservoir. It was constructed in 1963 to respond to the increasing demand for water supply. The dam incorporated an older dam in its upstream shoulder.

Prior to its construction the dam outlet facilities were model tested by BHRA in 1959. The design geometry tested incorporated a large radius at the connection between the bellmouth shaft and tunnel. The modelling work indicated that the system would 'gorge' at a flow of 7,769ft^3/s or 220m^3/s which was significantly above the maximum "catastrophic flood" requirement of 6,200ft^3/s (176m^3/s) at the time.

During construction both the shaft and tunnelling work proved more difficult than expected due to the behaviour of the soft shales. This resulted in significant modifications to the method of tunnelling and to the spillway shaft construction (Twort, 1964).

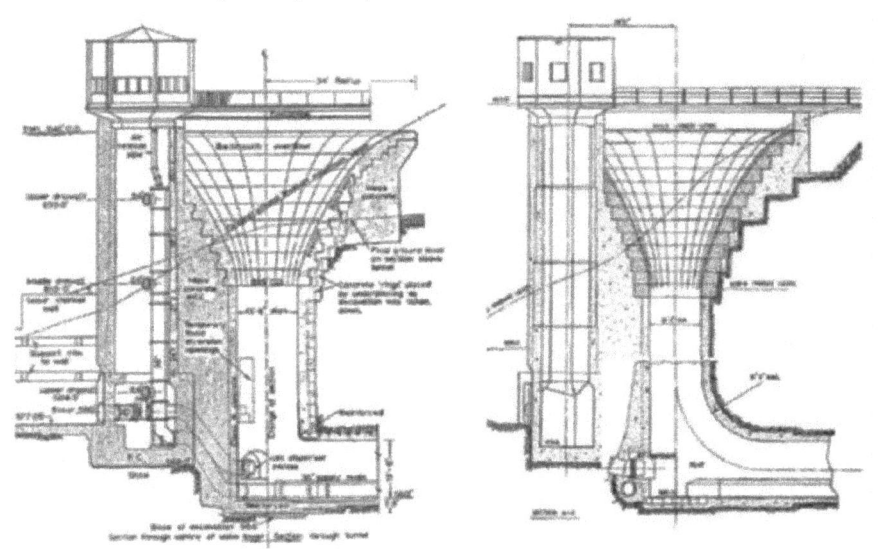

Figure 1(a). As-built Geometry   Figure1(b). BHRA Modelled
                                   Geometry used to derive rating curve

In 1993 the reservoir hydrology was reviewed by the undertaker prior to a statutory inspection. By this time the design flood had been increased to the

PMF, and the original BHRA spillway rating curve was used in reservoir routing to calculate a maximum spillway discharge during PMF of 222m³/s.

As part of STW's strategic approach to guarantee serviceability of its impounding reservoirs, a 'pre-inspection' two years ahead of any statutory inspection is commissioned (Hope 2012). This provides the opportunity for investigations, studies and options appraisal ahead of any mandated actions. The strategy is embodied in the company's vision of being the best water company in Great Britain and applies the ethos of continuous improvement referred to as "Safer, Better, Faster".

In 2012 All Reservoirs Panel Engineer (ARPE) Jim Claydon carried out a 'pre-inspection' of the reservoir to give early warning to STW of any work that may be required.

During review of the documents it was noted that the as-constructed geometry did not include the large radius at the connection of the shaft and tunnel as modelled by BHRA (Figure 1). It is likely this change was made during the difficult construction of the shaft and tunnel. It is not known if any resulting change to the spillway rating curve was considered given that the 'catastrophic flood' was significantly below the gorging point of the bellmouth.

Following the discovery that the rating curve was not based on the as-built geometry the ARPE recommended investigations to determine the effect of these changes on the rating curve and capacity of the spillway to pass the PMF.

The following methodology was recommended and was adopted for these investigations

- Obtain accurate as-built dimensions for the overflow.
- Undertake model testing (computer or physical) to check the performance and obtain a rating curve.
- Carry out a new flood study to the latest standards.
- Route the design flood (PMF) through the spillway.
- Consider the implications.

STW commissioned Mott MacDonald to model the spillway in order to revise the spillway rating curve and flood study and to propose options for the mitigation of any shortfall found.

This paper summarises the modelling work undertaken, the updates to the flood study and the evaluation of options arising from these.

DATA COLLECTION

Given that the spillway had been built with a somewhat modified geometry it was prudent to gather accurate and up-to-date as-built dimensions for the spillway. A 3D laser survey of the spillway bellmouth and shaft, tunnel and stilling basin was commissioned, the output of which was a detailed point cloud (Figure 2). A CAD model was then built around this point cloud for later modelling.

The laser scan was supplemented by standard detailed topographical survey, reservoir bathymetry and as-built drawings to construct the full 3D virtual model.

Figure 2. Laser Scan point-cloud (by North and Midland Construction Nomenca)

MODELLING

Selection of Modelling Method

In the early stages of the project an evaluation was undertaken comparing the advantages and disadvantages of both physical and numerical modelling.

Comparable quotations were obtained for both modelling options and based on the cost comparison and pros and cons of both methods the client selected a Computational Fluid Dynamics (CFD) modelling methodology.

The reasons for this were, in part, the durability of the numerical model and ability to be able to return to the model at a later stage; the accessibility of results; and the ability to undertake major modifications if necessary. In addition to these benefits, using a numerical model was considered acceptable as the results from the 1959 BHRA physical model were available to calibrate against.

CFD Modelling Methodology
Following the decision to proceed with CFD modelling, two phases of modelling were undertaken;

Validation modelling: Modelling of the physical model geometry to replicate results and determine appropriate modelling parameters

As-built modelling: Modelling of as-built geometry to predict spillway performance during flood events and to build a rating curve.

The following general methodology was adopted:

1. Information such as survey and data from as-built drawings was collected and a 3D virtual model built in AutoCAD.
2. Hand calculations were undertaken to evaluate the expected performance of the spillway based on empirical formula to gain an understanding of the different flow conditions likely in the spillway and tunnel.
3. Geometry was 'meshed', flow regions defined, boundary and initial conditions applied and other parameters, such as surface roughness, input as required.
4. Sensitivity runs were undertaken to evaluate appropriate mesh block size and other parameters.
5. Final runs were set up and simulations run to the degree needed to obtain the predictions required.
6. Results were processed and outputs generated and reported.

The CFD modelling was undertaken using the Flow-3D software package developed by Flow Science Inc which solves the three dimensional, transient Navier-Stokes equations on a structured grid. The software was selected due to its ability to track and define sharp fluid interfaces using the volume of fluid (VOF) method considered key for spillway flows and other civil hydraulic applications. Solid boundaries are defined by a fractional area-volume obstacle representation (FAVOR) method which resolved the geometries for the Tittesworth spillway well.

CFD modelling parameters and sensitivity
The CFD model was checked for sensitivity to various model parameters including mesh size, boundary conditions and surface roughness.

Model parameters
The Flow3D solver allows for the modelling of turbulent flow and various complex physical phenomena where they are anticipated to occur. The following specific parameters/models were adopted in the Tittesworth CFD study:

Turbulence: The RNG k-ε turbulence model was used in preference to the standard k-ε model as it has been found to be more suitable for turbulent flows such as those including hydraulic jumps.

Air Entrainment and Adiabatic bubble models: These were activated in the simulation to enable entrainment and the correct modelling of trapped air which was assigned density and compressibility.

Other model parameters such as momentum advection were selected based on the flows expected such as vortices, swirl within the shaft and tunnel and cavitation.

Model boundary conditions and sensitivity
The model boundary conditions were relatively straight forward. The upstream boundary was located within the reservoir and defined as a hydrostatic pressure and inflow boundary. Downstream boundaries were either simple outflow (where supercritical flow was expected) or pressure and flow boundaries where tail water levels where defined.

Initially, a large proportion of the reservoir, stretching more than 200m from the bellmouth, was incorporated into the model. This was later reduced to 48m for final runs given that the velocities at the boundaries were less than 0.1m/s.

The downstream boundary was also evaluated for sensitivity to low tail water levels which had been calculated using a 1D HEC-RAS model.

Surface roughness was added to the model run to calibrate from the physical model and to match the expectations of the ARP Engineer for the tunnel Manning's coefficient close to 0.016.

The model was run as a single phase model, disregarding the effect of air movement across the free surfaces which was considered reasonable for this application. Air was only modelled when entrained or trapped within the model, such as would occur during slug flow. An atmospheric boundary pressure was applied to the free surface.

Validation Modelling
The physical model, tested in 1959, was run for prototype design flows of 176m^3/s then tested to gorging point at 220m^3/s. The model was not tested for higher flows and the response beyond the gorging point was not evaluated as it involved flood flows much larger than required at the time.

Hand calculations were undertaken to attempt to confirm the gorging point and the spillway response above this. Equations for tunnel full, orifice and short tube control were developed and superimposed on the model weir performance curve. The short tube and tunnel control curves showed the best fit but required some adjustment to the head loss factors to match the predicted gorge point, highlighting the need for 3D modelling.

Calculated head loss was also predicted using Manning's equation and the tunnel roughness in the CFD model increased until the energy slope matched the calculations.

Once mesh dependency and sensitivity studies had been completed the CFD model runs showed good agreement with the 1959 weir rating curve (Figure 3).

The CFD model also replicated the hydraulic effects in the shaft, the connection to the tunnel and those resulting from the presence of the valve house and bridge pier as observed during physical modelling. These included jetting into the shaft caused by the valve house and bridge pier; formation of strong swirl in the bend at the foot of the shaft; vortex development; and flow separation at the shaft walls where the bellmouth curve terminates.

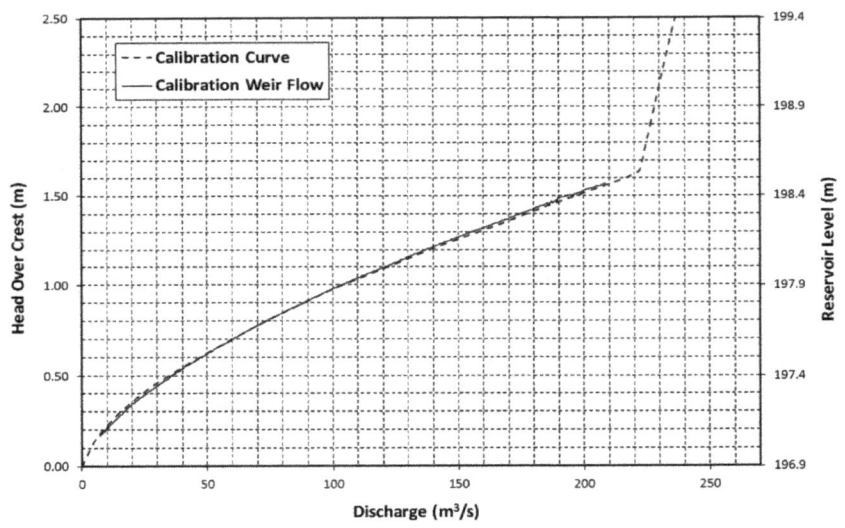

Figure 3. Physical modelling curve (Calibration Curve) and CFD model curve (Calibration Weir Flow)

As-built Modelling

As the validation modelling showed good agreement with the physical modelling results, the model parameters were then taken forward to the As-built Modelling phase.

All simulations carried out using Flow3D are transient and once as-built models were initialised they were run for thirty seconds of simulation time to enable natural fluctuations and oscillations in the flow regime to be observed. A rating curve was also generated to demonstrate the spillway performance over a full range of reservoir levels.

Revised Rating Curve
To derive a full rating curve and determine if the spillway capacity had changed as suspected, a transient rating curve simulation was run in which reservoir levels were gradually increased.

Weir flow in the as-built model closely matched the curve produced during physical and validation modelling. However, it was apparent that due to the modified bend geometry, the transition to pipe control (gorging) occurred at a lower head and flow giving the as-built geometry a lower maximum capacity when compared to the BHRA (1959) rating curve.

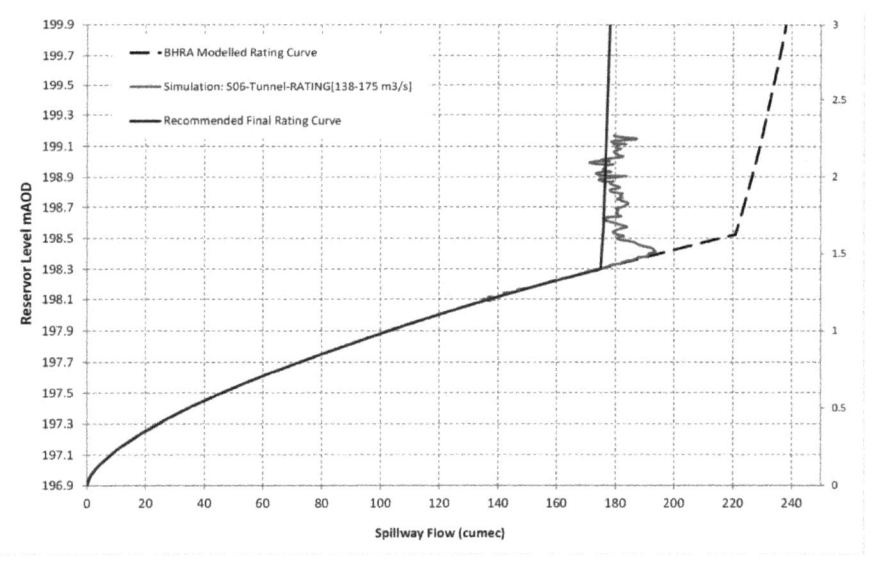

Figure 4. BHRA Modelled Rating Curve and transient model results for rising reservoir compared to recommended final rating curve.

Hydraulic control
The hydraulic performance of the spillway system was complex with multiple factors contributing to the final performance curve demonstrated in Figure 4.

The system is weir controlled until approximately 175m³/s but as the reservoir begins to spill, conditions develop within the shaft and tunnel that result in the final submergence of the weir.

Initially, water simply plunges to the base of the shaft almost completely dissipating the kinetic energy gained before flowing down a partially full tunnel. As flow increases to 15m³/s the depth of flow is enough to splash the roof of the tunnel and intermittently seal, although a significant proportion of this flow is entrapped air. Flows also begin to converge within the rather narrow shaft section.

At 50m³/s the tunnel initially flows completely full of the turbulent air water mixture and flows converge at the transition from bellmouth to shaft. Flow performance in the base of the shaft is influenced by the converging jets of flow created by the valve tower and bridge pier blocking a portion of the weir. The blockage enables air to be injected into the tunnel increasing air entrainment and the occasional entrapped larger slug of air.

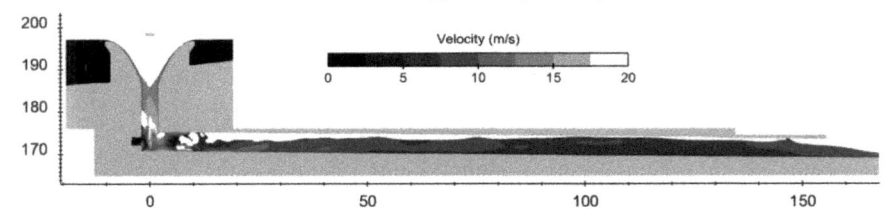

Figure 5. Flow at 15m³/s – 2D Long section Velocity (m/s)

Above 50m³/s the shaft chokes and the flow convergence begins to support a slug of water still well below the bellmouth crest. As flows increase the pressure below this slug increases and the head loss through the now full bend and tunnel begins to control the level of water in the bellmouth. The water rises to form an annular hydraulic jump behind the weir.

Finally, at ~175m³/s the weir is submerged as the head loss in the tunnel and shaft require an increasingly large driving head.

Weir control was seen to persist to higher flow rates temporarily in some simulations as transition from weir flow to tunnel control was dependant on oscillations in the annular hydraulic jump.

The system flow continued to oscillate once control had shifted and is through to be a function of intermittent vortex formation and strong but fluctuating swirl in the base of the shaft.

Hydraulic performance: Tunnel pressures
It has been assumed that, although not vented, the tunnel was not designed to flow full, or to develop hydraulic pressure on the liner. This is inferred from good practice but also from the fact that the tunnel pressures were not reported by BHRA and must not have been considered problematic at the time.

Simple 1D Manning's calculations indicated that the tunnel will run full under normal conditions for flows above 100m³/s, equal to the maximum design flood at the time of the BHRA modelling in 1959. However, the current modelling study has shown that due to turbulence and air entrainment at the base of the shaft, flows are significantly bulked and the tunnel flow depths will be greater and flow full much earlier than predicted by the hand calculations (at about 15-20m³/s).

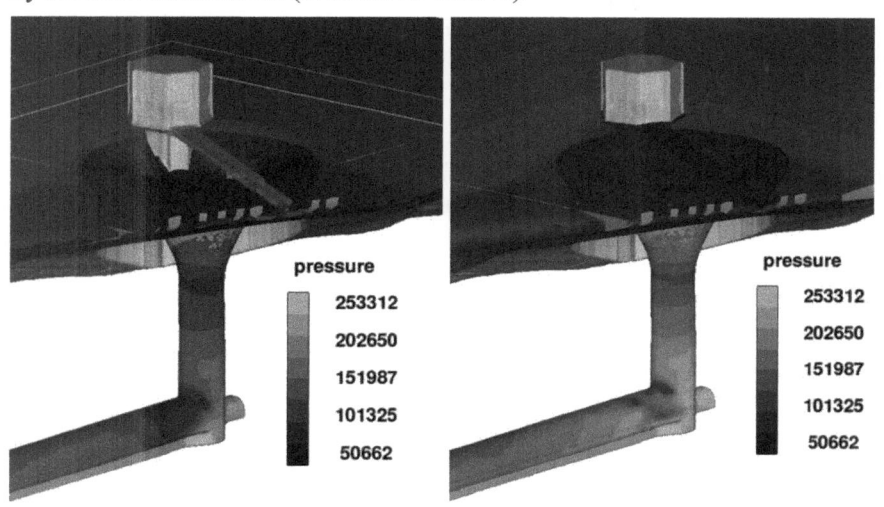

Figure 6. 3D view of tunnel and shaft pressure (in Pascal) for (a) 10,000yr Flood (138m³/s) and (b) PMF (175m³/s) – (101325 pa is atmospheric pressure)

The tunnel will definitely be flowing full for the PMF of 175m³/s calculated for the current study.

Low pressures are of more concern than high pressures as they act with the ground load and increase the overall load on the tunnel liner. As such an important output was the minimum and maximum tunnel pressures for various flows.

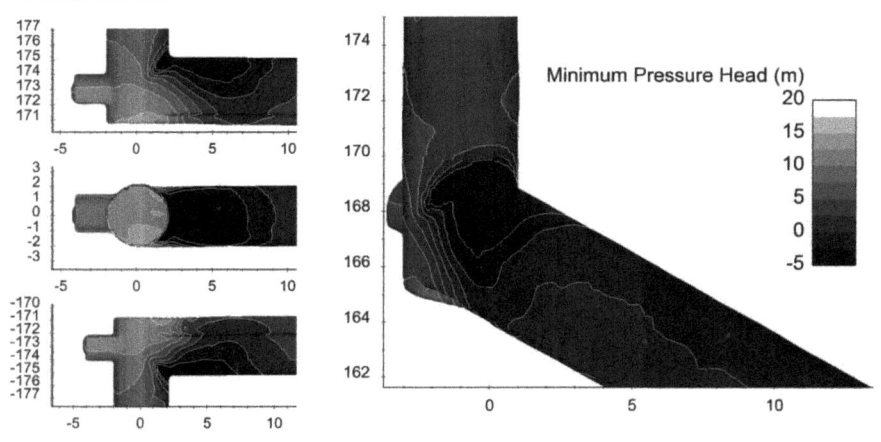

Figure 7. 10,000yr Flood [138m³/s] 3D View - Minimum Pressure (m)

During steady state full flow such as during PMF flow, pressures on the liner did not drop below negative 10m of head with the exception of a small cavitating region at shaft / tunnel connection. A slight contraction of the tunnel 140m from the shaft attracts low pressures as flow velocity increases but these are generally not more than negative 5m of head during PMF flows.

Hydraulic performance: Cavitation
As soon as the tunnel begins to flow full flow separation occurs at the connection between the shaft and the tunnel intrados. This becomes quite severe and not only restricts the flow through this connection but creates an area of very low pressure at the point of flow separation and immediately downstream, extending for up to 10m.

The modelled pressures were extracted from the model here and Figure 8 shows the region of low pressure flow that will cavitate and need protection or venting.

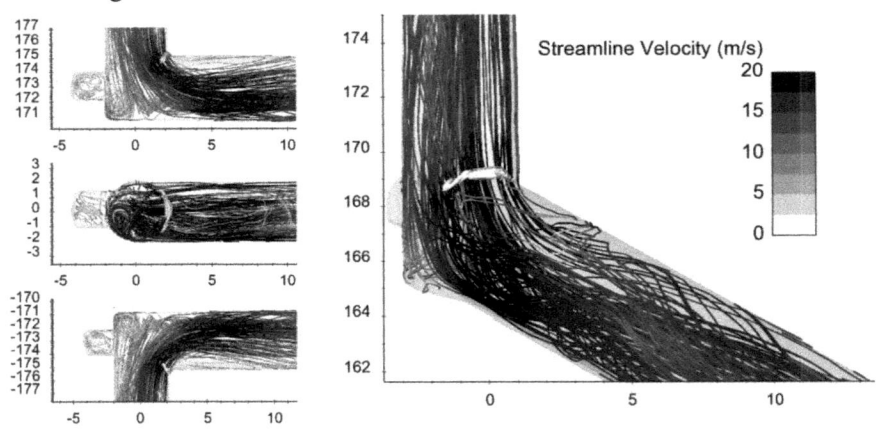

Figure 8. 10,000yr Flood (138m^3/s) 3D View – Cavitating region and velocity streamlines (m)

Hydraulic performance: Stilling Basin
The performance of the stilling basin was not evaluated in the 1959 model study and was included in the current study to confirm that the basin remained adequate for the flows it is expected deal with.

Model mesh resolution was increased within the basin area to resolve the basin teeth and to accurately capture the turbulent hydraulic jump.

The outputs as shown in Figure 9 demonstrated that the hydraulic jump was contained within the basin and was not particularly sensitive to tail water level. However, it was apparent that flows do inundate the area immediately downstream of the dam and turbulence generated by the stilling basin could be problematic and will need to be addressed.

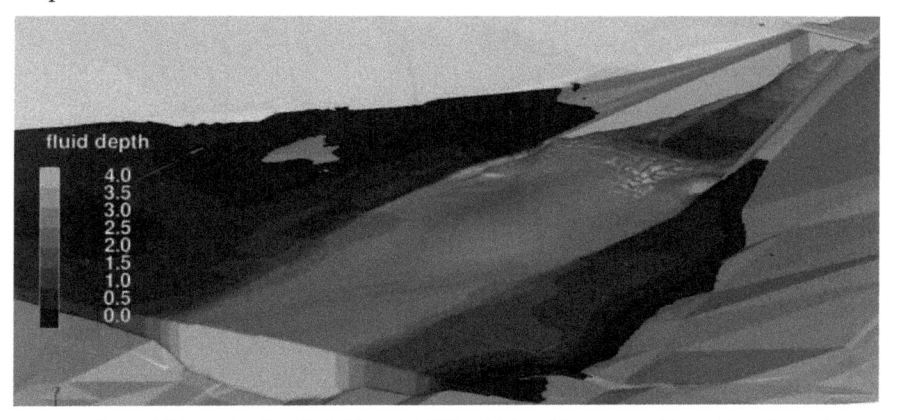

Figure 9. Stilling Basin PM Flood (175m³/s) – 3D view Flow Depth (m)

REVISED FLOOD STUDY

Once modelling had progressed sufficiently to generate a rating curve the flood study was revisited and inflows re-routed.

As Tittesworth is a Category A dam the appropriate standard is the Probable Maximum Flood (PMF) which was evaluated and routed through the reservoir as seen in Figure 10. The study also made assessments of the 1,000 and 10,000 year floods.

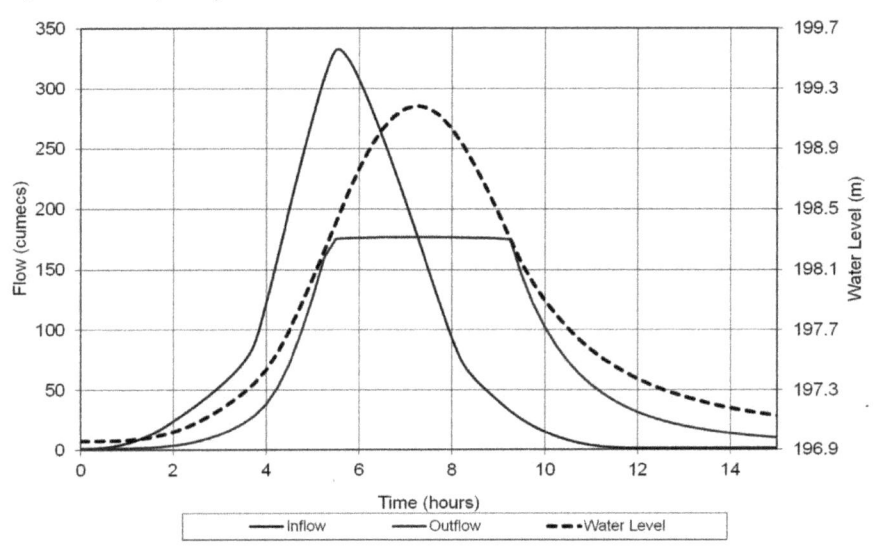

Figure 10. Tittesworth PMF Routing

As seen below, the gorging of the spillway dramatically affected the outflow leading to a decrease in the PMF outflow but an increase in flood attenuation and flood rise.

OPTIONS STUDY AND RECOMMENDATIONS

Following on from the model study, an options assessment was undertaken to address the issues that had been raised. The assessment looked to address the following areas;

- options to mitigate any negative effects of additional flood rise
- hydraulic conditions identified within the shaft and tunnel including cavitation and the potential for increased loading on the tunnel liner
- hydraulic conditions downstream of the dam

As a result of the options study, a number of remedial options have now been identified and will be carried out over the course of the next few years.

ENGAGEMENT WITH WARWICK UNIVERSITY

As part of its programme of engagement with aspiring engineers, STW concurrently sponsored the construction of a physical model of the spillway and tunnel at Warwick University. The authors provided inputs and support during the project and were invited to a demonstration of the completed project and contributed toward the assessment of the nine-strong team of Masters degree candidates.

CONCLUSIONS

The Tittesworth Spillway Study has illustrated the benefits of STW's pre-inspection strategy which has provided time to consider the options ahead of statutory drivers and provide opportunity for financial planning.

CFD modelling has managed to capture the flow conditions with control moving between inlet, shaft and tunnel and demonstrated the complex range of hydraulics possible in bellmouth spillways.

Outputs from the CFD model will now be available to inform the structural assessments that will be undertaken following the options study.

Considerable reliance was placed on the candid, post construction paper, (Twort 1964) which proved instrumental to the understanding of the conditions that lead to the changes made during construction. The valuable insight provided by this archive further emphasises the need for open and factual papers from current practitioners to better inform future generations of engineers.

REFERENCES

BHRA (1959) *RR 641, Model Tests on the Bellmouth Spillway for Tittesworth Reservoir*, BHRA, Cranfield, UK

FSI (2013) *Flow-3D User's Manual.* Flow Science, Inc., Sante Fe, NM

Hope I M (2012). *The implementation of Severn Trent Water's People Plan to be recognised as the best in Great Britain at Managing Reservoir Safety* ICOLD Kyoto 2012, Kyoto, Japan

Ho D K H and Riddette K M (2010). *Application of computational fluid dynamics to evaluate hydraulic performance of spillways in Australia*, Australian Journal of Civil Engineering, Vol 6 No 1, 2010

Khatsuria R M (2004). *Hydraulics of Spillways and Energy Dissipators.* CRC Press.

Kaheh M, Kashefipour S M and Dehghani A (2010). *Comparison of k-ε and RNG k-ε Turbulent Models for Estimation of Velocity Profiles along the Hydraulic Jump on Corrugated Beds*, 6th International Symposium on Environmental Hydraulics, Athens, Greece, June

Blaisdell F W (1958). *Technical Paper No 12, Series B: Hydraulics of closed conduit spillways.* St Anthony Falls Hydraulic Laboratory, University of Minnesota.

Severn Trent Water (1993). *Tittesworth Reservoir – Reservoir Safety Assessment Hydrological Analysis.* Severn Trent Water

Twort A C (1964). *The New Tittesworth Dam.* Journal of the Institution of Water Engineers, March (1964), pp. 125-179

USBR (1987) *Design of Small Dams, 3rd Edition.* United States Bureau of Reclamation, Washington DC, USA

Maintaining the Safety of our Dams and Reservoirs
ISBN 978-0-7277-6034-0

ICE Publishing: All rights reserved
doi: 10.1680/mdam.60340.494

Planning for emergencies and rehabilitation to improve operational safety at Spelga Dam, NI

J R BRADSHAW, URS, Belfast, UK
K J McCUSKER, URS, Belfast, UK
D A McKILLEN, URS, Belfast, UK

SYNOPSIS   Spelga Dam (Northern Ireland) is a 28m high concrete gravity dam located in the Mourne Mountains in County Down, with a storage capacity of 2.7Mm^3 and was originally constructed in 1957. In 2011 rehabilitation and maintenance work was undertaken to improve and maintain the operational safety of Spelga Dam. This work was relatively low cost but complex in nature with the works utilising innovative solutions. The paper will expand on the following elements of this work:

- The creation of a new passage from the downstream face of the concrete dam to access the longitudinal gallery including the studies undertaken; the alternative solutions considered and the final solution which involved cutting through the concrete dam using both hydro-demolition and diamond wire cutting technology. A review of the pros and cons of the methods of construction, a summary of the findings of an examination on vibration limits and international standards for vibration limits; and details of the challenges encountered on site are also covered

- The inspection works of the 12 No. air-regulated siphons constructed in 1974, undertaken by abseiling, resulting in the specification of a new paint system and concrete repairs to the internal faces of the siphons. The history of the siphons, the surveying methods used, the development of the paint system, the testing undertaken and how the paint system was applied are described.

- The replacement of the wave deflector along the crest of the dam where an Armco road crash barrier was utilised in the design, covering the design as well as the challenges faced installing the wave deflector and how a 'hanging' mobile scaffold, which could be moved along the dam crest, was utilised to facilitate the fitting of the barrier.

INTRODUCTION

Spelga Dam is a concrete gravity dam situated within the Mourne Mountains, Co. Down, Northern Ireland (Figure 1). It has a maximum height of 28m above ground level and a crest length of 305m. Construction of Spelga Dam was completed in 1957 at which stage the reservoir had a storage capacity of 2.7Mm^3. A Paper on the Construction of Spelga Dam was delivered by Frederick Forrest Poskitt, B.Sc., M.I.C.E and John Andrew Soye, B.Sc., A.M.I.C.E. to the Institution of Civil Engineers, Northern Ireland on 18th January 1960 [1].

Spelga Dam is owned and operated by Northern Ireland Water (NIW) which is a Government Owned Company set up in 2007 to provide water and sewerage services in Northern Ireland.

In 1974, the storage capacity of the reservoir was increased to 3.3Mm^3 with the construction of 12 siphons, comprising four banks each containing three barrels to raise the top water level of the reservoir by one metre.

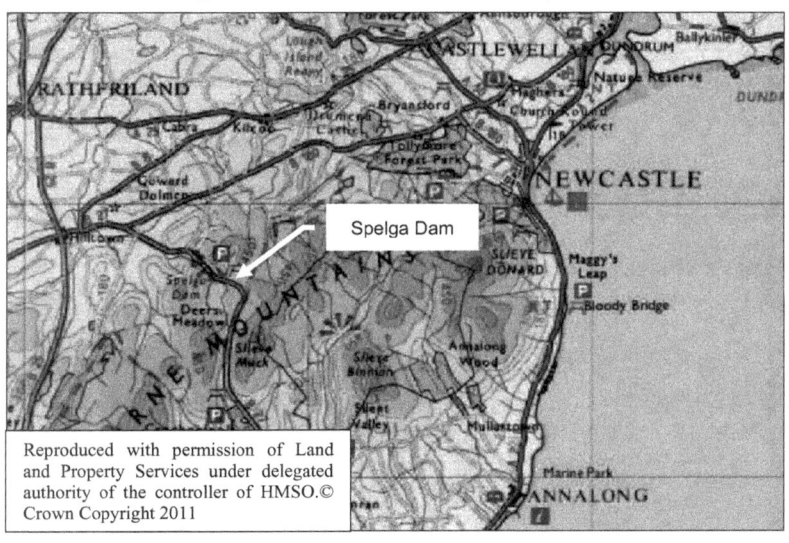

Figure 1. Location Plan, Spelga Dam

In 2008, URS was appointed as Designer, Project Manager and CDM Co-ordinator the rehabilitation and maintenance work to be undertaken at Spelga Dam in order to improve its operational safety. This work included:

- The creation of a new 'walk in' access from the downstream face of the concrete dam to access the longitudinal gallery.
- The replacement of the wave deflector along the crest of the dam.
- The inspection and repainting works of the 12 No. air-regulated siphons.

The contract was awarded to Graham Construction in August 2011 under the NEC Engineering and Construction Contract Option A (Priced Activity Schedule). The works began on site in September 2011 and were completed in February 2012. This Paper describes the design and construction of the above elements of work.

NEW 'WALK IN' ACCESS

Requirement for a 'walk in' access

In 2002 the dam owner considered the access arrangements to the internal longitudinal gallery of Spelga Dam. The longitudinal gallery runs through the centreline of the dam to enable inspection and provide access to various installed instrumentation. The only means of access available at the time was via the drawoff tower.

Achieving access to the gallery required using a number of vertical ladder sections with intermediate platforms and was considered a confined space for the purpose of inspection and maintenance.

Concerns were raised with regard to the practicality of removing a person from this area should they have a medical emergency. Previously consideration had been given to the installation of a lift facility. However, following further consideration of alternatives, it was agreed to proceed with the design of a new pedestrian 'walk in' access through the concrete of the downstream face into the longitudinal gallery.

Design of Access

The new access extends from the downstream face of the dam into the end of the Eastern section of the longitudinal gallery, a length of approximately 7m. The cross section dimensions of the new access are 2.0m high by 1.5m wide. The location of the proposed access is shown in Figure 2

Figure 2. Location of the new "walk in" access

The formation of the access was a difficult and significant construction procedure as it involved the removal of a volume of concrete without damaging the structure of the dam.

A structural analysis was carried out to ensure that creation of the opening through the dam into the existing gallery was acceptable in terms of dam stability. In addition, consideration was given to possible construction techniques and the impact on the dam during construction. As part of this work, construction techniques, including hydro-demolition and diamond wire cutting were investigated.

Concern was raised about the risk of excess vibration on the dam structure, which could potentially result in cracking or damage to the seals between the concrete bays. Consequently a review of the standards of vibration limits was undertaken.

Various standards and guidance papers were consulted prior to specifying vibration limits in the contract. The values of the vibration were set based on the dam being classified as a sensitive structure using DIN 4150 Part 3 - Structural Vibration - Effects of vibration on structures [2] – see Table 1.

Table 1: Guideline values of vibration velocity, vi, for evaluating the effects of short-term vibration on structures (from DIN 4150 Part 3).

Type of Structure	Values of Vibration Velocity (mm/s)		
	At Frequency (Hz)		
	<10	10 – 50	50 - 100
1. Commercial	20	20 to 40	40 to 50
2. Dwellings	5	5 to 15	15 to 20
3. Sensitive Structures	3	3 to 8	8 to 10

Other limitations placed on the Contractor included restricting overcutting to less than 100mm, prohibiting the use of explosives, and the requirement to drill a pilot hole prior to construction of the new access to accurately establish the location of the "walk in" access.

Construction of Access

Hydro demolition method
The initial chosen method of working was hydro demolition. The construction began with the drilling of a 70mm diameter pilot hole to accurately identify the location of the "walk in" access from the downstream face. The location of the pilot hole was determined based on the ventilation grille visible on the downstream face, the position of which was located at the eastern end of the longitudinal gallery.

Figure 3. Hydro demolition

A saw cut, 2.0m by 1.5m, was provided to mark the location of the new access on the downstream face. A hydro demolition lance was mounted on the arm of a tracked excavator (Figure 3) and once the hydro demolition reached the corners a hand lance was used.

However, once the Contractor reached the "hearting" concrete in the dam core, the hydro demolition rig experienced problems blasting the larger sized aggregate and progress was slowed to approximately 50mm in one week. As a result an Early Warning was raised and a risk reduction meeting held, at which it was agreed that the method of working should be changed to diamond wire cutting.

Diamond wire cutting method
Six additional holes were drilled in the four corners and at the mid point of the new "walk in" access. The diamond wire was threaded through the holes and the diamond wire cut along the perimeter of the proposed new access to form a large concrete block. Further diamond wire cutting was carried out to cut the block into three segments. A steel lifting eye was bolted into each block and the blocks were pulled using a tracked excavator. To ensure that the blocks would not become lodged, the separation cut was tapered. The method of cutting provided a smooth finish which required no further treatment of the final exposed surface (Figure 4).

Figure 4. First tapered concrete block removed

Vibration monitoring and completion
The vibration monitoring was carried out in accordance with British Standard BS 7385 Evaluation and Measurement for vibration in Buildings [3]. Vibration monitoring points were set up at three locations to measure vibration during construction. These were positioned on either side of the new access and within the longitudinal gallery.

Figure 5. Completed "walk in" access and steps

To complete the access, new reinforced concrete steps were formed against the downstream face of the dam with dowelled reinforcement to ensure a connection. In addition new security fencing and a heavy steel security door was installed (Figure 5).

SIPHON INSPECTION AND REPAINTING

History of the Siphons
In 1974 the capacity of the reservoir was increased by 25% though raising the top water level by one metre using a set of 12 No. air regulated siphons (Figure 6a). The more efficient siphon overflow arrangement enabled the design flow to be discharged at a reduced head and avoided the need to raise the crest of the dam [4].

Siphon Inspection
In order to assess the quality of the existing concrete and paintwork within the siphon barrels, a rope access survey was required. The rope access survey was carried out by a Chartered Civil Engineer trained in rope access, aided by a specialist rope access contractor (Figure 6b).

Figures 6a &6b - Existing siphons and the rope access survey

The internal inspections of the siphons concluded that paintwork repairs were required. The paint on the siphons serves two important functions. Firstly, to reduce the hydraulic friction losses through the siphons and secondly, to protect the reinforced concrete from high flow velocity.

Copies of the original paint specification were available. However, the original epoxy paint was no longer available and a specialised paint supplier was consulted to determine an alternative compatible product. The specification for preparing the internal faces of the siphons, the type of paint and testing was crucial to ensuring that the repainting works would provide the required 25 year life expectancy.

Paint specification and application
In order to facilitate the painting the contractor proposed erecting scaffolding against the downstream face of the dam, up to the siphons. At the top of the scaffolding a platform was cantilevered under the siphon barrels, in order to provide a safe working area (Figure 7). In addition, to facilitate the works, the water level in the reservoir was lowered to prevent waves from overtopping through the siphons.

The paint used for repainting of the siphons was Sikadur 31. As part of the paint works pull off tests were carried out to test the bond between the parent concrete and epoxy paint application. A number of tests were carried out at various locations on both the walls and the slope and over differing areas of concrete wear.

A pull-off value of $1.5 N/mm^2$ or greater was required based on BS-EN-1504 Products and Systems for the Protection and Repair of Concrete Structures [5].

Figure 7. Scaffolding to siphons

WAVE DEFLECTOR

Design of wave deflector
The previous wave deflector comprised coated corrugated iron bolted to galvanised steel support uprights fixed to the crest of the dam. During an inspection in 2006 by the All Reservoirs Panel Engineer, Andy Rowland, it

was identified that parts of the wave deflector were missing and/or in a state of disrepair, thereby allowing spray to overtop the crest. Therefore it was agreed that, in order to improve the safety for operatives, the wave deflector should be replaced to minimise spray over the crest.

The design of the new wave deflector utilised galvanised steel Armco Road Safety Barrier fixed to the existing supports. To prevent sagging, a 60mm diameter hollow section tube was used to support the Armco barrier. The existing uprights were prepared and repainted.

Erection of wave deflector
To enable the safe removal of the existing wave deflector and fitting of the new wave deflector, custom made mobile scaffolding was erected (Figure 8a). The scaffolding overhung the crest of the dam and was counter balanced by large tanks. This enabled the scaffolding to be moved along the crest of the dam quickly and efficiently, thereby eliminating the requirement to frequently dismantle scaffolding.

Figure 8a & 8b. Scaffolding used to erect the wave deflector and the new fitted wave deflector.

SUMMARY
The addition of a new "walk in" access was achieved without any damage to the structure of the dam. In addition, work to the siphons and wave deflector required careful consideration in terms of providing safe access during inspection and construction of the improvements.

ACKNOWLEDGMENTS

The authors would like to thank Northern Ireland Water and Graham Construction for information on Spelga Dam.

REFERENCES

1. Poskitt F F and Soye J A (1960). *The Design and Construction of the Spelga Dam.*, The Institution of Civil Engineers, London, UK

2. German Standard DIN 4150-3 (1999-2002), Structural vibration - Effects of vibration on structures

3. BSI (1993). *British Standard BS 7385: Part 2: 1993, Evaluation and Measurement for vibration in Buildings: Guide to damage levels from ground-borne vibration.* British Standards Institution, London, UK

4. Poskitt, F F and Elsawy Dr E M (1976) *Air Regulated Siphon Spillways at Spelga Dam*, The Institute of Water Engineers and Scientists, London, UK.

5. BSI (2005; 2013). *British Standard BS-EN-1504, Products and Systems for the Protection and Repair of Concrete Structures.* British Standards Institution, London, UK

6. Molyneux J D, Zhou J, Bradshaw J and Hogan J (2013). *New Passages in Old Dams.* ICOLD 2013 International Symposium, Seattle, 2013.

7. BRE (1995). *Damage to structures from ground-borne vibration, Building Research Establishment Digest 403.* BRE, Watford, UK

8. BSI (1992). *British Standard BS6472-1992, Evaluation of human exposure to vibration in buildings (1-80 Hz).* British Standards Institution, London, UK.

9. Australian Standard AS 2187.2 – 1993, Explosives - Storage, transport and use. Part 2 Use of explosives

10. Swedish Standard SN 640 312:1978, For steady-state vibration, from machines, traffic and construction in buildings

Refurbishment of Woodburn Reservoirs Eduction Towers and Scour Pipework

G BRIGGS, URS Infrastructure and Environment UK Ltd
G A COOPER, URS Infrastructure and Environment UK Ltd
D BELL, URS Infrastructure and Environment UK Ltd

SYNOPSIS The Woodburn Reservoir system in the hills above Carrickfergus, Northern Ireland, comprises seven impounding reservoirs in two cascades. These water supply reservoirs are managed by Northern Ireland Water (NIW) and they supply a wide area of North Belfast and South Antrim. The structures date from 1868 and include designs by both Macassey and Bateman.

The most recent Section 10 Inspection Reports recommended investigations into the operation and condition of the reservoir scour pipework and highlighted concerns regarding the access arrangements within the associated towers.

This paper outlines the findings of the resulting investigation works and the subsequent refurbishment works at the first two sites with particular reference to the lessons learned in the refurbishment of historic structures and pipework. It will also comment on the programming challenges associated with temporarily removing reservoirs from service and the environmental issues associated with emptying reservoirs to provide safe access for repairs

INTRODUCTION

The Woodburn reservoir system is located approximately 12 miles North of Belfast, in the vicinity of the historic town of Carrickfergus. The system comprises seven reservoirs which supply Dorisland Water Treatment Works. The reservoirs are in parallel series, with Upper South Woodburn, Middle South Woodburn and Lower South Woodburn to the West, (North Woodburn feeding into Middle South Woodburn) and Lough Mourne and Copeland to the East. Both series feed into Dorisland Reservoir, which acts as a header reservoir to the water treatment works. The layout of the system is shown in Figure 1

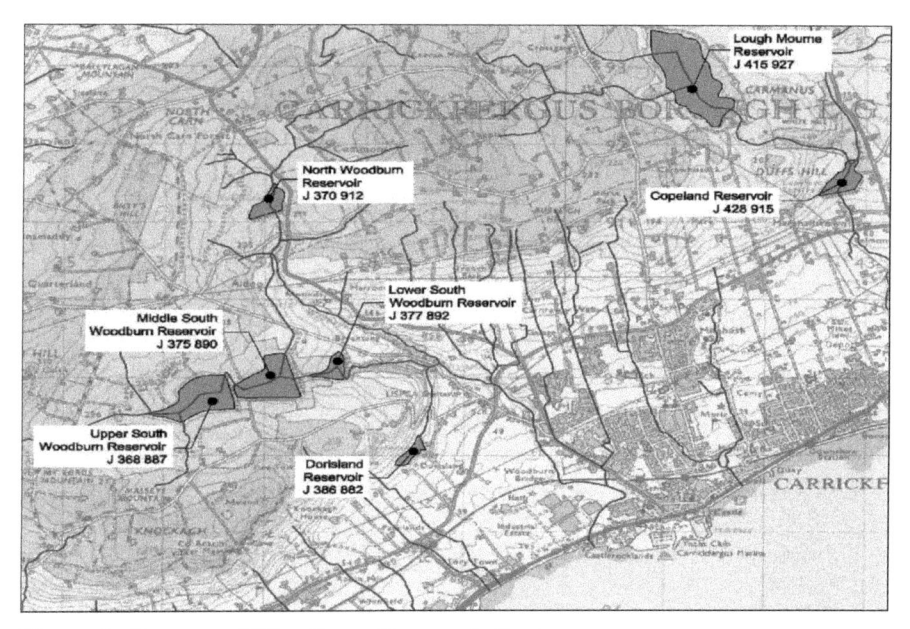

Figure 1. Layout of Woodburn Reservoir System

Table 1 outlines the scale of each site. The reservoirs are all classified as Category A dams under the guidance within the Floods and Reservoir Safety Handbook where a breach could endanger lives in a community

Table 1 Dam Characteristics

Name	Storage (Ml)	Max Height (m)
North Woodburn	372	13
Upper South Woodburn	1669	22
Middle South Woodburn	2135	26
Lower South Woodburn	489	19
Lough Mourne	2620	5
Copeland	607	19
Dorisland	300	13

History

The reservoir system was constructed between 1868 and 1881 by the Belfast Water Commissioners to supply water to the greater Belfast region. The Middle and Lower reservoirs on the South Woodburn River together with the North Woodburn reservoir on the North Woodburn River were based on a scheme prepared by J F Bateman working in conjunction with Sir Charles Lanyon.

The Middle South and North Woodburn earth embankment dams were completed in 1868. This work was followed by the construction of a further five embankment dams designed by Macassey, between 1876 and 1881, to complete the development of the Woodburn catchments. These were at Upper South Woodburn, Lower South Woodburn, Dorisland, Lough Mourne and Copeland.

The dams are all constructed of earth with puddle clay cores. The cast iron standpipe drawoffs, which are located within masonry towers, lead to brick lined outlet tunnels, cut through the basalt below the cut offs.

In the last 40 years the spillways, drawoffs and embankments have all been brought up to currently acceptable safety standards where necessary. Works have included the installation of weighted filter layers and rockfill berms were installed at some of the sites to improve stability and to control seepage.

With the exception of Dorisland the eduction towers are all wet towers, with cast iron standpipes connected to downstream tunnels and supported within masonry structures. The associated scour and drawoff valves and pipework are only fully accessible with the water level drawn down.

Current Use
The sites were designed to supply approximately 50Ml/d through Dorisland Water Treatment Works. However, since the development of sources from Lough Neagh, the average daily supply has dropped to around 25Ml/d, although the surplus yield is often used to supplement peak week demands in the Belfast area. During the winter of 2010 the abstraction was increased to 50Ml/d due to the significant increase in demand resulting from network failures following the winter thaw.

FINDINGS OF THE 2006 SECTION 10 INSPECTIONS
While the 1975 Reservoirs Act does not apply in Northern Ireland, NIW operates and maintains its dam structures generally within the spirit of the Act. It has in place a strict monitoring regime requiring monthly, bi-annual and annual inspections from designated staff members and the systematic recording of findings within a prescribed form set out in Section 1703 of NIW's 'Water Supply – Catchment Management Manual'.

For all other duties under the Act, NIW employs a Qualified Civil Engineer, from the All Reservoirs Panel, where appropriate. It also complies with the requirement for Section 10 Inspections and the implementation of any recommendations under Section 10(6) forms part of NIW's business plan and becomes a reported Regulator target.

The last Section 10 Reports for the reservoirs within the Woodburn system were completed by Mr Alan Cooper OBE and issued in October 2006.

Summary of report findings
In general the findings of the Section 10 reports were that the reservoirs were in a satisfactory condition in terms of the stability of the overall structural condition of the dams. There were no indications of significant structural deterioration since the previous (1996) Section 10 reports.

The monitoring records over the 10 years had not identified any unusual behaviour in terms of movement or leakage to give any cause for concern.

The reports all noted concerns about the condition of the control valves, which were critical for the drawdown of the reservoirs in an emergency. Therefore measures to be taken in the interests of safety included a recommendation that a full inspection of each of the eduction towers and all control valves within the complete system should be carried out.

Consultant appointments
In 2010, NIW appointed URS to carry out detailed inspections of the eduction towers and pipework and subsequently prepare feasibility reports and cost estimates for any recommended refurbishment works.

INITIAL INVESTIGATIONS AND FEASIBILITY STUDIES
The site inspections and testing followed the completion of a detailed review of all available drawings and records of each site. In general the available records were incomplete and not necessarily representative of what was identified on site.

It was not possible to have the reservoirs fully drained for the inspections, although the inspection team worked closely with NIW to facilitate partial lowering into supply where possible, to allow an appreciation of the internal structure of each tower to be gained.

The site inspections of the towers and pipework were completed between March and June 2010 by URS assisted by W A Bamford Mechanical Engineers who had previous detailed knowledge in their maintenance and repair. In general the inspections determined that:

- Some of the scour valves and the intermediate drawoff valves could not be operated from crest level
- Some intake screens were damaged, missing or could not be moved
- Valve headstocks were damaged with spindle threads sheared in a number of instances
- Spindles, where visible above the water level, showed signs of corrosion and/or deformation
- Internal pipework metalwork needed painting
- Access arrangements did not meet current safety standards

With over 8000ML of storage in seven reservoirs located above a large population, the inability to draw all the reservoirs down in an emergency would present a serious risk to NIW and the downstream communities. The feasibility reports therefore recommended that each reservoir should be drained to allow access for the complete refurbishment of the valve mechanisms and for repairs to be carried out.

Development of Cost Estimates

While the full scope of work could not be fully determined, cost estimates were prepared for NIW to allow the works to be procured under its capital investment programme. It was agreed at this stage that the scope of any resulting contract should also include any maintenance and refurbishment works that may be required below top water level in order to take full advantage of the drawdown of the reservoirs.

The cost estimate for the works totalled £1.4 million, including an assessed optimism bias risk percentage of 62% due to the nature of the work and its unknown extent

Approach to procurement

The procurement of a contractor for works which are not fully defined, and that may be subject to considerable change can pose significant cost risks to clients. In this case, not only was the scope not fully defined but the available existing site information could not be fully relied on. Subsequently any construction contract was likely to involve a number of compensation events and require a high level of project management.

The options available for the procurement of a contractor from one of NIW's existing frameworks included:

- NEC3 ECC Option A – effectively a lump sum contract based on an activity schedule.
- NEC3 ECC Option B – effectively a lump sum contract based on a Bill of Quantities.
- NEC3 ECC Option C – a target cost contract based on an activity schedule
- NEC3 ECC Option E – a time and materials contract target with fixed profit and overhead percentages

It was agreed that without draining the reservoirs prior to the contract, the scope of work could not be sufficiently defined for either an Option A or Option B form of contract, as this would put excessive risk onto the contractor and client respectively. While the use of a time based contract had the advantage of reducing the risk to the contractor, it was felt that a target cost contract would share the risk more appropriately encouraging contractor involvement in the decision making and value engineering processes. The contract for the construction works was advertised in November 2012 and awarded to Graham Construction in April 2013.

PROGRAMMING CHALLENGES AND OPPORTUNITIES

The requirement to drain each of the reservoirs while the Woodburn system remained in service resulted in the need for a considered approach to programming of the contract using a defined sequence of drawdowns.

While the average daily yield from the system was approximately 50% of the safe yield from the complete reservoir system, the interconnection between the individual sites presented a number of challenges. For example, water can only be abstracted from the North, Upper, Middle and Lower South Woodburn series through Dorisland Reservoir. If the Dorisland Reservoir was out of service all flow must come from the Copeland / Lough Mourne series, and the associated raw water transfer main could not deliver more than 25ML/d. There was also a requirement from NIW to have a minimum level of storage available between November and February each year to provide resilience against a major freeze-thaw event within the supply network.

A series of programming workshops were held with NIW operations staff and a programming model was developed to demonstrate how programming decisions would impact on the water available for use within the system, see Figure 2. These models were also used to assess the impact of drought events, or periods of prolonged wet weather on the programme.

Where possible the sequencing and programme was developed to allow NIW to lower water levels to at least the lowest operable drawoff into supply, so as to minimise the amount of water sent to waste. Following the modelling of a number of scenarios a preferred programme was selected and approved by NIW's Operations Manager and the 32 month programme was incorporated within the works information in the NEC3 contract.

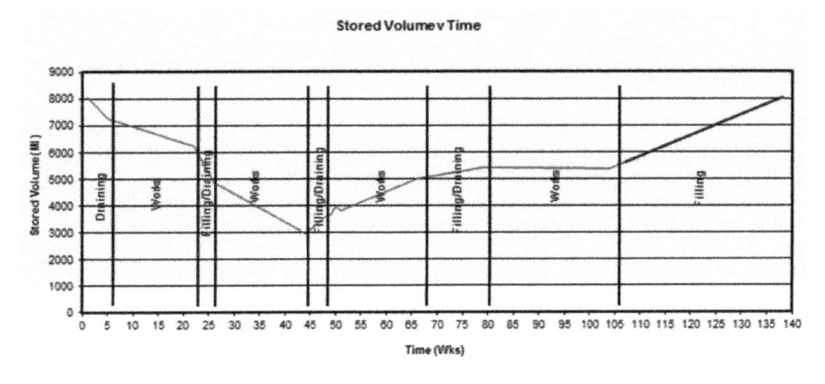

Figure 2. Example of programming sequencing charts

ENVIRONMENTAL CHALLENGES AND OPPORTUNITIES
Following the decision to draw the reservoirs down to facilitate the proposed works, it was recognised that a number of environmental issues existed that needed to be addressed. These issues were compounded by the fact that the Woodburn catchment lies within an Area of Special Scientific Interest (ASSI) and discharges into Belfast Lough Special Protection Area.

Early discussions were held with the Water Management, Water Framework Directive and Natural Heritage Departments of the Northern Ireland Environment Agency to agree on the required studies and applications that would be required to facilitate the planned draw downs and discharges.

The discussions regarding the discharge consents for the site quickly identified doubts as to the relevant regulatory applications required. This was complicated further as there was no overriding legislation governing the public safety requirement for the works and they were initially treated like any other developer application.

The agreed process involved the production of the following reports.

- EU Water Framework Directive 2000/60/EC Article 4.6 Exemption – while typically Article 4.6 exemptions are used after an event to justify why an objective set within the River Basin Management Plan was not achieved, this submission was developed and issued to cover the potential impact arising from the releases of water into the Woodburn Heavily Modified Water Body

- Water and Sewerage Services (NI) Order Consent under Section 226/227 - Under the Water (Northern Ireland) Order 1999, the consent of the Department of Environment is required to discharge any trade or sewage effluent into our waterways or underground strata. This includes any potentially polluting matter (including site drainage liable to contamination) from commercial, industrial or domestic premises to waterways or underground strata. This section includes details of how to apply for a Discharge Consent for both domestic and industrial discharges.

- Habitats Regulatory Assessment (HRA) – HRAs were required to assess the impact on the downstream Natura 2000 site, Belfast Lough SPA which was designated under Article 4.2 of EC Directive 79/409 for the conservation of wild birds.

- Flood Risk Assessment (FRA) – an FRA was required for DARD Rivers Agency approval to demonstrate that the works would not have a flood impact downstream.

Each of these applications and reports had to outline and assess the effectiveness of the proposed mitigation procedures for the works and for every application a detailed Scour Management Plan (SMP) was prepared for each of the proposed drawdowns.

Scour Management Plans

The scour management plans were prepared as site specific and standalone documents for NIW to apply to any discharge from their reservoir. Discharges were categorised within each SMP as either:

- Emergency releases
- Planned maintenance releases where the total time of discharge was less than 30 minutes
- Planned maintenance releases where the total time of discharge was greater than 30 minutes

The plan then outlined, where possible by a series of flowcharts, the statutory notification and applications, approval timescales, mitigation measures and records required for each type of release.

Depending on the nature of the release the typical mitigations could include.

- Flow restrictions and restrictions on use of drawoff valves
- Stratification risk assessments
- Sediment sampling and testing
- Installation of upstream control of sediment (hydrodams)
- Installation of geotextile weirs within the downstream watercourse at defined Mitigation and Monitoring Points
- Restrictions on the release during adverse weather
- Monitoring requirements and trigger levels for additional controls

The development of the SMPs was completed in conjunction with on-going consultation with NIEA and NIW's operational teams. The SMPs were approved by NIEA, alongside the other applications for the Woodburn sites, in May 2013 and were included as part of the works information for the construction contract.

TECHNICAL CHALLENGES AND OPPORTUNITIES

Application of the Scour Management Plans

While only Upper South, Middle South and Dorisland Reservoirs have been drained to date, the application of the SMPs has generally been a success with no pollution incidents reported. Where possible the existing scour valves have been forced partially open to allow clearance of the sediment from the area immediately upstream of the valve, with any sediment released collected at the downstream mitigation point prior to entry into the Woodburn River. Figure 3 below shows an aerial photograph of Middle South Woodburn drained and Upper South Woodburn refilling.

Figure 3. Aerial photograph of Upper, Middle and Lower South Woodburn during the works.

There have been issues with the levels of sediment within the interconnecting pipework and tunnels between the reservoirs. However generally this sediment has flowed over the course of a few days to naturally clear the tunnels without the need for mechanical intervention. Figure 4 shows the outlet of the tunnel between Upper South Woodburn and Middle South Woodburn immediately after draw down of Middle South Woodburn and a week later.

Figure 4. Photographs of outlet of the tunnel between Upper and Middle South Woodburn

The design of the scour intake tunnel within Middle South Woodburn meant that cleaning the sediment from the tunnel was not straightforward. Works completed in the early 1980s reduced the large diameter jacket scour valve to twin 250mm valves. While the reduced valve size was adequate to discharge the required scour flows, it appears to have reduced the velocity within the scour intake tunnel such that manual excavation was required to remove the sediment.

Figure 5. Middle South Woodburn Tower during refurbishment and Scour Intake

To gain access required the construction of a causeway out to the intake point, and man entry excavation works within the tunnel. This resulted in considerable Health and Safety issues and required management of the natural inflows to the reservoir.

Access Arrangements within Towers
The nature of the works and the size of the existing structures posed a number of significant access issues for the contractor. In many cases modifications were required to the tower structure to facilitate safe access and, where possible, these were incorporated into the completed works to improve future access. This included the provision of new flooring platforms, the installation of davit socket winch systems and the provision of fall arrest systems on ladders where necessary.

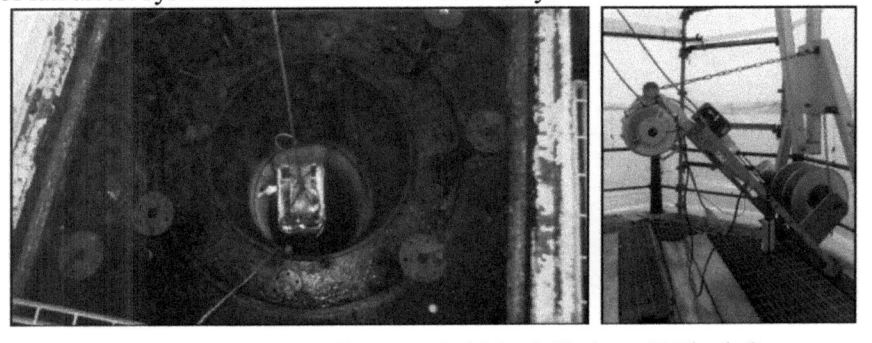

Figure 6. Middle South Woodburn Refurbished Chair and Winch System

Condition of Valves Arrangement

Valve Bodies

Following draw down of each reservoir, the detailed inspections and testing of the existing drawoff and scour valves was carried out. In general the condition of the penstock type valves was reasonable. There was considerable corrosion evident between the valve body and guides but in most cases, the valve could be forced open.

The valve bodies were sand blasted and painted, all bolts were replaced and new stainless steel guides and retaining slips were installed. Figure 7 shows the typical before and after condition of a valve door.

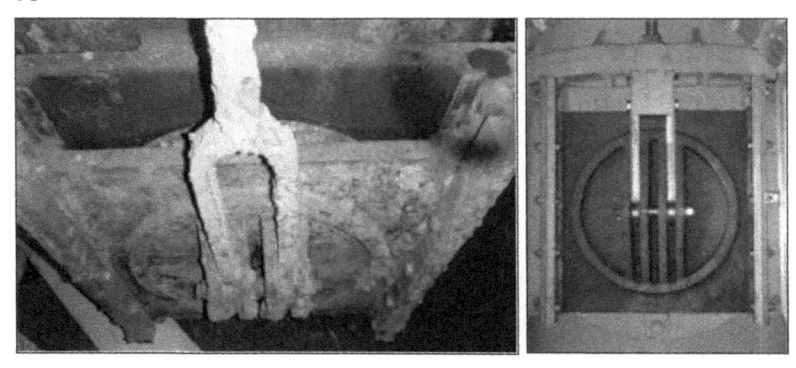

Figure 7. Upper South Woodburn Refurbished Valve Door showing before and after condition

Valve Spindles

In most cases the valve spindle, rather than the valve body was identified as the main cause of failure or weak link within the mechanism. In many cases heavy corrosion was evident between the spindle and the spindle support bracket arrangement, resulting in a seizure at these supports. Subsequent overstressing at the headstock then caused further damage to the spindle.

Figure 8. Middle South Woodburn Spindle Support bracket showing before and after condition

The design of each spindle was checked and all existing spindles replaced by a stainless steel spindle, of appropriate diameter, supported by stainless steel support brackets fitted with nylon bushes. This arrangement should prevent corrosion in the future and provide a free turning spindle arrangement.

Valve headstocks
Similar to the spindle arrangement, the valve headstocks were also found to have caused a number of problems, although these defects were generally identified prior to drawdown during the feasibility stage. The headstocks were all non-rising and a number had seized or had sheared threads due to overloading. To reduce the risk of future seizing, we replaced added lubrication to all headstocks and provided a shear pin to provide mechanism to prevent overloading.

Figure 9. Refurbished headstock before and after condition

Structural repairs
The drawdown of the reservoirs has provided an opportunity to complete a detailed inspection of the structure of the valve towers, the associated access bridges, tunnels and revetments, including the part of the downstream slope which is normally submerged within the downstream reservoir. While the works to date have shown that the stone clad valve towers are generally in excellent condition (especially when compared to cast iron towers of a similar age) a number of the bridge supports have required attention.

During the draining of the Upper South Woodburn reservoir the opportunity was taken, using a V Notch weir to monitor seepage flow from toe drains at the embankment seat of the upper dam, as the Upper South Woodburn refills.

Figure 10. Upper South Woodburn tower structure

CONCLUSIONS AND LESSONS LEARNED

There are a number of lessons learned on the project that we believe can influence future similar works, namely;

- Respect the environmental and other interested parties concerns – early engagement of the environmental stakeholders has allowed the works to progress with minimal long term environmental impact and has ensured that all statutory requirements were met. A site specific Scour Management Plan is be a useful addition to the management plan for reservoirs.

- Generally old cast iron valves and penstocks were found to be in good condition and in need of minimal repair. The majority of issues which prevented the operation of the valves related to spindles and spindle support brackets. Modern materials and manufacturing can be used to put systems in place which should require minimal future maintenance.

- The importance of regular testing and maintenance of valves cannot be underestimated.

ACKNOWLEDGMENTS

The authors would like to thank: Northern Ireland Water, Graham Construction, Mr David Gawn, Mr WA Bamford, NIEA and DARD Rivers Agency for information on the Woodburn Reservoirs.

Maintaining the Safety of our Dams and Reservoirs
ISBN 978-0-7277-6034-0

ICE Publishing: All rights reserved
doi: 10.1680/mdam.60340.517

Design of a new grout curtain for Wimbleball Dam

J G PENMAN, CH2M Hill
M J PALMER, CH2M Hill
A C MORISON, CH2M Hill
D K MASON, South West Water
J J WELBANK, Wessex Water

SYNOPSIS Wimbleball Dam was constructed in the late 1970s. The reservoir is a strategic water resource asset operated jointly by South West Water and Wessex Water. It is a 50m high buttress dam on a tributary of the River Exe in Somerset. The dam has a single line grout curtain drilled from a concrete plinth at the upstream toe of the dam. Leakage beneath the south abutment of the dam was observed during first filling. Despite undertaking remedial grouting at the time of construction and again in 2003, leakage flows have continued to rise. In 2010 the Inspecting Engineer judged the increasing flows to be a risk to the long term safety of the dam and recommended that measures be put in place to reduce the leakage. Implementation of remedial measures was complicated by a requirement that the reservoir could not be drawn down during construction works.

This paper describes the development of the design of a new grout curtain from an initial concept of drilling grout holes from a tunnel beneath the upstream toe to the adopted solution of drilling an inclined grout curtain from ground level from the areas between the buttress webs. It will also cover finite element analysis undertaken to identify the location of tensile zones in the dam foundations and to assess the impact on stability of moving the grout curtain downstream, evaluation of instrumentation data and design of a replacement pressure relief system.

INTRODUCTION
Wimbleball dam near Dulverton in Somerset is a 50m high concrete diamond headed buttress approximately 300m long and 50m high. The dam was constructed between 1974 and 1978 and stores 21.5Mm³ of water [1]. A view of the south flank from downstream is shown in Figure 1.

The reservoir is strategic water resource for South West Water (SWW) and Wessex Water (WW). SWW is able to release water into the River Exe in dry summers to support abstractions at its water treatment works for

517

Tiverton and Exeter (population of c150,000). WW pumps directly from the reservoir to the treatment works which serves Taunton and central Somerset (population served c100,000).

Figure 1. View of the south flank of the dam from downstream

Historical records [2] and a ground investigation undertaken in 2011 revealed that the dam is founded on Upper Devonian rocks comprising an interbedded sequence of medium strong to strong sandstones, weak to medium strong siltstones and very weak to weak mudstones. The strata are weathered to the full depth (c40m) of the existing grout curtain, with clay infilling many joints. The rock mass is typically 'blocky' with sub-horizontal bedding joints and sub-vertical joint sets. During construction, 'crush zones' and 'shattered areas' resulted in additional excavation up to 24m below original ground level on the south hillside to reach an acceptable foundation.

To provide bearing onto the weaker foundation, the concrete foundation blocks were widened to form a continuous foundation plinth and the spaces between the buttresses backfilled with spoil to add to the gravity loading. A longitudinal section of the dam is shown in Figure 2. There is a shallow concrete cut-off at the upstream toe and, beneath this, a single-line curtain [3].

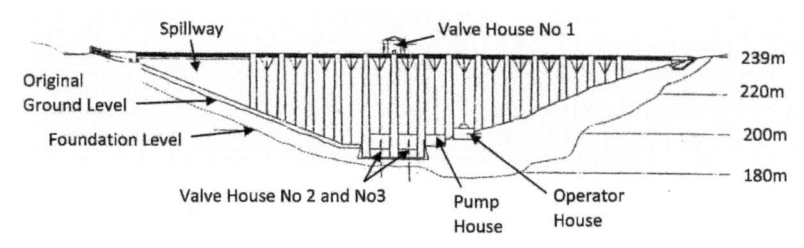

Figure 2. Longitudinal section of the dam

There is a history of leakage through the foundation beneath the dam and a limited programme of grouting was undertaken in 2003 in the rock beneath the south abutment in an attempt to reduce the under-seepage in this area. This work produced some reduction in seepage although monitoring over subsequent years indicated steadily increasing flows from this lower baseline. In 2010 the Inspecting Engineer judged the increasing flows to be a risk to the long term safety of the dam and recommended that measures be put in place to reduce the leakage. Early studies attributed the observed leakage to deterioration of the grout curtain. This led to the development of proposals for re-grouting the curtain to reduce leakage.

The importance of the reservoir for public water supply for both companies led to the common view that a long term solution was required for the deterioration of leakage and the decision that the reservoir could not be drawn down during construction.

The cost of the construction of the dam in the 1970s was funded jointly by the predecessor authorities and the cost of maintenance of the dam is shared approximately equally between SWW and WW. The joint nature of the funding of any remedial works meant that close cooperation was required throughout the project, including development of the business case to secure the funding for the project through the regulatory settlement and agreement on the project objectives, design criteria and proposals.

HISTORY OF LEAKAGE

There are three systems collecting leakage from the dam south abutment:

- the foundation pressure relief drainage system
- the backfill drain
- collection and measurement of the hillside spring flow downstream of the dam, known as the fish trap springs.

The foundation pressure relief system comprises a pipe network collecting flows from the individual pressure relief holes drilled through the dam into the foundation. The system discharges through connector pipes and manholes beneath the backfill. Groundwater within the south side backfill flows downhill towards the centre of the dam under gravity. The water can pass through the buttress webs by way of low level drain holes in Buttresses 12 to 15. There are typically five 300mm diameter drains through each web. At Buttress 11 there are no through-web drains so the water ponds behind the buttress web. It then exits via a pipe linking a drain hole through a retaining wall downstream of Buttress 11 to a chamber outside the entrance to the Pump Hall. This drain is called the south abutment backfill drain

The south abutment backfill drain thus collects:

- foundation seepage flows emerging at the downstream toe of the dam and leakage around the south abutment;
- leakage from the dam pressure relief drainage system;
- natural groundwater flow emerging from the south valley side plus rainfall and runoff percolating through the backfill.

Based on the comprehensive records maintained by the dam's operational team it is evident that flows into the pressure relief drainage system have risen steadily since construction. The flow reached 15l/s in 2002 and then levelled off at around 10l/s following the 2003 remedial grouting, but with flows diverted into the backfill drain. However this overall figure hides local continuing increases in individual drains and reductions in others due to silting up with washed-in clay. Backfill drainage flow has increased by 300% since 1990, to approximately 20l/s. Figure 3 shows the total groundwater inflows into the pressure relief wells and backfill drains.

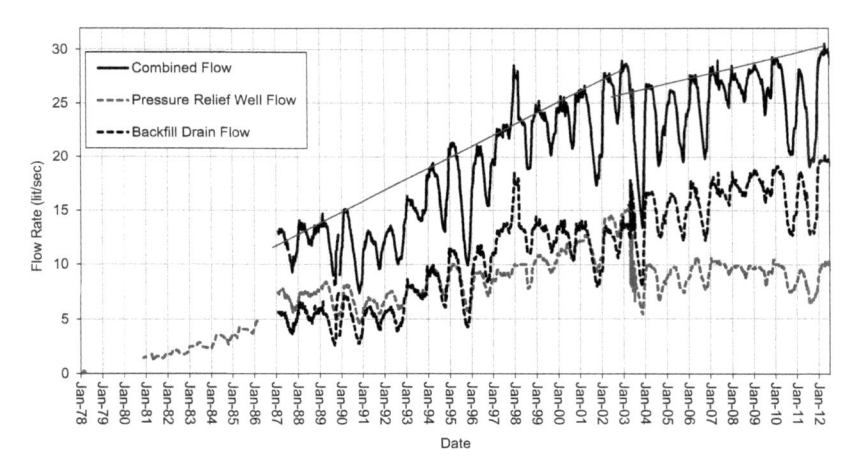

Figure 3. Variation in flow rates in the drainage system with time

Inflows into both the backfill and pressure relief well systems are influenced by the reservoir level. The effect of reservoir level on the inflow quantities can be assessed by plotting reservoir level against flow. The results for the backfill drain inflow (Figure 4) show a year on year increase with time. It is notable that flow rates generally show 'step changes' rather than a gradual increase with time. At other occasions there appears to be a degree of hysteresis during the refilling cycle following a drawdown.

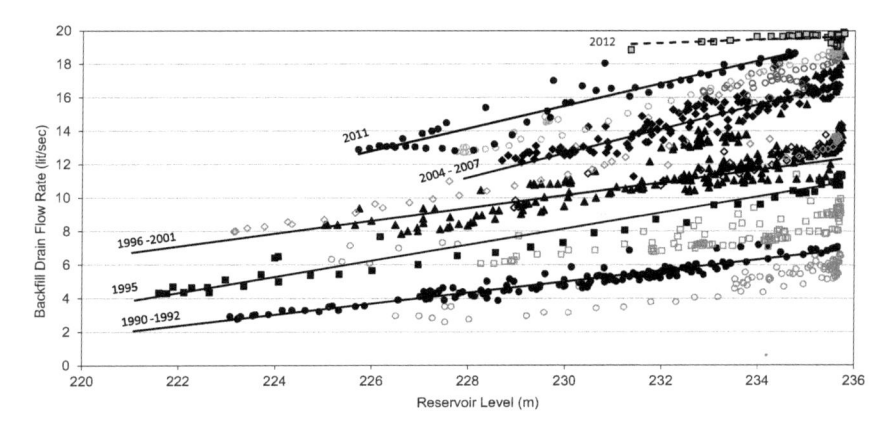

Figure 4. Variation in flow in the backfill drain with reservoir level and time

GROUND INVESTIGATIONS

A ground investigation was undertaken in 2011 to allow preliminary design and refinement of the chosen option. The works included rotary cored boreholes, geophysical televiewer and flow logging and *in situ* hydrotests. Locally, difficulties with borehole stability were experienced.

Inspection of the cores and laboratory testing assisted estimates of the rock mass deformation modulus. These were compared to previous estimates based on seismic velocity profiling prior to dam construction, *in situ* plate loading tests during construction, and observations of dam performance including pendulum and laser deformation monitoring.

In situ hydrotests indicated that the existing grout curtain showed little evidence of the low permeabilities reported to have been achieved during construction. Permeabilities of between 10^{-6} m/s and 10^{-4} m/s were instead recorded. Similar permeabilities were measured within and below the zone of blanket grouting beneath the downstream foundation area of the dam. Downhole geophysical flowmeter surveys indicated the existence of multiple, probably sub-horizontal, flow paths.

The reservoir water has a negative Langellier index which makes it undersaturated with respect to calcium carbonate, and therefore has the potential to dissolve calcium carbonate.

STABILITY ANALYSIS

In order to better understand the stresses and location of tensile zones in the dam and foundation, stability and stress analyses were carried out. Two forms of analysis were used:

- a rigid block gravity stability analysis on the concrete foundations, similar to the type of analysis that was used in the original design;
- two-dimensional linear elastic finite element stress analyses of the dam and the rock foundation, with a joint modelled at the concrete-rock foundation interface.

Gravity Analysis

The gravity stability analysis produced factors of safety well in excess of the minimum values recommended for stability and shear friction [4] even considering worst case assumptions for uplift pressures and cracking in the buttress heads. The analysis indicated that the buttress head would be in compression, the whole buttress section area would be available to resist sliding and propagation of cracking was unlikely.

Finite Element Analysis

The finite element analysis was carried out using SAP 2000 v14, with plane-stress elements for the dam and plane-strain elements for the foundation. Each buttress was modelled separately taking account of the actual 3-d section by varying the stiffness of the dam elements. The foundation profile at each buttress was taken from the as-constructed drawings. The contact between the concrete and the rock foundation was modelled using link elements, which could either be fixed, with stress in the link or, if in tension, released and allowed to separate with reservoir pressure applied in the joint. By progressively releasing links in tension, the development of a crack at the foundation interface could be simulated.

The finite element model was used to consider:

- three water level cases: normal, flood and drawdown (50% depth);
- alternative uplift pressure profiles;
- the effect of the backfill and of its possible removal;
- the potential for crack development along the foundation contact and the possibility of hydraulic fracture beneath the dam;
- sensitivity of the results to the rock mass deformation modulus.

The model output (Figures 6 and 7) for each of the load cases included:

- horizontal and vertical displacement of the section;
- maximum and minimum principal stresses;
- stresses and openings in the foundation links at each iteration of crack release (usually taking up to 5 iterations to fully release tensions).

The finite element stress analysis revealed conditions in the original grout curtain which would not have been recognised in the original design.

The key conclusions drawn from the finite element analyses included:
- with the reservoir full, a tension zone exists in the foundation in the region of the original grout curtain;
- should a crack develop in the tension zone, it can extend a significant distance downstream along the dam-foundation interface;
- the dam remains stable with such a tension crack in place. In most cases the extent of crack development is limited but it provides a possible leakage path through the grout curtain;
- the peak tensile stresses calculated locally at the upstream heel of the dam are high enough to cause local cracking of mass concrete;
- removal of backfill from downstream of the dam between the buttresses has only a minor effect on peak tensile stresses in the dam body and at the foundation contact, but results in cracking at the foundation interface extending further downstream;
- installing a new grout curtain downstream of the existing curtain may cause an increase in uplift pressures beneath the upstream heel of the dam, but does not result in a significant reduction in dam stability.

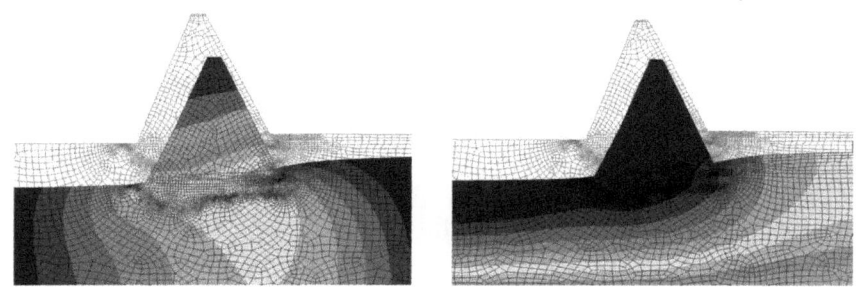

Figure 6. Finite Element analysis - Buttress 10 - contours of horizontal and vertical displacements

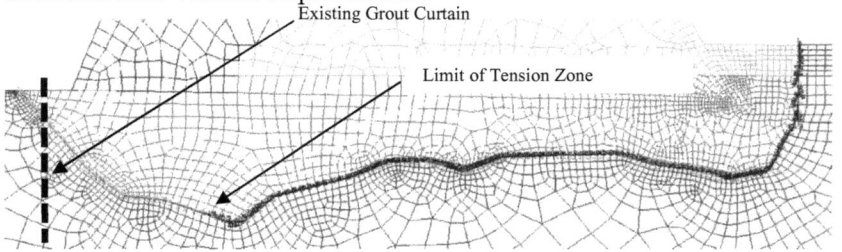

Figure 7. Finite element analysis – stresses in foundation links

SELECTION OF PREFERRED SOLUTION

Design and construction of the new curtain had to address the following project specific key issues:

- the curtain had to be constructed without drawing down the reservoir;
- zones of high permeability in the vicinity of the curtain were expected to connect with the reservoir and have significant leakage flow paths, or 'flowing features', to be grouted;
- the original grout curtain was located in an area of possible high tensile stress which may have been contributing to development of the leakage;
- the foundation included zones of closely jointed rock, which could cause difficulty in drilling and alignment control.

Initial studies focused on attempting to renew the existing vertical grout curtain at the upstream toe of the dam. Options considered included grouting from a tunnel driven along the line of the curtain from either a shaft on the south abutment or from access tunnels downstream of the dam passing beneath the dam foundation. However, this strategy was changed when the stability analyses indicated that the original grout curtain was within a tension zone and that there was opportunity to move the grout curtain downstream without adversely impacting the stability of the dam.

Moving the curtain downstream allowed an inclined curtain to be constructed on the line of the downstream edge of the buttress head with the curtain inclined upstream at a gradient of 1V:0.4H.

Options for constructing the curtain then included:

- Option A: driving galleries through the buttress webs and backfill accessed from a tunnel leading to the downstream toe;
- Option B: lowering the backfill level between the buttresses and driving galleries though the buttresses at the lowered ground level.

Option B was the cheapest option in terms of capital cost, long term O&M costs and construction time. It had significant safety advantages over the other options including avoiding underground works and confined spaces. Although cased drill holes would be required to reach the grout curtain area, this was offset by being able to use larger, higher-head drilling rigs, providing better drilling rates and directional accuracy compared with the smaller rigs required for work underground. The construction of a new curtain from the downstream side of the dam constrains the alignment of the grout curtain to be parallel to the face of the dam and so inclined across the plane of the existing grout curtain.

DETAILED DESIGN OF GROUT CURTAIN
After a review of potential access arrangements for Option B it was concluded that there was no benefit in the excavation of backfill material from between the buttresses as it would be possible to construct access

tracks into each bay at levels close to existing. The additional drilling lengths through the backfill material were considered acceptable.

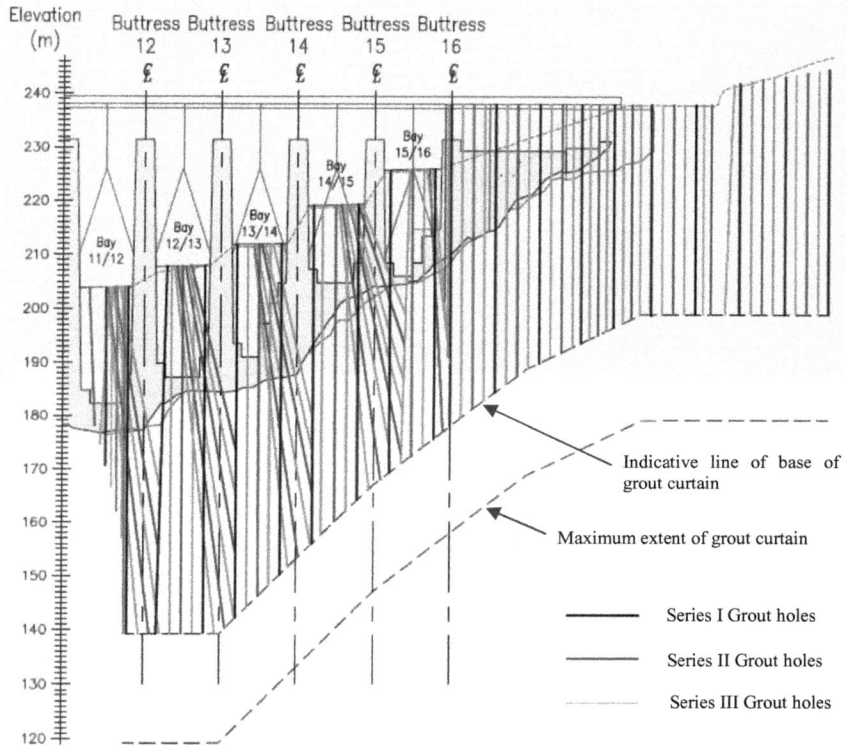

Figure 8. Longitudinal section of proposed grout curtain

The initial concept of drilling parallel grout holes along the full length of the curtain required forming galleries through the buttresses, which would have been time consuming. The arrangement was therefore developed into a combination of parallel and inclined holes such that the sections of curtain beneath the buttresses were covered by inclined holes drilled from the adjacent bays. The final arrangement is shown in Figures 8 and 9. At each end of the new curtain there are transitions back to the original curtain.

GROUTING CONCEPTS

The preferred solution was to create a new section of grout curtain by a planned systematic grouting procedure. In order to develop the best possible construction methodology, a series of grouting trials were planned for the start of construction to enable detailed design, plant commissioning, materials selection, testing and refinement of grouting mixes.

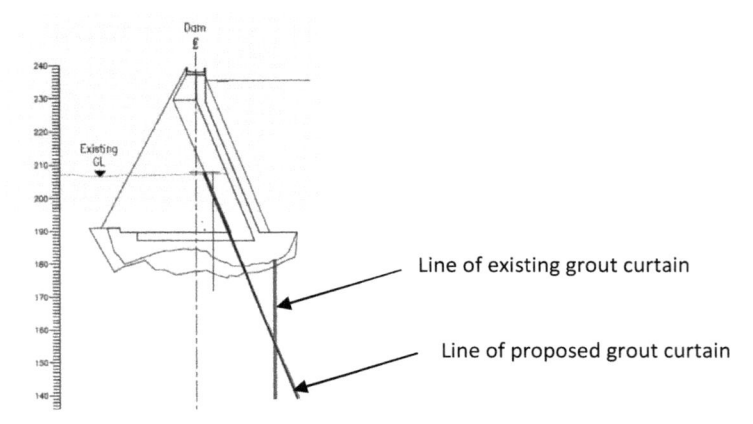

Figure 9. Cross section through proposed grout curtain

The development of the grouting design was directed towards addressing the project-specific key issues and maximising overall cost effectiveness by:

- optimising the use of modern grout materials to produce thixotropic, stable, low shrinkage, durable grouts of low permeability when injected into in a wide range of fissure apertures;
- setting a low target permeability for the grout curtain to minimise the opportunity for groundwater to come into contact with the grout, thereby increasing durability of the grout curtain;
- considering the use of cementitious materials which are resistant to water with a negative Langellier index;
- using regulated set grouts to achieve final set in shorter times than would be achieved by more conventional primary grout;
- using borehole sleeving systems to maintain hole stability in fractured rock;
- utilising grouts and grout control procedures that allow not only the filling of wider fissures under controlled conditions (*i.e.* with prescribed limits on the volume of grout injected in each stage), but also the maximum use of pressure to achieve penetration in fine fissures thereby allowing the required ultimate low permeability to be achieved.

Inherent in the proposed approach was the use of the split-spacing principle for the layout of boreholes and sequence of injection. The initial design anticipated three series of grout holes in a single row, with the intention that the design would be subject to progressive review as the trials progress. Contingency planning was undertaken to allow the design to be developed at any stage of the trials if found necessary.

The grout mix design is a critical element as it needs to be optimised to ensure stability (*i.e.* minimal bleed, separation under pressure, and shrinkage in the hardened state), good penetrability through discontinuities with apertures of less than 100micron, acceptable 'pot life', and predictable gel characteristics.

Modern grout materials using microfine (*i.e.* with a D_{95} <30microns) or ultrafine cements (D_{95} <15microns) with super/hyperplasticisers and silica fume were specified. Mixes using these materials produce stable grouts that are thixotropic, do not intermix readily with water, do not experience significant pressure filtration, and have the required durability and penetration under a wide range of aperture.

Microfine and ultrafine cements are readily available in the UK, sometimes through agents for European producers. The effectiveness of these materials can vary between suppliers and early identification and pre-qualification of products is necessary to allow time for comprehensive material selection trials. Laboratory tests alone do not allow technical and commercial material selection to be made, and trial injections are fundamental to the selection process. It was envisaged that grout mixes and the trial grouting approach would be developed collaboratively with a specialist grouting contractor during an Early Contractor Involvement (ECI) project phase prior to construction.

Experience indicated that control procedures based on the Grouting Intensity Number (GIN) [5] approach were appropriate to the requirements at Wimbleball Dam. In general, the approach advocates the use of a single, stable, fine-grained, cement grout mix of relatively low water:cement ratio. A steady, low to medium rate of injection is maintained, which results in a progressively increasing injection pressure. The rate of pressure increase varies depending on the fissures being injected, the rate being higher the smaller the volume of grout required.

The overall composite GIN curve is a pressure *v.* volume injection control line that comprises three components, *viz.* an upper pressure limit (P_{max}), an upper volume limit (V_{max}), and a hyperbolic curve (*i.e.* constant (pressure x volume) – the 'GIN curve') linking these. The composite curve provides the control limits for the injection process. Use of the GIN curve prevents combinations of high pressure and large volumes occurring that can result in uncontrolled hydraulic jacking and disruption of the rock mass, while at the same time allowing the use of high pressure at low volumes to ensure penetration of fine fissures.

The grout plant and equipment was an important part of the grouting design and the ability to have computer controlled injections was considered necessary to achieve the objectives using production grouting methods.

For most conventional projects where the GIN approach is used, such as a dam foundation, the observational approach is used. The observational method proposed for Wimbleball Dam can be summarised as follows:

- initial hydrotests, televiewer inspection and geophysical logging of ungrouted boreholes to give information on permeability, fissure frequency and fissure apertures, and to allow zones of 'homogeneous' ground with similar grouting properties to be established;
- representative trial curtain segment length constructed using the selected grout materials and an initial estimate of GIN and split-hole spacings;
- from each Series to the next, the grout take should be seen to diminish, while the final pressure will progressively increase. Increasing pressure from one borehole Series to the next indicates that the first Series has sealed the wider fissures and with the following Series sealing the remaining finer fissures; the grouting records therefore form an important part of the validation of the grouting works;
- the overall result is confirmed by hydrotests in validation holes on the completed segment and the adjacent ground to demonstrate the achieved reduction in permeability and the width of curtain achieved.

NEW PRESSURE RELIEF SYSTEM

It was accepted that the new grout curtain arrangement would conflict with the existing system of pressure relief wells and that this could lead to the existing wells being grouted up. Provision was therefore made for a replacement pressure relief system.

The new permanent pressure relief system could not replicate the original system as drains could not be drilled or laid within the dam foundations. The initial concept was to install the new pressure relief well system as fans of holes drilled from levels which would afford a similar degree of pressure relief as the original system. The arrangement required drilling sub-horizontal and inclined holes from between buttresses downslope of the area which the new pressure relief wells targeted. The replacement system could not be installed until after all grouting was completed.

In case the dam foundation pressure relief drainage was disrupted while the grouting works were in progress, provision was also made for temporary pressure relief wells to be installed, if required, during construction of the grout curtain. The proposed arrangement for the temporary pressure relief wells was to install vertical drains to foundation level between the buttresses, which would allow groundwater to discharge into the backfill, thus providing pressure relief to the foundations. The decision on whether or not to install temporary pressure relief wells was to be based on monitoring data from new vibrating wire piezometers installed just below foundation level in each of the bays.

CONCLUSIONS

The adopted solution to the increased leakage at Wimbleball Dam is to construct a new grout curtain by drilling from the top of the existing backfill on the downstream side of the buttress heads. Construction is to use modern grouting materials including microfine cement with silica fume. The grouting will be undertaken using the GIN principle. The new grout curtain is expected to compromise the existing pressure relief drainage system and a new pressure relief system will be constructed.

ACKNOWLEDGEMENTS

The authors wish to thank South West Water and Wessex Water for their permission to publish this paper and Dr Peter Mason for his invaluable contributions as specialist advisor to the project team. Special thanks are also due to Stewart Shapland of the SWW operations team for his diligence in maintaining the dam's instrumentation records.

REFERENCES

1. Battersby D, Bass K T, Reader R A and Evans K W (1979). *Promotion, Design and Construction of Wimblebal.* IWES, London, pp329-428

2. Bass K T and Isherwood C W (1978). *The Foundations of Wimbleball Dam.* IWES, London, pp187-197

3. Bruce D A and George C R F (1982), *Rock Grouting at Wimbleball Dam.* Geotechnique, VolXXXII, No4

4. Kennard M F, Owens C L and Reader R A (1996). *Engineering Guide to the safety of concrete and masonry dam structures in the UK. CIRIA Report 148.* CIRIA, London, UK

5. Lombardi G (2003), *Grouting of Rock Masses.* Keynote Address, Proc 3rd Int Conf on Grouting and Ground Treatment, New Orleans.

Maintaining the Safety of our Dams and Reservoirs
ISBN 978-0-7277-6034-0

ICE Publishing: All rights reserved
doi: 10.1680/mdam.60340.530

Index

Ackers, J. C.	336, 361	Mann, R.	271
Ambler, T.	426	Marchant, L.	35
Ashworth, J. R.	16	Mason, D. K.	517
Atyeo, M.	309	Matos, D. S.	457
		McCusker, K. J.	175, 494
Bell, D.	504	McKillen, D. A.	48, 494
Blower, T.	322	Morison, A. C.	517
Bradshaw, J. R.	494	Moura, G.	457
Briggs, G.	504	Mulreid, G.	101
Brinded, P.	348		
Brown, A. J.	469	Neeve, D. E.	35
Brown, C.	234	Noble, M. A.	186
Brown, M.	309		
Bruggemann, D. A.	125	Palmer, M. J.	517
		Panzeri, M.	295
Cambridge, M.	243, 414	Parks, C. D.	101, 402
Chesterton, O. J.	221, 480	Pawson, J. R.	148
Clarke, L.	75	Penman, J. G.	517
Claydon, J. R.	480	Peters, A.	135, 348
Coombs, M.	195	Pickles, A.	75
Cooper, G. A.	48, 175, 504	Pickles, G.	369
Courtnadge, A. P.	426	Porter, D. N.	24
Coutts, H.	59	Porter, S.	35
		Prisk, D. M.	361
Deakin, A.	3	Pryce, S. A.	135, 336
Digby, R. J.	186		
		Rebollo, D.	369
Eddleston, M.	113, 257	Rigby, P. J.	88
Edmondson, M.	101	Rose, C.	257
Elder, A. J.	469	Ross, A. W. D.	450
		Rundle, S.	3
Francis, O. J.	125	Russell, E.	148
		Russell, S.	381
Gallagher, E.	257		
Gardiner, K. D.	234	Sandham, R.	75
Gauldie, R. L.	221	Saunders, G. J.	283
Gethin, D. A.	336	Scott, T. A.	336
Gilbert, R.	348	Simm, J.	75
Goff, C. A.	295	Smith, A. D.	295
Gosden, J. D.	426	Sugden, P.	257
Grosfils, R.	309		
		Taylor, H.	113
Harvey, P. J.	161, 243	Terrell, R. J.	361
Hill, T. J.	221, 243	Thomas, H. V. H.	16
Hope, I. M.	221, 257, 480	Thompson, A. N.	88
Hughes, A. K.	135, 391, 438	Thomson, D.	101, 113
		Tietavainen, M. T.	207
Jackson, T.	381	Tudhope, J.	336
Jones, D. E.	88, 402		
		Walker, J. P.	207
Karunaratne, G.	336	Wanner, T. R.	59
Kelham, P.	309, 348	Welbank, J. J.	517
Kirby, A. M.	322, 480	Wheeler, M.	135, 336
		Williamson, T.	438
Lewis, R. I.	3	Windsor, D. M.	195
Lockett, S.	271		
Malia, J.	271		